VOLUME SIXTY NINE

ADVANCES IN
BOTANICAL RESEARCH
Genomes of Herbaceous Land Plants

ADVANCES IN BOTANICAL RESEARCH

Series Editors

Jean-Pierre Jacquot
Professor, Membre de L'Institut Universitaire de France, Unité Mixte de Recherche INRA, UHP 1136 "Interaction Arbres Microorganismes", Université de Lorraine, Faculté des Sciences, Vandoeuvre, France

Pierre Gadal
Honorary Professor, Université Paris-Sud XI, Institut Biologie des Plantes, Orsay, France

VOLUME SIXTY NINE

ADVANCES IN BOTANICAL RESEARCH

Genomes of Herbaceous Land Plants

Volume Editor

ANDREW H. PATERSON

Plant Genome Mapping Laboratory
Department of Crop and Soil Sciences,
Department of Plant Biology, and Department of Genetics
University of Georgia
USA

ELSEVIER

AMSTERDAM • BOSTON • HEIDELBERG • LONDON
NEW YORK • OXFORD • PARIS • SAN DIEGO
SAN FRANCISCO • SINGAPORE • SYDNEY • TOKYO
Academic Press is an imprint of Elsevier

Academic Press is an imprint of Elsevier
32 Jamestown Road, London NW1 7BY, UK
Radarweg 29, PO Box 211, 1000 AE Amsterdam, The Netherlands
The Boulevard, Langford Lane, Kidlington, Oxford, OX5 1GB, UK
225 Wyman Street, Waltham, MA 02451, USA
525 B Street, Suite 1800, San Diego, CA 92101-4495, USA

First edition 2014

ISBN: 978-0-12-417163-3
ISSN: 0065-2296

For information on all Academic Press publications
visit our website at store.elsevier.com

DEDICATION

To my sons

CONTENTS

Andrew H. Paterson
Plant Genome Mapping Laboratory, Department of Crop and Soil Sciences, Department of Plant Biology, and Department of Genetics, University of Georgia, USA

Daniel G. Peterson
Institute for Genomics, Biocomputing & Biotechnology, and Department of Plant & Soil Sciences, Mississippi State University, Mississippi, USA

Michael D. Purugganan
Center for Genomics and Systems Biology, NYU Abu Dhabi Institute, New York University Abu Dhabi, Abu Dhabi, United Arab Emirates, and Department of Biology, Center for Genomics and Systems Biology, New York University, New York, USA

Marnie E. Rout
Department of Forensics and Investigative Genetics, University of North Texas Health Science Center, Fort Worth, Texas, USA

Hiroaki Sakai
National Institute of Agrobiological Sciences, Kannondai, Tsukuba, Ibaraki, Japan

Takuji Sasaki
Tokyo University of Agriculture, Sakuragaoka, Setagaya-ku, Tokyo, Japan

Robert J. Schmitz
Department of Genetics, University of Georgia, Athens, Georgia, USA

James C. Schnable
Plant & Microbial Biology, University of California, Berkeley, California, and Donald Danforth Plant Science Center, St. Louis, Missouri, USA

Tsuyoshi Tanaka
National Institute of Agrobiological Sciences, Kannondai, Tsukuba, Ibaraki, Japan

Haibao Tang
FAFU and UIUC-SIB Joint Center for Genomics and Biotechnology, Fujian Agriculture and Forestry University, Fuzhou, Fujian Province, China, and J. Craig Venter Institute, Rockville, Maryland, USA

Jingpeng Wang
Center for Genomics and Computational Biology, and College of Life Sciences, Hebei United University, Tangshan, PR China

Xiyin Wang
Plant Genome Mapping Laboratory, University of Georgia, Athens, Georgia, USA; Center for Genomics and Computational Biology, and College of Life Sciences, Hebei United University, Tangshan, PR China

Xiaoyu Zhang
Department of Plant Biology, University of Georgia, Athens, Georgia, USA

CHAPTER ONE

A Backdrop: Principles of Plant Genome Organization

Andrew H. Paterson[1]
Plant Genome Mapping Laboratory, Department of Crop and Soil Sciences, Department of Plant Biology, and Department of Genetics, University of Georgia, USA
[1]Corresponding author: e-mail address: paterson@plantbio.uga.edu

Contents

Abstract

In this volume, leading practitioners of plant genomics and experts in study of many of the world's most important crop plants and botanical models update the state of knowledge of the genomes of herbaceous angiosperms (flowering plants), also necessarily touching on a few woody plants such as poplar and grape that have become important as genomic models or contribute to important inferences about genome evolution. Within the diversity of taxa, concepts, and scientific approaches addressed in this volume, a few central features of plant genome organization recur as both constraints (i.e. to experimental design and execution) and questions (i.e. as to their basis). Some of these include the nearly 2000-fold variation in genome size among angiosperms, the nature and evolution of repetitive DNA sequences that comprise most of this genome size variation, and the tendency of angiosperm lineages to survive genome duplication much more frequently than most if not all other branches of the tree of life. While the richness of diversity in the angiosperm genome sequences now available is great, as detailed herein, the richness of as yet undiscovered botanical diversity may remain still greater. Ongoing opportunities in cataloguing and studying the genomes of herbaceous plants and their associated microbiomes are expected to shed further light on the nature of the biodiversity that has permitted angiosperms to become the world's dominant vegetation and to sustain humanity by providing a wide range of ecosystem services. One chapter (Chapter 13) not only illustrates seminal concepts in genome biology but also provides a 'tutorial' for those intrigued by this field and wishing to learn how to explore new questions.

Advances in Botanical Research, Volume 69
ISSN 0065-2296
http://dx.doi.org/10.1016/B978-0-12-417163-3.00001-9

1. BACKGROUND FOR THIS VOLUME

Angiosperms (flowering plants) are the Earth's dominant vegetation, with nearly global distribution from the oceans to the deserts, from the tropics to the polar regions, and from sea level to 20,000 feet in altitude (Younghusband, 1926). The 'domestication' of a small subset of angiosperms is arguably one of the greatest human achievements and inarguably was essential to human cultural advancement and population growth (Raven, Evert, & Eichhorn, 1992), sustaining humanity by providing 'ecosystem services' including oxygen, food, feed, fibre, fuel, medicines, spirits, erosion and flooding control, soil regeneration and remediation, urban cooling and greenspace, wildlife habitat, and other benefits (Paterson, Freeling, Tang, & Wang, 2010).

While humanity has been dependent upon angiosperms for thousands of years, we are very early in the process of learning their 'parts list(s)' and how these fit together to make a whole. At the most elemental level, the 'parts' are genes, and revealing their nature, evolution and function is an area of vigorous scientific activity.

A prerequisite for learning how an angiosperm works is to have a complete 'parts list(s)', specifically a genome sequence that reveals the identity and arrangement of virtually every nucleotide in its genome. A few short years ago, this was a daunting task—only 4 years before this writing, the author and others reviewed the nature and features of all six(!) angiosperm genome sequences that had been published at that time (Paterson et al., 2010). With the reduction to practice of sequencing methodologies able to provide deep coverage of large genomes in single experiments costing tens of thousands of dollars or less, more than 30 angiosperm genomes now have at least one 'reference' sequence, and, in a few cases, hundreds of additional genotypes have been sequenced to reveal variations on such reference sequences. The authors' prediction (in 2006) that '. . . the sequencing of 200 domesticated plants would take a relatively short 14 years' (Paterson, 2006), at the time considered audacious by many colleagues, is beginning to look unduly conservative.

In this volume, leading practitioners of plant genomics and experts in study of many of the world's most important crop plants and botanical models update the state of knowledge of the genomes of herbaceous plants (also necessarily touching on a few woody plants such as poplar and grape that have become important as genomic models or contribute to important inferences about genome evolution). The analysis of plant genomes enjoys a rich history that predates the knowledge that DNA was the underlying genetic material (Chapter 2) but has recently experienced a dramatic acceleration, thanks in particular to massively parallel techniques for sequencing of nucleic acids (Chapter 3). The first plant genome sequence (*Arabidopsis thaliana*, Chapter 4) and first monocot genome sequence (*Oryza sativa*, Chapter 5) have now been joined by multiple representatives of their respective lineages, permitting well-informed deductions to be made about the nature of the common ancestors that spawned all modern eudicots (Chapter 6) and cereals (Chapter 7). Comparison of the members of these lineages to one another and consideration of additional taxa offer insight into the nature of the angiosperm common ancestor (Chapter 8) and into the history of this important lineage prior to the evolution of angiosperms (Chapter 9).

Through modern genomic approaches, we have gained appreciation that heritable mechanisms other than the information encoded by the DNA sequence have important consequences for plant biology and explore the additional layer of complexity added to a genome by the 'epigenome' (Chapter 10).

Likewise, it is now clear that 'individual higher organisms' are actually *ecosystems of host and associated microbial cells (the 'microbiome') working in concert* to produce functionality that was previously attributed to the host alone, and we explore early knowledge of the role of the 'microbiome' in angiosperm biology, reaching far beyond well-studied cases such as nitrogen-fixing legumes (Chapter 11).

The ability to sequence not only individuals but also large samplings of the members of populations has accelerated the pace of discovery of the mechanisms of plant genome evolution and begun to address fundamental questions in population genomics and molecular evolution (Chapter 12).

The broad spectrum of angiosperm taxa for which we now have at least one well-groomed sequence to study provides new insight into the nature of biodiversity at its most elemental level, specifically evolutionary novelty encoded within DNA. The central role of genomic redundancy in the origin of this evolutionary novelty is highlighted, and a 'tutorial' is provided for those intrigued by this field and wishing to learn how to explore new questions (Chapter 13).

Finally, our growing insight into the history of plant evolution combined with recent technological advances is beginning to suggest exciting new

Cooke, & Delseny, 2002; Simillion, Vandepoele, Saeys, & Van de Peer, 2004; Vandepoele, Simillion, & Van de Peer, 2002). This controversy was eventually resolved, using improved techniques and additional sequence information to show that one could deduce eudicot–monocot correspondence although deviations from colinearity are frequent (Tang, Bowers, Wang, & Paterson, 2010) (see also Chapter 8).

Their more rapid genome structural evolution than vertebrates appears to be due largely to the propensity of angiosperms for chromosomal duplication and subsequent gene loss (Coghlan, Eichler, Oliver, Paterson, & Stein, 2005), fragmenting ancestral linkage arrangements across multiple chromosomes. It has long been suspected that many angiosperms were paleopolyploids (Stebbins, 1966). A host of recent analyses of genome sequences and ESTs now make a compelling case that virtually all angiosperms are paleopolyploids (Paterson et al., 2010). An attractive but still speculative hypothesis suggests that genome duplication in angiosperms may not be merely episodic but truly cyclic, imparting various fitness advantages that erode over time, favouring new polyploidizations (Chapman, Bowers, Feltus, & Paterson, 2006).

Like genome size variation, polyploidy has a range of practical and fundamental consequences for genome studies. With regard to sequencing per se, paleopolyploidy is only infrequently a practical problem as most paleoduplicated sequences have accumulated sufficient divergence that modern sequencing technology readily distinguishes between them. Intriguing exceptions to this generalization are cases in which specific genomic regions or individual genes have experienced one or more forms of concerted evolution (Holliday, 1964; Ohta, 1984), that is, that duplicated genes have preferentially maintained similarity to one another (Wang, Tang, Bowers, Feltus, & Paterson, 2007). For example, one region duplicated in the common ancestor of cereals appears to have experienced this phenomenon independently in different cereals over the past 70 million years (Wang, Tang, & Paterson, 2011). Recent evidence suggests that such cases may be of great fundamental importance, perhaps contributing to novel features that distinguish some polyploids from their progenitors (Paterson et al., 2012). The possibility of convergent evolution of duplicate loci may need to be given special consideration in the sequencing of polyploids that are 'relatively recent', that is, with origins in the past few million years.

Neopolyploidy, formed very recently in nature or even by humans (see Chapter 14), may pose a considerable challenge to high-quality sequence assembly. In 'allopolyploids' such as wheat or cotton, formed from genomes that experienced a few million years of divergence prior to being reunited in a common polyploid nucleus, the duplicated chromosomes normally do not pair and the sequences of gene pairs are usually distinguishable (again noting that recent evidence of concerted evolution may need to be considered; Paterson et al., 2012). However, 'autopolyploids' such as sugarcane and alfalfa contain multiple chromosome sets that pair and recombine in many if not all possible combinations (albeit to varying degrees; Ming et al., 1998). These species generally have the additional complication of being intolerant of inbreeding, containing heterozygosity (sequence polymorphism) within individuals. This complication has the consequence that sequence assembly must resolve not only variations among members of a family of DNA sequences but also alleles within such members (loci). Thus, one is faced both with assembly of a 'reference sequence' and simultaneously with a 'resequencing' project to capture allelic variation in the reference sequence. To date, this author is not aware of any reference sequences published for an autopolyploid per se. An attractive solution to this challenge has been to sequence diploid members of genera that include autopolyploid crops, using the diploid 'reference genome' as a platform to study allelic variations, thus separating the tasks of reference sequencing and resequencing. The sequences for diploid members of several genera that include autopolyploid crops (e.g. potato (Xu et al., 2011) and banana (D'Hont et al., 2012)) are being employed in this manner. In other cases, closely related genera provide useful reference genomes (e.g. sorghum as a reference for sugarcane; Paterson et al., 2009).

2.3. Fundamental consequences of polyploidy

Genome duplication is a punctuational event in the evolution of a lineage, with permanent consequences for all descendants—if the lineage survives. Most higher organisms pass through different ploidy levels at different stages of development (Galitski, Saldanha, Styles, Lander, & Fink, 1999; Hughes et al., 2000) and continuously produce aberrant unreduced gametes at low rates. However, the extreme rarity of genome duplications in the evolutionary history of extant lineages, usually surviving only once in many millions of years, shows that the vast majority quickly go extinct.

Classical views suggest that genome duplication is potentially advantageous as a source of genes with new functions (Ohno, 1970; Stephens, 1951). Some polyploids appear to realize these and other benefits (Comai, 2005), with genome duplication thought to be central to the evolution of morphological complexity (Freeling & Thomas, 2006). Numerous mechanisms could account for the evolution of new genes, for example, neo- or subfunctionalization of existing genes (Lynch & Force, 2000) or transposon-based juxtaposition of previously unrelated exons (Jiang, Bao, Zhang, Eddy, & Wessler, 2004). However, analysis of whole genome sequences shows that following genome duplication, most genes are restored to singleton status, reducing the possibility of such adaptive divergence. Many genes for which duplicated copies do survive may be functionally altered relative to their ancestors. Classical ideas about one copy diverging to new function (neofunctionalization—Ohno, 1970; Stephens, 1951) have more recently been tempered by findings that mutations in genes or their *cis*-regulatory elements may result in subdivision of ancestral functions between duplicated copies (subfunctionalization—Lynch & Force, 2000), which then become interdependent upon one another, and favour the preservation of both duplicated copies. Subfunctionalization may be a stepping stone to neofunctionalization (He & Zhang, 2005).

The angiosperms are an outstanding higher eukaryote model in which to elucidate consequences of genome duplication, both due to their relative propensity for this macroevolutionary event and in view of population genetic considerations. Seminal findings from yeast (Christoffels et al., 2004; Gu et al., 2003; Scannell, Byrne, Gordon, Wong, & Wolfe, 2006) and *Paramecium* (Aury et al., 2006) are shedding valuable light on consequences of genome duplication *in microbes*. However, these consequences are expected to be very different in higher eukaryotes with small effective population sizes such as angiosperms and mammals (Lynch, 2006; Lynch, O'Hely, Walsh, & Force, 2001). For example, neofunctionalization (evolution of new gene functions) is more probable in large populations, which contain more targets for mutations conferring new beneficial functions. In contrast, subfunctionalization (subdivision of ancestral gene function between duplicated genes) is improbable in large populations since a partially subfunctionalized allele (the first step in the process) is likely to be silenced by secondary mutations before reaching fixation by drift (Lynch et al., 2001).

Several chapters in this volume contribute insights into the fundamental consequences of genome duplication, and the editor envisions this area as being one with particularly great learning opportunities in the near future.

REFERENCES

Aury, J. M., Jaillon, O., Duret, L., Noel, B., Jubin, C., Porcel, B. M., et al. (2006). Global trends of whole-genome duplications revealed by the ciliate Paramecium tetraurelia. *Nature, 444*(7116), 171–178.

Bennett, M. D., & Smith, J. B. (1991). Nuclear DNA amounts in angiosperms. *Philosophical Transactions of the Royal Society of London B, 334*, 309–345.

Bennetzen, J. L. (2002). Mechanisms and rates of genome expansion and contraction in flowering plants. *Genetica, 115*(1), 29–36.

Chapman, B. A., Bowers, J. E., Feltus, F. A., & Paterson, A. H. (2006). Buffering crucial functions by paleologous duplicated genes may impart cyclicality to angiosperm genome duplication. *In: Proceedings of the National Academy of Sciences of the United States of America, 103*, 2730–2735.

Christoffels, A., Koh, E. G. L., Chia, J. M., Brenner, S., Aparicio, S., & Venkatesh, B. (2004). Fugu genome analysis provides evidence for a whole-genome duplication early during the evolution of ray-finned fishes. *Molecular Biology and Evolution, 21*(6), 1146–1151.

Coghlan, A., Eichler, E. E., Oliver, S. G., Paterson, A. H., & Stein, L. (2005). Chromosome evolution in eukaryotes: A multi-kingdom perspective. *Trends in Genetics, 21*(12), 673–682.

Comai, L. (2005). The advantages and disadvantages of being polyploid. *Nature Reviews Genetics, 6*(11), 836–846.

Davies, T. J., Barraclough, T. G., Chase, M. W., Soltis, P. S., Soltis, D. E., & Savolainen, V. (2004). Darwin's abominable mystery: Insights from a supertree of the angiosperms. *In: Proceedings of the National Academy of Sciences of the United States of America, 101*(7), 1904–1909.

D'Hont, A., Denoeud, F., Aury, J. M., Baurens, F. C., Carreel, F., Garsmeur, O., et al. (2012). The banana (Musa acuminata) genome and the evolution of monocotyledonous plants. *Nature, 488*, 213–217.

Freeling, M., & Thomas, B. C. (2006). Gene-balanced duplications, like tetraploidy, provide predictable drive to increase morphological complexity. *Genome Research, 16*(7), 805–814.

Galitski, T., Saldanha, A. J., Styles, C. A., Lander, E. S., & Fink, G. R. (1999). Ploidy regulation of gene expression. *Science, 285*(5425), 251–254.

Grant, D., Cregan, P., & Shoemaker, R. C. (2000). Genome organization in dicots: Genome duplication in Arabidopsis and synteny between soybean and Arabidopsis. *In: Proceedings of the National Academy of Sciences of the United States of America, 97*(8), 4168–4173.

Greilhuber, J., Borsch, T., Muller, K., Worberg, A., Porembski, S., & Barthlott, W. (2006). Smallest angiosperm genomes found in Lentibulariaceae, with chromosomes of bacterial size. *Plant Biology, 8*, 770–777.

Gu, Z. L., Steinmetz, L. M., Gu, X., Scharfe, C., Davis, R. W., & Li, W. H. (2003). Role of duplicate genes in genetic robustness against null mutations. *Nature, 421*(6918), 63–66.

He, X. L., & Zhang, J. Z. (2005). Rapid subfunctionalization accompanied by prolonged and substantial neofunctionalization in duplicate gene evolution. *Genetics, 169*(2), 1157–1164.

Holliday, R. (1964). Mechanism for gene conversion in fungi. *Genetical Research, 5*(2), 282–304.

Hughes, T. R., Roberts, C. J., Dai, H. Y., Jones, A. R., Meyer, M. R., Slade, D., et al. (2000). Widespread aneuploidy revealed by DNA microarray expression profiling. *Nature Genetics, 25*, 333–337.

Ibarra-Laclette, E., Lyons, E., Hernández-Guzmán, G., Anahí Pérez-Torres, C., Carretero-Paulet, L., Chang, T. H., et al. (2013). Architecture and evolution of a minute plant genome. *Nature, 498*, 94–98. http://dx.doi.org/10.1038/nature12132.

CHAPTER TWO

Evolution of Plant Genome Analysis

Daniel G. Peterson[*,†,1]
[*]Institute for Genomics, Biocomputing & Biotechnology, Mississippi State University, Mississippi, USA
[†]Department of Plant & Soil Sciences, Mississippi State University, Mississippi, USA
[1]Corresponding author: e-mail address: peterson@igbb.msstate.edu

Contents

Abstract

Here, I present a brief history and technical overview of key genome analysis methods developed prior to massively parallel DNA sequencing. Disciplines discussed include cytology/cytogenetics, genome size analyses, DNA reassociation kinetics, genetic mapping, and physical mapping. These tools have been crucial in advancing our knowledge of plant genomes and have been useful in guiding and validating whole genome sequencing efforts. While some of these methods will likely fade into obscurity, others may become even more useful as focus shifts from DNA sequence to higher-order genome (chromatin) structure.

1. INTRODUCTION

After publication of the human genome in 2003, I began to hear some of my colleagues use the phrase *postgenomics era* (or *postgenome era*) to describe the state of biomolecular research. I immediately disliked this new catch phrase. My primary objection to the expression was that sequencing a handful of genomes (yes, even a human genome) was only a minute step forward in understanding genome structure, function, and evolution. Saying we had

Advances in Botanical Research, Volume 69
ISSN 0065-2296
http://dx.doi.org/10.1016/B978-0-12-417163-3.00002-0

entered the *postgenomics era* was, in my opinion, analogous to saying that the Apollo 11 moon landing ushered us into the 'postgalactic era'.

Clearly, humans have a strong desire to fit things into nice, neat categories. However, few things fit into such 'boxes'. Today, a wide variety of new and old techniques are being utilized separately or together to provide insight into the nature of plant genomes. If we are in the *postgenomics era*, then we are simultaneously in the *genomics era* and the *pregenomics era*.

This chapter discusses the evolution of genome analysis with emphasis on investigation methods that do not rely upon whole genome sequences. These include cytology/cytogenetics, genome size studies, DNA reassociation kinetics, genetic (linkage) mapping, and physical mapping. The majority of these methods were developed as a means of characterizing and comparing genomes either prior to the development of DNA sequencing techniques or at least prior to the availability of massively parallel sequencing (MPS), that is, *next-generation sequencing*.[1] All of these tools have been crucial in the development of our understanding of genomes, most are still utilized for the novel information they provide, and all have been employed in the validation, correction, and exploitation of DNA sequence data. While MPS and (dare I say it) *postgenomics* techniques have diverted attention away from these older methods, they have not made them any less valid. As I will discuss, in some situations, the *old ways* are still the *best ways*. Moreover, some of these methods will likely become even more valuable as we begin to study genomes as they are actually found in living plants (i.e. packaged into highly organized chromatin domains within the dynamic and highly complex architecture of the nucleus).

2. EVOLUTION OF THE TERM GENOME

I propose the expression Genom *for the haploid chromosome set, which, together with the pertinent protoplasm, specifies the material foundations of the species.*
Hans Winkler, 1920, as translated by Lederberg and McCray (2001)

When Hans Winkler coined the term *genome* in 1920, it already had been shown that genes reside on chromosomes and that the linear order of genes

[1] The phrase 'next-generation sequencing' is as scientifically meaningless as 'postgenomics era'. However, these terms, which may have originated in the marketing departments of biotech companies, may well have value when selling products.

on chromosomes is conserved within species (Morgan, 1910). However, the molecular constituents of chromosomes had yet to be determined. In 1933, Jean Brachet demonstrated that chromosomes contained DNA (Alexandre, 1992), but it was widely believed that DNA was 'too simple' to contain complex genetic information; proteins, another known component of chromosomes, seemed the more obvious molecular substance of genes.[2] It would be another eleven years before DNA was definitively shown to be the molecule of heredity (Avery, Macleod, & McCarty, 1944) and an additional nine years until the structure of DNA would be elucidated (Watson & Crick, 1953) (see Fig. 2.1). Of importance here, however, is that the term *genome* has evolved and will likely continue to do so. Of note, the term has become both more specific as understanding of the biochemical nature of genes has increased and more generalized as the term has been adopted by those studying noneukaryotic organisms. Moreover, the study of genomes began several decades before the term was coined (Fig. 2.1). At present, *genome* has the following meanings:

1. For eukaryotes, *genome* is now used to indicate the DNA content (in both mass and sequence) of the nucleus of an unreplicated gamete regardless of ploidy. Proteins and other non-DNA constituents of chromosomes are not included in any modern *genome* definitions. For a particular species, one genome's worth of DNA is referred to as its 1C DNA content. The mass of a genome is traditionally provided in picograms or in millions or billions of base pairs, that is, megabases (Mbs) or gigabases (Gbs), respectively.
2. Though Winkler's original definition of *genome* only referred to eukaryotes, the term has been adopted by prokaryote scientists to describe the hereditable information transmitted from bacteria and archaea to their respective progeny. When first used to describe prokaryote DNA, *genome* referred only to the (usually) circular 'chromosomal' DNA molecule of prokaryotes and not their 'extrachromosomal' elements (e.g. plasmids). However, this definition fell into question as some species/lines were found to possess multiple (sometimes linear) chromosomes and/or plasmids with housekeeping genes (Gregory & DeSalle, 2005;

[2] Once Phoebus Levene determined the structures of the four nucleotide components of DNA, he hypothesized that the nucleotides were found in equal proportions. Indeed, Levene suggested that each DNA molecule was a tetramer composed of the four nucleotides in an alternating purine–pyrimidine arrangement. Levene's hypothesis, which he called the 'tetranucleotide theory' was widely accepted, and in Levene's own words, '. . .conclusively demonstrated by experimental evidence' (Levene, 1919). Even when DNA molecules were shown to be much longer than simple tetramers, the idea of DNA as a simple repeating polymer still predominated (Olby, 1974).

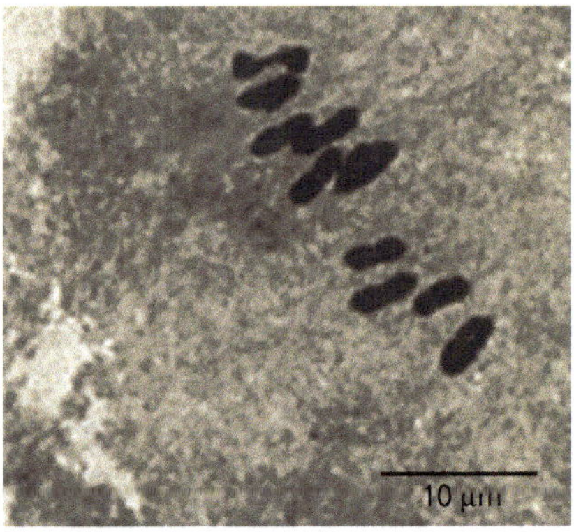

Figure 2.2 Light micrograph of early anaphase I chromosomes of *Sorghum bicolor* ($2n = 20$) stained with aceto-orcein. Each chromosome is composed of two homologues attached by chiasmata. Homologues are just beginning to be pulled towards opposite ends of the cell. *Photo by Daniel G. Peterson.* (See colour plate.)

the rest of the cellular protoplasm. However, the *coloured body* moniker reflects the observation that when squashed or otherwise disrupted cells are exposed to basophilic stains (e.g. aceto-orcein, basic fuchsine, or acetocarmine), the stains preferentially bind to chromatin allowing the chromosomes to be differentiated (by light microscopy) from other cellular materials (Fig. 2.2).

It is difficult to say when chromosomes were first observed. They were certainly observed long before they were given a name. In 1842, Carl Nägeli ostensibly published the first drawings of chromosomes (Posner & Skutil, 1968), which he called 'transitory cytoblasts'.[3]

In the latter half of the nineteenth century, significant advances in chromosome cytology were made by Walther Flemming and August Weismann (see Fig. 2.1). Flemming made numerous contributions including discovering and naming chromatin and mitosis (Olins & Olins, 2003). From his observations, Flemming concluded that all nuclei are derived from existing nuclei (Flemming, 1878). His work established the importance of the nucleus and hinted at the importance of chromatin/chromosomes. Though

[3] Interestingly, Nägeli (whose last name is sometimes spelled Naegeli) corresponded with both Gregor Mendel and Charles Darwin. Though a strong proponent of Darwin's work, Nägeli did not trust or properly understand Mendel's findings and famously encouraged Mendel to abandon his work on pea plants (advice which Mendel took). In a paper cowritten by Walter Naegeli, the great-grandson of Carl Nägeli (Naegeli & Wiedemann, 1993), the authors note, '[Darwin], who found the hybrid problem fascinating and had himself worked with *Pisum*, never, tragically, heard anything of Mendel. What an opportunity Naegeli missed here! He could easily have put the 2 men in touch with each other'.

meiosis had been previously described, Weismann was the first to postulate that the two back-to-back cell divisions characteristic of meiosis were essential in transforming a single diploid cell into four haploid cells; only through this means could ploidy be maintained after fertilization (Conklin, 1915).

A tremendous breakthrough in chromosome research came in the first few years of the twentieth century soon after the rediscovery of Mendel's work. While a student at Columbia University, Walter Sutton published two papers (Sutton, 1902, 1903) in which he presented observations supporting the premise that chromosomes behave according to the principles of Mendelian genetics. In the last paragraph of his 1902 paper, Sutton noted: 'I may finally call attention to the probability that the association of paternal and maternal chromosomes in pairs and their subsequent separation during the reducing division as indicated above may constitute the physical basis of the Mendelian law of heredity'. These were the only two papers that Sutton published—he left Columbia without finishing his PhD, worked for a couple of years in the oil fields, went on to medical school, and had a distinguished career in medicine until his death from a ruptured appendix at the age of 39 (Crow & Crow, 2002). Sutton's hypothesis was widely heralded and supported by the findings of others including Theodor Boveri who independently came to many of the same conclusions as Sutton based on his observations of fertilization and development in sea urchins (Baltzer, 1967). In an act of political astuteness, E. B. Wilson, Sutton's graduate advisor and a close friend of Boveri, referred to the concept that chromosomes are the units of inheritance as the 'Sutton–Boveri Theory'. Whether the assignment of cocredit to Boveri was warranted is still a subject of debate (Martins, 1999). Regardless, research by geneticist extraordinaire, Thomas Hunt Morgan, soon convinced all but the most stubborn of sceptics that the Sutton–Boveri Theory was indeed a reality. Based on Frans A. Janssen's insightful description and interpretation of chiasmata (Koszul, Meselson, Van, Vandenhaute, & Zickler, 2012) and Morgan's own revolutionary research on fruit fly genetics/cytology, Morgan went on to develop the concept of genetic linkage (Morgan, 1910). Morgan's student, Alfred Sturtevant, produced the first gene linkage map shortly thereafter (Sturtevant, 1913). The field of cytogenetics (i.e. the union of cytology and genetics) had been born.

Genes were on chromosomes ... indeed, genomes were contained within chromosomes. The study of inheritance and genetic processes was now in the cross hairs of the light microscope and, starting in the late

1930s, the electron microscope. With regard to light microscopy, a number of stains were developed to help visualize and differentiate chromosomes (Singh, 2003), and considerable strides were made in studying chromosome behaviour in numerous plant systems. Heitz (1928) coined the terms *heterochromatin* and *euchromatin* based upon observations of moss chromosomes and several years later demonstrated that specific bands of fly polytene chromosomes contained specific genes (Rubin & Lewis, 2000). Using *Drosophila* as a model, H.J. Muller showed that X-rays cause mutations and chromosome aberrations and described the inactivation of euchromatic genes translocated into heterochromatic regions (i.e. position effect variegation) (Carlson, 2013; Dillon, 2004). Using barley as a model, Stadler (1928) independently showed that X-rays and radium can cause mutations. In 1937, Dobzhansky (1937) described chromosomal aberrations as a source of mutations and touted their potential importance in speciation. Meanwhile, Sax (1938, 1940) developed the plant *Tradescantia* as a model for exploring radiation and its role in producing chromosome aberrations, work that ultimately was of tremendous value in advancing understanding of cancer and DNA repair.

The electron microscope played a pivotal role in helping advance understanding of chromosome structure. Serial sectioning and reconstruction of nuclei provided information on the state of chromatin in interphase including the association of chromatin with the nuclear matrix (Nickerson, 2001). Various nuclear spreading techniques revealed that the simplest state of chromatin is the so-called 11 nm fibre (also known as the 10 nm fibre), which has a 'beads on a string' appearance and represents DNA wrapped around histone octamers; the resulting *nucleosomes* are connected by stretches of naked DNA (Olins & Olins, 2003). Electron microscopy showed that the 11 nm fibre is often coiled into a 30 nm fibre, which, in cross section, is a solenoid with six nucleosomes per complete turn. Interestingly, whether the 30 nm fibre is representative of chromatin *in vivo* is still a subject of considerable debate and research (Grigoryev & Woodcock, 2012; Maeshima, Hihara, & Eltsov, 2010). Sections and spreads of mitotic chromosomes, along with various differential staining techniques, revealed the presence of protein-based chromosome cores (Stack, 1991). The electron microscope shed new light on meiosis revealing structures such as the synaptonemal complex (SC), a tripartite proteinaceous scaffold that joins homologous chromosomes along their entire lengths during pachytene (a substage of prophase I) of meiosis (Moses, 1956, 1958). A diagram of two homologues

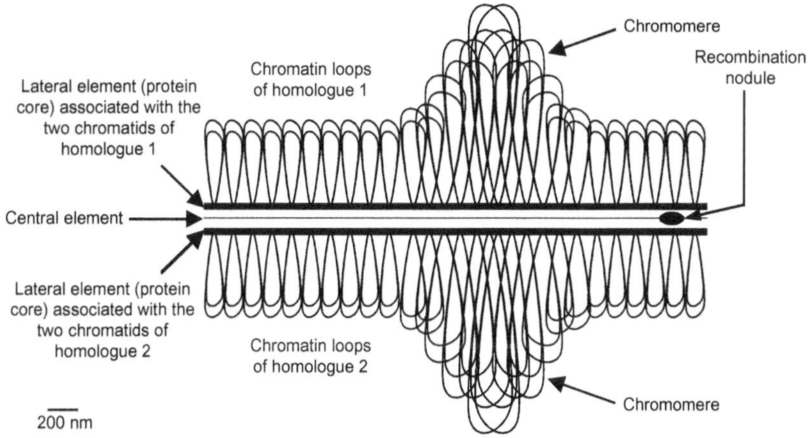

Figure 2.3 Model of a pachytene chromosome. Two homologous chromosomes (each composed of two identical sister chromatids) are joined along their entire lengths by a proteinaceous synaptonemal complex (SC). The SC itself is composed of lateral elements (each associated with the chromatin of one homologue) and a central element/region. Loops of chromatin extend from each lateral element. Chromatin is differentially packaged along the length of the pachytene chromosome with heterochromatin-rich areas sometimes visible as dark-staining chromomeres. In normal synapsis, chromomeres of one homologue are aligned with corresponding chromomeres on their homologue. Recombination nodules, ellipsoidal structures sometimes associated with the central element of the SC, mark the sites of crossovers (i.e. evidence indicates that recombination nodules are involved in mediating meiotic recombination).

joined together by an SC is shown in Fig. 2.3, while electron and light micrographs showing SCs and normal and abnormal chromosome synapsis are shown in Fig. 2.4. Just as Janssen, Morgan, Barbara McClintock, and others used light microscopy and genetic analyses to reveal that chiasmata, seen in diplotene and diakinesis, are the cytological manifestations of crossing over, electron microscopy led to the discovery of recombination nodules (Carpenter, 1975), protein structures associated with SCs during pachytene (e.g. Figs. 2.3 and 2.4B). Recombination nodules are now divided into 'early nodule' and 'late nodule' subgroups (Anderson & Stack, 2005). Of note, late nodule distributions are correlated with the sites of crossovers (Anderson et al., 2004; Anderson & Stack, 2005; Sherman & Stack, 1995), and proteins known to be involved in recombination have been found in both early and late nodules (Anderson, Offenberg, Verkuijlen, & Heyting, 1997; Lhuissier, Offenberg, Wittich, Vischer, & Heyting, 2007).

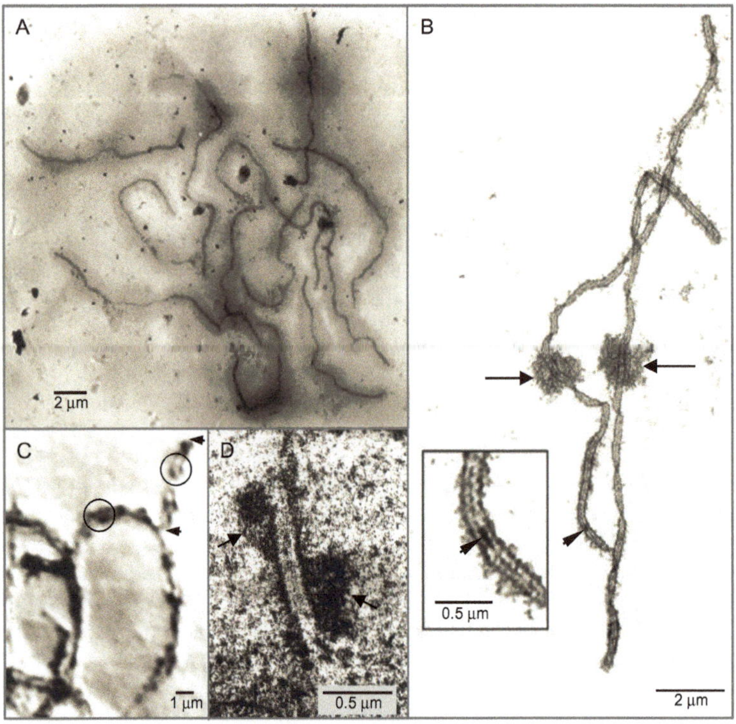

Figure 2.4 Pairing and synapsis during meiosis. (A) Electron micrograph of pachytene chromosomes of *Solanum lycopersicum* (tomato; $2n=24$). Each chromosome is composed of two homologous chromosomes held together by a proteinaceous synaptonemal complex (SC). The chromatin appears as a lighter staining halo around a darker-stained, centrally located SC. While chromomeres are not visible in this preparation, the chromatin halo around each SC varies in width indicative of differential chromatin packaging. (B) Electron micrograph of two tomato pachytene chromosomes from which the chromatin has been largely stripped away. SCs appear as tripartite ribbon-like structures (see Fig. 2.3 for details). Kinetochores are visible as ball-like structures (arrows). A recombination nodule (e.g. see arrowhead in full image and in inset) is visible as an ellipsoidal, dark-staining structure on the central element of the left SC. (C) Light micrograph showing pachytene chromosomes in a haploid *Hordeum vulgare* (barley) plant. A normal diploid barley plant would enter meiosis with $2n=14$ chromosomes and a 4C DNA content. In this haploid plant, meiosis began with $2n=7$ and a 2C DNA amount. As the chromosomes do not have homologues with which to pair, they have synapsed in a nonhomologous fashion as evidenced by mismatched chromomeres (e.g. circled areas). The region demarcated by the two arrowheads is a portion of one chromosome that has folded back and synapsed with itself. (D) Longitudinal section through a haploid *Hordeum vulgare* pachytene chromosome. While SC structure appears normal, the chromomeres on opposite sides of the SC are clearly mismatched (arrows) as expected in nonhomologous synapsis. *Panels A and B were taken by the author. They have been published previously, with his permission, in a* PLoS Biology *article (Hey, 2004) and in Wikipedia (article: 'Synaptonemal complex'). Panels C and D are previously unpublished and have been generously provided by Stephen M. Stack.*

In 1969, Mary Lou Pardue and Joseph Gall demonstrated that single-stranded, radiolabeled DNA probes could be hybridized to denatured chromosome preparations, and the sites where reassociation occurred could be visualized by autoradiography. The resulting technique was deemed chromosomal *in situ* hybridization and its implications were enormous (Pardue & Gall, 1969). In practice, *in situ* hybridization to chromosome preparations of plants turned out to be rather difficult compared to experiments using animal chromosomes. However, considerable advances have been made. Specifically, digestion of cell walls with enzymes has allowed preparations of cleaner squashes/spreads. Also, the move from autoradiography to fluorescence-based chromosome visualization and hybridization detection has made *in situ* hybridization easier and less hazardous. In fluorescence *in situ* hybridization (FISH), chromosome preparations are hybridized with a probe that contains hapten-linked nucleotides. Haptens are small, inert molecules that can be attached to nucleotides and incorporated into DNA without disrupting base pairing or synthesis/transcription reactions. Two of the most common haptens are biotin and digoxigenin. The hapten-containing probes are hybridized to the chromosomes, and sites of probe hybridization are visualized using an antibody or other molecule that preferentially binds to the hapten. The antibody is conjugated to a fluorescent tag. After counterstaining the chromosomes, the sites of hybridization are visualized using a fluorescence microscope. Examples of FISH to plant chromosomes are shown in Fig. 2.5. Note that FISH can be used to explore the locations of both single-copy sequences and repeat sequences. FISH has proven highly useful in the integration of genetic maps with actual chromosome structure (Islam-Faridi et al., 2002; Peterson, Lapitan, & Stack, 1999). More recently, FISH has become a powerful tool in guiding genome assembly (Stack et al., 2009; Tomato Genome Consortium, 2012). Specifically, FISH appears to be complementary and often superior to linkage map-based techniques in properly positioning and orienting sequence scaffolds on chromosomes (Hoskins et al., 2002; Satou et al., 2008; Stack, Shearer, Szinay, van de Belt, & de Jong, 2013). Although both molecular linkage mapping and FISH tend to correctly position scaffolds in areas of euchromatin, the latter is far superior in determining the arrangement of scaffolds in heterochromatic regions (Hoskins et al., 2002; Stack et al., 2013). Also in species where cytological techniques are fairly advanced, scaffolds/contigs that cannot be anchored into chromosome sequence assemblies can be positioned with high accuracy; not surprisingly, such unanchored scaffolds tend to fall within gaps in chromosome assemblies (Stack et al., 2013).

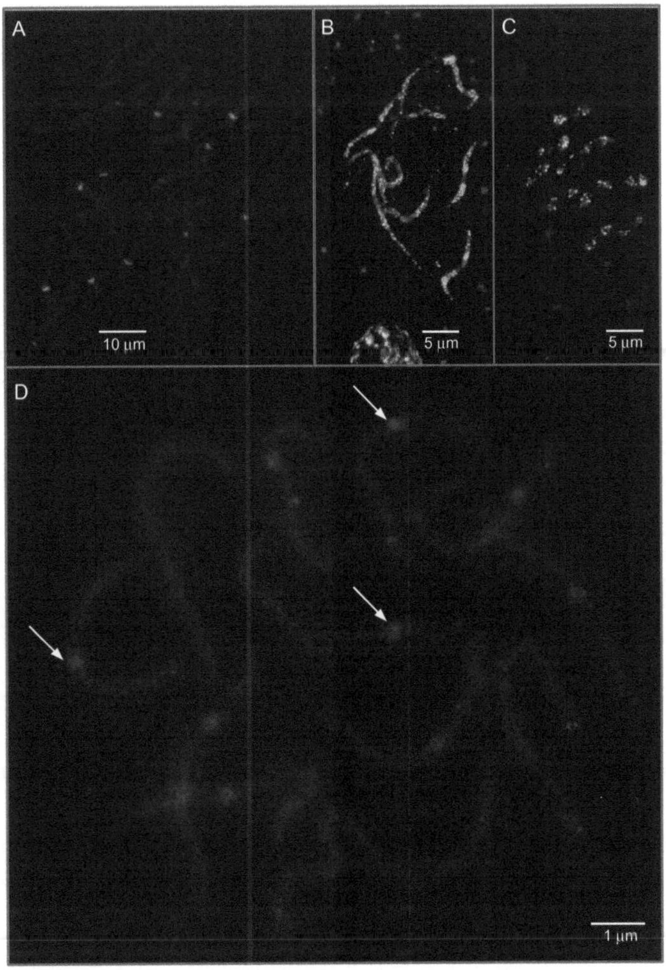

Figure 2.5 Fluorescence *in situ* hybridization (FISH) to plant chromosomes. All images are previously unpublished. (A) FISH to mitotic metaphase chromosomes of *Pinus taeda* (loblolly pine; $2n = 24$). The chromosomes have been hybridized with a 45S rDNA probe (green/yellow) and a dispersed repeat sequence (red). Image courtesy of Nurul Islam-Faridi. (B) FISH to pachytene chromosomes and (C) anaphase I chromosomes of *Sorghum bicolor* (sorghum; $2n = 20$). The pink regions indicate hybridization of a centromeric probe, while the green/yellow signals show hybridization of a pericentromeric repeat sequence to the chromosomes. Note the considerable difference in length between the pachytene and anaphase I chromosomes. (Images by Daniel G. Peterson.) (D) FISH localization of a single-copy sequence (pink signal) on tomato pachytene chromosome *2*. Note that a kinetochore is visible on each chromosome (e.g. arrows) and that chromatin is visible as a halo around each SC. FISH has greatly facilitated assembly of the tomato genome (Stack et al., 2009). *(Image courtesy of Stephen M. Stack.)* (See colour plate.)

Cytology/cytogenetics predated MPS by roughly 100 years. However, both cytogenetics and MPS are methodologies still in their infancies. Elucidation of genome sequences is likely to become extremely cheap and fast, and there is no reason to believe it will not continue to accelerate at a tremendous rate. Cytogenetics, a much more difficult 'art', will continue on at a slower pace. Ultimately, however, all the sequence in the world will not be enough to truly understand plant genomes. *In vivo*, DNA is always associated with numerous proteins. It is packaged into chromatin fibres, which themselves are highly coiled and folded into higher-level chromatin loops. It is within the context of chromosomes and nuclear chromatin that genes function and evolve in living cells. Cytology and cytogenetics will become increasingly important as we try and study genomes in their natural environment.

4. GENOME SIZE DETERMINATION

The relativity principle in connection with the basic Maxwellian equations demands that the mass should be a direct measure of the energy contained in a body; light transfers mass.

Albert Einstein as quoted in Bell (1996)

Einstein's theory of relativity established the interconnection and interconvertibility of mass and energy and in all physical sciences, including biology, mass and energy are two of the most readily measured and compared features of systems. For example, long before sequencing DNA was a reality, genome size (i.e. mass) was utilized to compare species and gain insight into molecular evolution (Gregory, 2005).

While eukaryotes exhibit a 17-fold difference in transcriptome size (Cavalier-Smith, 2005), variation in 1C DNA content is as great as 66,000-fold (Pellicer, Fay, & Leitch, 2010). In flowering plants, genome sizes vary by at least 2000-fold (Fig. 2.6). The observation that C-value is not correlated with structural complexity was originally deemed the *C-value paradox* (Thomas, 1971). For almost 20 years, the basic assumption of most biologists was that differences in eukaryotic genome sizes reflected differences in gene number and/or gene lengths. The reasons why genomes exhibit such enormous variation in mass have been elucidated; while polyploidy and gene duplication account for some variation in C-value, mobile element and other repeat sequence amplification events account for the vast majority of differences in genome sizes (Bennett & Leitch,

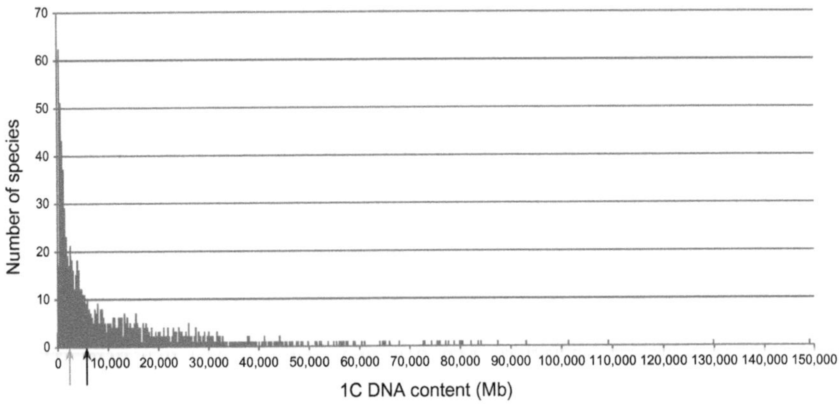

Figure 2.6 Genome size (1C) distribution of the 7541 angiosperm species in the Plant C-values Database (Bennett & Leitch, 2012). The mean genome size of this relatively small sampling of angiosperms is 5666 Mb (black arrow), while the median is 2161 Mb (grey arrow). The largest reported genome belongs to *Paris japonica* (1C=148,852 Mb), while the smallest belongs to *Genlisea margaretae* and *G. aurea* (both 1C=63 Mb).

2005; Oliver, Petrov, Ackerly, Falkowski, & Schofield, 2007). Thus, the *C-value paradox* is now more properly called the *C-value enigma* (Gregory, 2005).

With regard to plants, the collection and analysis of genome size data has been one of the principal undertakings of a small but highly dedicated group of scientists. The central figure in this group is Dr. Michael D. Bennett whose research group at Kew Royal Botanic Gardens generated many genome size estimates, published extensively on the biological implications of genome size, helped develop standards and 'best practice' procedures for determining genome sizes, and oversaw creation of the Plant C-values Database (Gregory et al., 2007), a highly curated online resource featuring prime C-value estimates for numerous plant taxa.[4] As of 10 June 2013, there are genome size estimates for 7541 different flowering plants, 355 gymnosperms, 128 pteridophytes, and 232 bryophytes. Although Dr. Bennett retired in 2006, he is still active in various aspects of plant genome research. Fortunately, the Plant C-values Database has been left in the care of Dr. Ilia Leitch, one of Bennett's highly successful *protégées* and collaborators. Other major contributors to the study of plant C-values include Johann Greilhuber

[4] The animal equivalent of the Plant C-Values Database is the Animal Genome Size Database (www.genomesize.com). The curator of this resource is T. Ryan Gregory (University of Guelph, Canada).

(University of Vienna), Jaroslav Dolezel (Institute of Experimental Botany, Czech Republic), H. James Price (Texas A&M), J. Spencer Johnston (Texas A&M), Thomas Cavalier-Smith (University of Oxford), Richard B. Flavell (Ceres, Inc.), A. Lane Rayburn (University of Illinois), J. S. 'Pat' Heslop-Harrison (University of Leicester), Brian G. Murray (University of Auckland), David W. Galbraith (University of Arizona), Jeffrey Bennetzen (University of Georgia), and Jonathan F. Wendel (Iowa State University).

There are a number of different means for estimating genome size. These include chemical extraction, Feulgen densitometry, flow cytometry (FCM), and reassociation kinetics.

The first genome size estimates were made using a chemical extraction approach. In short, nuclei were isolated from a large quantity of tissue via acid digestion. The total number of nuclei was estimated by sample counts using a hemocytometer or similar device. The nuclear DNA was then isolated from the nuclei and quantified. The quantity of DNA was divided by the estimated number of nuclei to yield a genome size value. In studies utilizing cattle, pig, and guinea pig tissues, Boivin, Vendrely, and Vendrely (1948) and Vendrely and Vendrely (1948) used the chemical extraction approach to show that there is '. . .a remarkable constancy in the nuclear DNA content of all the cells in all the individuals within a species' (translated from French to English by Gregory (2005)). Based on their findings using the chemical extraction approach, the husband and wife team of Roger and Colette Vendrely argued that genome size constancy was evidence that DNA, not protein (which shows considerable variation in quantity from tissue to tissue), was the molecule of heredity. However, the chemical extraction approach was difficult and required considerable amounts of tissue (e.g. 35 kg of cow pancreas was used to get a suitable amount of DNA) (Gregory, 2005) and thus was not practical for species in which tissue was limited. New approaches were needed.

In 1924, Robert Feulgen and Heinrich Rossenbeck published a procedure for preferentially staining DNA in chromosomes fixed to microscope slides (Feulgen & Rossenbeck, 1924). The procedure, which has come to be known as Feulgen staining, utilizes hydrochloric acid to depurinate the DNA followed by treatment with Schiff's reagent (fuchsin–sulphurous acid). The Schiff's reagent turns a purple-red colour when it interacts with the aldehyde groups exposed by depurination. The staining procedure is quantitative (staining intensity is directly related to DNA amount), although considerable care must be taken not to over- or underexpose the chromosomes to HCl. Also, the polyphenolic compounds present in many plant

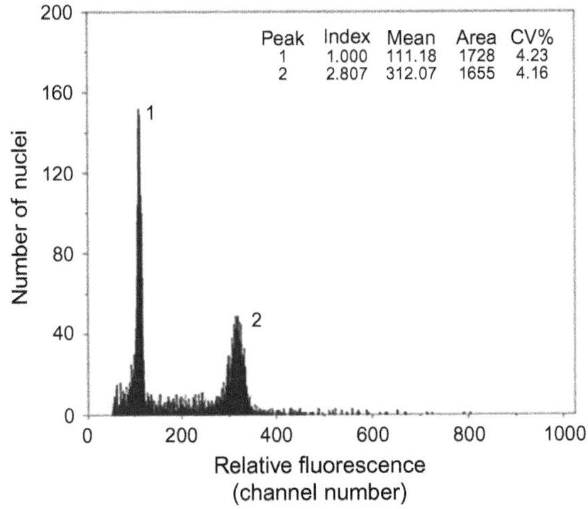

Figure 2.8 Histogram showing G_0/G_1 (i.e. 2C) peaks for *Trillium sessile* (peak 1) and *Paris japonica* (peak 2). *P. japonica* has the largest known genome of any plant species (1C = 148,852 Mb). *Image from Pellicer et al. (2010). Used with permission of John Wiley & Sons and the authors.* (See colour plate.)

2. Nuclei should be prepared according to Galbraith et al. (1983). The procedure, which involves maceration of fresh leaf tissue in a cold buffer, is simple and works well for most plants. Of note, the buffer contains magnesium ions that stabilize nuclear membranes and Triton X-100, a detergent that (when used at a proper concentration) can preferentially lyse chloroplast and mitochondrial membranes while leaving nuclear membranes intact. Additionally, the technique includes elimination of RNA with RNase A. This is important as double-stranded RNA will bind propidium iodide. Alternate buffers are reviewed in Dolezel and Bartos (2005);
3. There is some debate as to what standards should be utilized in FCM-based genome size determinations. In general, it is agreed that plant nuclei should be used as standards if possible (Dolezel & Bartos, 2005; Johnston et al., 1999). Ideally, one could isolate the sample and standard nuclei in a single experiment; that is, mix tissues of the standard and sample together and isolate nuclei from the mixture so as to minimize experimental variation (Dolezel & Bartos, 2005). However, this seems to be less common than adding the standard to the sample after nuclear isolation but prior to staining. Both Johnston et al. (1999) and Dolezel et al. (1998) developed lists of suggested plant standards; of note, the two lists share some plants in common, and the genome sizes calculated for the shared plants are highly similar between the two groups (Dolezel & Bartos, 2005).

Reassociation kinetics (also known as Cot analysis), which is discussed in detail in a later section, can be used to make genome size estimates. However, a Cot analysis is a labour-intensive process requiring relatively large quantities of highly purified DNA, relatively precise mechanical shearing of DNA, several dozen reassociation reactions, and complex calculus-based analysis of the data. The sheer number of steps in a Cot analysis makes it a highly suspect means for obtaining anything other than a 'ballpark' estimate of genome size. This statement is based on my own personal experience with DNA reassociation studies (Liu, Thummasuwan, Sehgal, Chouvarine, & Peterson, 2011; Peterson, Pearson, & Stack, 1998; Peterson, Schulze, et al., 2002; Peterson, Wessler, & Paterson, 2002) and the opinions of some of the central players in animal and plant genome size research (Bennett & Leitch, 1995; Hardie et al., 2002).

In theory, whole genome sequencing would provide the most accurate genome size determinations. However, eukaryotic genome assemblies are typically far from complete and contain gaps (especially at centromeres and in other repetitive regions), often of unknown size. Based on 'whole' genome assembly, The Arabidopsis Genome Initiative (2000) produced a 1C genome size estimate for *Arabidopsis thaliana* of 125 Mb (115.4 Mb in sequence assemblies plus 10 Mb in unsequenced centromeric and rDNA regions). This 125 Mb value was widely cited despite the fact that earlier estimates based on Feulgen densitometry and FCM produced considerably higher 1C values (150–180 Mb). Bennett, Leitch, Price, and Johnston (2003) conducted a study to explore the discrepancies between the sequence-based value and other reported values and convincingly showed that the *Arabidopsis* genome is about 157 Mb (in line with previous FD and FCM data). Until considerable advances are made in sequencing and assembly of genomes, Feulgen densitometry and FCM should be viewed as the most accurate means of determining genome sizes.

5. RENATURATION KINETICS

Roy was interested in genomes more than in particular genes and in DNA more than in particular nucleotide sequences. Underlying all this – again beginning in his time at DTM (Department of Terrestrial Magnetism) when he and his colleagues carried out some of the first interspecific comparisons of genomes by hybridization (1963) – was an abiding interest in physical clues that would explain evolution of the genome.

Davidson (2012)

As discussed in the section on cytology earlier, a picture is worth at least 10^5 words. However, there are many things that are not amenable to photography (at least at present). Long before whole eukaryotic genomes could be sequenced, Roy Britten and his colleagues were using sophisticated DNA reassociation studies to 'visualize' the structure of genomes. Their method involved no cameras, but the 'images' produced during their research revolutionized understanding of the molecular structure of genomes. It was Britten's group that discovered that eukaryotic genomes contain numerous repetitive DNA sequences (Britten & Kohne, 1968). The enormity of this discovery is difficult to overstate. During the period between elucidation of the genetic code and the publication of Britten's experiments, it was widely believed that all DNA coded for proteins and that genomes consisted of different, unrelated genes. The enormous observed variation in eukaryote genome sizes made little intuitive sense, and a number of complex theories arose to explain why some relatively 'simple' life forms needed more genes to maintain their mundane lifestyles. Dr. Britten's work effectively showed that eukaryotic genomes possess considerable amounts of repetitive DNA sequences, some of which occur in thousands to millions of copies per 1C genome. Subsequent work by Britten and others confirmed that differences in C-values among eukaryotes are primarily due to differences in repetitive DNA content. The molecular nature of the C-value paradox had been determined without any DNA sequencing, although the origin and function of repeats was far from clear at the time.[5]

Britten produced 'images' of genomes using a technique now known as *Cot analysis*. Cot is a neologism derived from the mathematical equation C_0t where C_0 is DNA concentration in moles per litre at time zero and t is reassociation time in seconds.[6] In a Cot analysis, sheared DNA in solution is heated until the molecular forces holding complementary base pairs together are disrupted and the two strands of the double-helix dissociate (denature). If the DNA is then cooled to a temperature roughly 25 °C below its melting point, the single-stranded sequences will begin to reassociate with complementary strands. A series of genomic DNA samples are reassociated to different Cot values, and, for a given Cot value, the fraction of genomic

[5] I had the pleasure of corresponding with Dr. Britten on several occasions (via phone and e-mail). He graciously provided me with advice on how to conduct Cot analyses. Dr. Britten died in 2012 at the age of 92.

[6] An additional factor accounting for the cation concentration of the buffer is required if reassociation is carried out in a solution other than 0.12 M sodium phosphate buffer (Britten, Graham, & Neufeld, 1974).

DNA that has reassociated is elucidated by separating the double- and single-stranded DNA by hydroxyapatite chromatography and quantifying the amount of DNA in both fractions (usually by spectrophotometry). When the percentage of DNA remaining single-stranded is graphed against the logarithm of Cot for Cot values ranging from ~0% reassociation to nearly complete reassociation, the resulting profile is called a *Cot curve*.

In initial studies of bacterial genomes, Britten et al. found that the DNA for a given species had a sigmoid-shaped Cot curve in which 80% of the sequences reannealed within a 'two Cot decade region' centred at the curve's Cot½ value (i.e. the Cot value at which 50% of a sample has reassociated). For a particular bacterial species, the Cot½ of its genome was shown to be proportional to its genome size (see Fig. 2.9A).

Eukaryotic genomes are so much larger than those of prokaryotes that Britten et al. predicted that eukaryotic genomes would not exhibit any reassociation unless samples were reassociated to extremely high Cot values. However, when they actually conducted experiments on eukaryotes, they found a considerable fraction of the DNA reassociated at a rate a million times faster than predicted. Britten hypothesized that this quickly reannealing DNA was composed of repeat sequences. In their seminal paper, Britten and Kohne (1968) showed that eukaryotic genomes were composed of complex mixtures of sequences of different iterations. Through study of Cot curves of different eukaryotic species, it was discovered that eukaryotic genomes tend to be composed of several distinct kinetic components—specifically, highly repetitive, moderately repetitive, and single-/low-copy DNA (Fig. 2.9B and C). Of note, Britten's group found that genome size could be predicted from a Cot curve of a eukaryote by comparing the Cot½ value of its slowest reassociating (single-copy) component to the genome size and Cot½ value of *E. coli* (Peterson, 2005).

For roughly a decade, Cot analysis was the primary means of characterizing genomes at the molecular level. When molecular cloning and DNA sequencing technologies emerged in the late 1970s, Cot analysis, a notoriously difficult technique, was quickly abandoned. However, the information about genome structure gained from Cot analysis was considerable, and the principles of DNA reassociation elucidated by Britten and his colleagues form the basis of virtually all subsequent molecular techniques (e.g. PCR, primer-based DNA sequencing, microarrays, macroarrays, and Southern blotting).

In the early part of the twenty-first century, Cot analysis experienced something of a revival. Using a Cot curve as a guide, hydroxyapatite

markers is calculated by analysis of the genotypes/phenotypes of the progeny. Figure 2.10 shows generation of a linkage map from a 'three-point cross', a technique developed in Sturtevant's work and still the basis of most linkage mapping today (Johnson, 1995).

The first molecular markers were genes associated with visually assayable traits. In his map of the fruit fly X chromosome, Sturtevant used an eye colour gene, a gene associated with wing shape, and a gene for body colour as markers (Sturtevant, 1913). However, it is difficult to find enough

A Three linked markers X, B, and Q with dominant alleles *X*, *B*, and *Q* and recessive alleles *x*, *b*, and *q*

B Parent 1 genotype *xxbbqq* — Test cross X — Parent 2 genotype *XxBbQq*

Potential gametes produced by Parent 1: *xbq*

Potential gametes produced by Parent 2: *XBQ*, *XbQ*, *XBq*, *xBQ*, *xbQ*, *xbq*, *Xbq*, and *xBq*

C

Possible Genotypes	*XxBbQq*	*XxbbQq*	*XxBbqq*	*xxBbQq*	*xxbbQq*	*xxbbqq*	*Xxbbqq*	*xxBbqq*
Corresponding Phenotypes	XBQ	XbQ	XBq	xBQ	xbQ	xbq	Xbq	xBq
Observed frequency	110	850	6	60	6	110	60	850

D 2 most common phenotypes, XbQ and xBq, are deemed *parentals*
2 least common phenotypes, XBq and xbQ, are *double crossovers*

E Comparison of the *double crossover* class XBq with its most similar *parental* counterpart xBq reveals that only the X has changed between the *parental* and the *double crossover*. Since a double crossover will only result in a change in the middle locus, X likely occurs between B and Q. Comparison of the other *double crossover* with the other *parental* phenotype (xbQ and XbQ) again implicates X as the middle marker. Thus the marker order is BXQ.

F

	Parental	*Double crossover*	*Single crossover* between B & X		*Parental*	*Double crossover*	*Single crossover* between X & Q	
Phenotype	bXQ	BXq	BXQ	bxq	Bxq	bxQ	BxQ	bXq
Frequency	850	6	110	110	850	6	60	60

G The single crossover group that differs at only one marker from a parental is identified (e.g., BXQ differs at only one marker from bXQ while BxQ differs at only one marker from Bxq). Note that the difference between BXQ and bXQ is at the B locus. As the loci are in order, a crossover between B and X is responsible for this single crossover class. The same would be true of its reciprocal, bxq. As shown in the table above, these single crossover classes have been grouped together with their appropriate parental and double crossover classes. Conversely, BxQ differs from Bxq at the Q locus suggesting a crossover between X and Q, and its reciprocal, bxq, would also have been produced by a single crossover between X and Q.

H Recombination between B & X: Sum four classes that have crossovers between B and X, i.e., single-crossover classes BXQ and bxq and the double-crossover classes BXq and bxQ. Adding together the frequencies of these four classes gives us 232 offspring that have crossovers between B and X genes. There are 2052 offspring total, so there is (232/2052 =) 0.11 or 11% recombination between B and X.

Recombination between X & Q: Sum four classes that have crossovers between X and Q, i.e., bxQ, BxQ, and bXq, and divide by total progeny number [i.e., (60 + 60 + 6 + 6)/2052 = 0.06]. There is 6% recombination between X and Q.

I
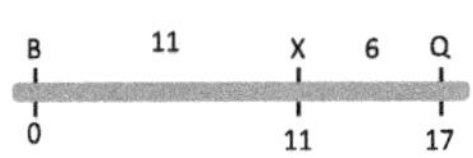

Figure 2.10 Construction of genetic map using three-point cross method. (A) Markers utilized. (B) Test cross. (C) Expected genotypes and phenotypes and their observed frequency (2052 offspring total). Note that in this case each genotype has a unique phenotype. (D) Identification of parental and double crossover classes. (E) Determination of marker order. (F) Phenotypic frequencies of offspring rewritten to reflect relative gene order. (G) Classification of single crossover classes. (H) Calculation of linkage distances. (I) Linkage map showing relative order and recombination distance between markers.

nonlethal morphological markers to make a linkage map of any density. Moreover, such maps can only be constructed using traits under the control of a single gene.

For roughly 50 years, linkage maps were constructed solely using morphological markers. However, in the late 1950s, Clement Markert and Freddy Moller discovered '. . .different variants of the same enzyme having identical functions and present in the same individual' (Markert & Moller, 1959). These variants, which they deemed *isozymes*, could be separated from each other by gel electrophoresis and isozyme bands could be visualized by colorimetric reactions centred on the enzymatic properties of the isozyme class being studied. In their original definition, the term *isozyme* included variants in proteins coded by a single locus (i.e. alleles) and variants produced by different loci (i.e. paralogous genes). However, allelic variants are now referred to as *allozymes*, while the term *isozymes* is used to describe similar proteins produced by different loci (Zeidler, 2000). As electrophoresis methods became more sophisticated, allozymes that differed by even a single amino acid residue could be differentiated by electrophoresis, and consequently allozymes were adopted as a new form of codominant marker and the first true molecular markers (Berg & Hamrick, 1997). Though allozymes are still utilized, the number of genes for which allozymes can be identified is too small to generate high-density molecular maps (Schlotterer, 2004).

Logically, the highest resolution maps could be obtained by comparing DNA sequences directly. However, until recently, such direct comparisons were not affordable. Consequently, a number of indirect DNA-based molecular mapping techniques have been developed. As these have been the subjects of numerous reviews (Avise, 2004; Duran, Appleby, Edwards, & Batley, 2009; Duran et al., 2010; Henry, 2012; Paterson, 1996; Xu, 2012), they will only be mentioned briefly in the succeeding text.

In 1980, David Botstein and his collaborators described construction of a linkage map based upon DNA polymorphisms in restriction endonuclease sites (Botstein, White, Skolnick, & Davis, 1980). The markers, known as restriction fragment length polymorphisms or RFLPs, allowed mapping at polymorphic loci without any prerequisite knowledge about the function of the loci and/or the gene products, if any. In RFLP mapping, random DNA and/or cDNA clones were hybridized to blots of gels of restriction digested DNA from numerous related and unrelated individuals. If the pattern of hybridization for a particular clone differed in a manner consistent with a change in a single locus, the clone could be utilized as an RFLP marker. As codominant markers, RFLPs were as informative as allozymes. RFLPs have been especially useful in plant genetic mapping (Paterson, 1996).

With the advent of PCR, new DNA-based markers were developed including randomly amplified polymorphic DNAs, AFLPs, and simple sequence repeats (SSRs). SSRs, also called microsatellites, are highly conserved, short (2–6 bp), tandem repeat sequences. Many SSRs are preferentially associated with genes. Primers are designed to amplify SSRs and polymorphisms are differentiated by gel electrophoresis. Because microsatellites are codominant and easy to assay, they have become particularly popular markers in forensics, positional cloning, kinship testing, etc. (Richard, Kerrest, & Dujon, 2008).

Eventually, advances in DNA sequencing made direct sequence-based genetic mapping possible. The single-nucleotide polymorphism or SNP has become the molecular marker of choice. SNP markers are much more numerous than other markers allowing for generation of extremely dense linkage maps. SNPs have provided depth sufficient for high-quality mapping of quantitative trait loci and association mapping studies (Duran et al., 2010).

Genetic mapping is unlikely to disappear anytime soon. While whole genome resequencing may eventually become the means by which maps are made, this is unlikely to occur until accurate ultralong read DNA sequencing is both possible and affordable. More importantly, maps, by their very nature, are simplified representations of more complex things. A map with too much detail is as worthless as a map without enough detail. Whatever the source of data used to construct genetic maps, the maps themselves need to serve as a tractable guide through a genome, not the genome itself.

7. PHYSICAL MAPPING

Since DNA sequencing was developed in the 1970s (Maxam & Gilbert, 1977; Sanger, Nicklen, & Coulson, 1977), the idea of sequencing whole genomes, including the often vast spaces between markers/genes, became a goal. At the time, shotgun sequencing was far too expensive (and too low throughput) to be considered as an affordable path to a genome sequence. Physical mapping, the identification of relatively large pieces of partially overlapping DNA collectively spanning a chromosome or part of a chromosome, was deemed a practical way of assembling a genome in a fashion that would allow eventual complete sequencing. The first eukaryotic genomes to be sequenced were sequenced using a physical mapping approach. Physical mapping is highly resource-intensive, but those genome sequences based on physical maps are still unparalleled with respect to quality and accuracy. A highly simplified outline of steps in building a physical map is as follows:

1. Large-insert genomic libraries are prepared for the organism of interest. The first large-insert libraries to be used in physical mapping were yeast artificial chromosomes or YACs. YACs can carry pieces of insert DNA up to 3 Mb, but roughly 50% of the time YAC inserts are chimeric (i.e. derived from two or more fragments of DNA). In 1992, bacterial artificial chromosomes (BACs) were introduced as alternatives to YACs (Shizuya et al., 1992). BACs are derived from bacterial F-factors. They have several features that make them exceptional cloning vectors. Specifically, they can be isolated like any other bacterial plasmid, they contain partitioning genes that prevent more than one BAC from inhabiting a cell at a time, and they are maintained in only one copy per cell (the latter two features help curb chimerism). BACs can carry inserts of 100 kb or longer (up to 350 kb), only 1/10 as much as can be carried in a YAC. However, the simplicity of BAC manipulation and the lack of chimerism have made BACs the vector of choice for physical mapping (Peterson, Tomkins, Frisch, Wing, & Paterson, 2000).[7]
2. BAC clones are digested with restriction enzymes and the fragments are visualized (conventionally on agarose gels but more recently on acrylamide that permits resolution of greater numbers of fragments and, in principle, increases statistical power) to produce BAC fingerprints.
3. Automated Sanger cycle sequencing with vector-specific primers is used to generate end sequences of BAC clones. The distance between end sequences is estimated by pulse-field gel electrophoresis.
4. BAC libraries are gridded onto high-density nylon macroarrays, and macroarrays are probed with molecular genetic markers (e.g. RFLPs), cDNAs, etc. The BACs recognized by each probe are determined.
5. BAC end sequences are used to design oligonucleotide probes that are then used to screen the macroarrays and find partially overlapping clones.
6. Using computational methods and relational databases, probe hybridization, fingerprint, BAC end data, and genetic maps are integrated to

[7] P1-derived artificial chromosomes (PACs) and fosmids can also be used in physical mapping. PACs can carry inserts of similar size to those of BACs, but have not been as widely used. Fosmids can only carry inserts of ca. 40 kb limiting their use in large-scale physical mapping projects.

Britten, R. J., Graham, D. E., & Neufeld, B. R. (1974). Analysis of repeating DNA sequences by reassociation. *Methods in Enzymology, 29*, 363–418.
Britten, R. J., & Kohne, D. E. (1968). Repeated sequences in DNA. *Science, 161*, 529–540.
Brown, T. A. (2007). *Genomes 3* (3rd ed.). New York, NY: Garland Scientific Publishing.
Carlson, E. A. (2013). H.J. Muller's contributions to mutation research. *Mutation Research, 752*, 1–5.
Carpenter, A. T. C. (1975). Electron microscopy of meiosis in *Drosophila melanogaster* females: II: The recombination nodule—A recombination-associated structure at pachytene? *Proceedings of the National Academy of Sciences of the United States of America, 72*, 3186–3189.
Cavalier-Smith, T. (2005). Economy, speed and size matter: Evolutionary forces driving nuclear genome miniaturization and expansion. *Annals of Botany, 95*, 147–175.
Conklin, E. G. (1915). August Weismann. *Proceedings of the American Philosophical Society, 54*, iii–xii.
Crow, E. W., & Crow, J. F. (2002). 100 years ago: Walter Sutton and the chromosome theory of heredity. *Genetics, 160*, 1–4.
Davidson, E. H. (2012). Retrospective. Roy J. Britten (1919-2012). *Science, 335*, 1183.
Dillon, N. (2004). Heterochromatin structure and function. *Biology of the Cell, 96*, 631–637.
Dobzhansky, T. Y. (1937). *Genetics and the origin of species*. New York, NY: Columbia University Press.
Dolezel, J., & Bartos, J. (2005). Plant DNA flow cytometry and estimation of nuclear genome size. *Annals of Botany, 95*, 99–110.
Dolezel, J., Greilhuber, J., Lucretti, S., Meister, A., Lysak, M. A., Nardi, L., et al. (1998). Plant genome size estimation by flow cytometry: Inter-laboratory comparison. *Annals of Botany, 82*(Suppl. A), 17–26.
Dolezel, J., Sgorbati, S., & Lucretti, S. (1992). Comparison of three DNA fluorochromes for flow cytometric estimation of nuclear DNA content in plants. *Physiologia Plantarum, 85*, 625–631.
Duran, C., Appleby, N., Edwards, D., & Batley, J. (2009). Molecular genetic markers: Discovery, applications, data storage and visualisation. *Current Bioinformatics, 4*, 16–27.
Duran, C., Eales, D., Marshall, D., Imelfort, M., Stiller, J., Berkman, P. J., et al. (2010). Future tools for association mapping in crop plants. *Genome, 53*, 1017–1023.
Feulgen, R. J., & Rossenbeck, H. (1924). Mikroskopisch-chemischer Nachweis einer Nucleinsäure vom Typus der Thymonucleinsäure und die darauf beruhende elektive Färbung von Zellkernen in mikroskopischen Präparate. *Hoppe-Seyler's Zeitschrift fur physiologische Chemie, 135*, 203–248.
Flemming, W. (1878). Zur Kenntniss der Zelle und ihrer Theilungs-Erscheinungen. *Schriften des Naturwissenschaftlichen Vereins für Schleswig-Holstein, 3*, 23–27.
Galbraith, D. W., Harkins, K. R., Maddox, J. M., Ayres, N. M., Sharma, D. P., & Firoozabady, E. (1983). Rapid flow cytometric analysis of the cell cycle in intact plant tissues. *Science, 220*, 1049–1051.
Goldberg, R. B. (1978). DNA sequence organization in the soybean plant. *Biochemical Genetics, 16*, 45–68.
Gregory, T. R. (2005). Genome size evolution in animals. In T. R. Gregory (Ed.), *The evolution of the genome* (pp. 3–87). Boston, MA: Elsevier Academic Press.
Gregory, T. R., & DeSalle, R. (2005). Comparative genomics in prokaryotes. In T. R. Gregory (Ed.), *The evolution of the genome* (pp. 585–673). Boston, MA: Elsevier Academic Press.
Gregory, T. R., Nicol, J. A., Tamm, H., Kullman, B., Kullman, K., Leitch, I. J., et al. (2007). Eukaryotic genome size databases. *Nucleic Acids Research, 35*, D332–D338.
Greilhuber, J. (2008). Cytochemistry and C-values: The less-well-known world of nuclear DNA amounts. *Annals of Botany, 101*, 791–804.

Grigoryev, S. A., & Woodcock, C. L. (2012). Chromatin organization—The 30 nm fiber. *Experimental Cell Research, 318*, 1448–1455.

Haiminen, N., Feltus, F. A., & Parida, L. (2011). Assessing pooled BAC and whole genome shotgun strategies for assembly of complex genomes. *BMC Genomics, 12*, 194.

Hardie, D. C., Gregory, T. R., & Hebert, P. D. (2002). From pixels to picograms: a beginners' guide to genome quantification by Feulgen image analysis densitometry. *Journal of Histochemistry & Cytochemistry, 50*, 735–749.

Heitz, E. (1928). Das heterochometin der Moose. *Jahrbiicher fur Wissenschaftliche Botanik, 69*, 762–818.

Henry, R. J. (2012). *Molecular markers in plants*. Hoboken, NJ: Wiley-Blackwell.

Hey, J. (2004). What's so hot about recombination hotspots? *PLoS Biology, 2*, e190.

Hoskins, R. A., Smith, C. D., Carlson, J. W., Carvalho, A. B., Halpern, A., Kaminker, J. S., et al. (2002). Heterochromatic sequences in a *Drosophila* whole-genome shotgun assembly. *Genome Biology, 3,* research0085-0085.16.

Islam-Faridi, M. N., Childs, K. L., Klein, P. E., Hodnett, G., Menz, M. A., Klein, R. R., et al. (2002). A molecular cytogenetic map of sorghum chromosome 1. Fluorescence in situ hybridization analysis with mapped bacterial artificial chromosomes. *Genetics, 161*, 345–353.

Johnson, G. B. (1995). *How scientists think*. Dubuque, IA: William C. Brown Publishing.

Johnston, J. S., Bennett, M. D., Rayburn, A. L., Galbraith, D. W., & Price, H. J. (1999). Reference standards for determination of DNA content of plant nuclei. *American Journal of Botany, 86*, 609–613.

Kiper, M., & Herzfeld, F. (1978). DNA sequence organization in the genome of *Petroselinum sativum* (Umbelliferae). *Chromosoma, 65*, 335–351.

Koszul, R., Meselson, M., Van, D. K., Vandenhaute, J., & Zickler, D. (2012). The centenary of Janssens's chiasmatype theory. *Genetics, 191*, 309–317.

Lamoureux, D., Peterson, D. G., Li, W., Fellers, J. P., & Gill, B. S. (2005). The efficacy of Cot-based gene enrichment in wheat (*Triticum aestivum* L.). *Genome, 48*, 1120–1126.

Lederberg, J., & McCray, A. T. (2001). Ome sweet omics—A genealogical treasury of words. *The Scientist, 15*, 8.

Levene, P. A. (1919). The structure of yeast nucleic acid. IV. Ammonia hydrolysis. *Journal of Biological Chemistry, 40*, 415–424.

Lhuissier, F. G., Offenberg, H. H., Wittich, P. E., Vischer, N. O., & Heyting, C. (2007). The mismatch repair protein MLH1 marks a subset of strongly interfering crossovers in tomato. *Plant Cell, 19*, 862–876.

Liu, W., Thummasuwan, S., Sehgal, S. K., Chouvarine, P., & Peterson, D. G. (2011). Characterization of the genome of bald cypress. *BMC Genomics, 12*, 553.

Maeshima, K., Hihara, S., & Eltsov, M. (2010). Chromatin structure: Does the 30-nm fibre exist in vivo? *Current Opinion in Cell Biology, 22*, 291–297.

Markert, C. L., & Moller, F. (1959). Multiple forms of enzymes: Tissue, ontogenetic, and species specific patterns. *Proceedings of the National Academy of Sciences of the United States of America, 45*, 753–763.

Martins, L. A. C. P. (1999). Did Sutton and Boveri propose the so-called Sutton-Boveri chromosome hypothesis? *Genetics and Molecular Biology, 22*, 261–272.

Maxam, A. M., & Gilbert, W. (1977). A new method for sequencing DNA. *Proceedings of the National Academy of Sciences of the United States of America, 74*, 560–564.

Mirsky, A. E., & Ris, H. (1951). The desoxyribonucleic acid content of animal cells and its evolutionary significance. *Journal of General Physiology, 34*, 451–462.

Morgan, T. H. (1910). Sex limited inheritance in *Drosophila*. *Science, 32*, 120–122.

Moses, M. J. (1956). Chromosomal structures in crayfish spermatocytes. *The Journal of Biophysical and Biochemical Cytology, 2*, 215–218.

Moses, M. J. (1958). The relation between the axial complex of meiotic prophase chromosomes and chromosome pairing in a salamander (Plethodon cinereus). *The Journal of Biophysical and Biochemical Cytology*, *4*, 633–638.

Naegeli, W., & Wiedemann, H. R. (1993). Charles Darwin and other great men in correspondence with Carl Wilhelm von Naegeli. *American Journal of Medical Genetics*, *46*, 236–243.

Nickerson, J. A. (2001). Experimental observations of a nuclear matrix. *Journal of Cell Science*, *114*, 463–474.

Nierman, W., Eisen, J. A., & Fraser, C. M. (2000). Microbial genome sequencing 2000: New insights into physiology, evolution and expression analysis. *Research in Microbiology*, *151*, 79–84.

Nikolaitchik, O. A., Dilley, K. A., Fu, W., Gorelick, R. J., Tai, S. H., Soheilian, F., et al. (2013). Dimeric RNA recognition regulates HIV-1 genome packaging. *PLoS Pathogens*, *9*, e1003249.

Olby, R. (1974). DNA before Watson–Crick. *Nature*, *248*, 782–785.

Olins, D. E., & Olins, A. L. (2003). Chromatin history: Our view from the bridge. *Nature Reviews Molecular Cell Biology*, *4*, 809–814.

Oliver, M. J., Petrov, D., Ackerly, D., Falkowski, P., & Schofield, O. M. (2007). The mode and tempo of genome size evolution in eukaryotes. *Genome Research*, *17*, 594–601.

Pardue, M. L., & Gall, J. G. (1969). Molecular hybridization of radioactive DNA to the DNA of cytological preparations. *Proceedings of the National Academy of Sciences of the United States of America*, *64*, 600–604.

Paterson, A. H. (1996). Making genetic maps. In A. H. Paterson (Ed.), *Genome mapping in plants* (pp. 23–39). Georgetown, TX: R.G. Landes Company and Academic Press, Inc.

Pellicer, J., Fay, M. F., & Leitch, I. J. (2010). The largest eukaryotic genome of them all? *Botanical Journal of the Linnean Society*, *164*, 10–15.

Peterson, D. G. (2005). Reduced representation strategies and their application to plant genomes. In K. Meksem & G. Kahl (Eds.), *The handbook of genome mapping: Genetic and physical mapping* (pp. 307–335). Weinheim: Wiley-VCH Verlag GmbH & Co.

Peterson, D. G., Lapitan, N. L. V., & Stack, S. M. (1999). Localization of single- and low-copy sequences on tomato synaptonemal complex spreads using fluorescence *in situ* hybridization (FISH). *Genetics*, *152*, 427–439.

Peterson, D. G., Pearson, W. R., & Stack, S. M. (1998). Characterization of the tomato (*Lycopersicon esculentum*) genome using *in vitro* and *in situ* DNA reassociation. *Genome*, *41*, 346–356.

Peterson, D. G., Schulze, S. R., Sciara, E. B., Lee, S. A., Bowers, J. E., Nagel, A., et al. (2002). Integration of Cot analysis, DNA cloning, and high-throughput sequencing facilitates genome characterization and gene discovery. *Genome Research*, *12*, 795–807.

Peterson, D. G., Tomkins, J. P., Frisch, D. A., Wing, R. A., & Paterson, A. H. (2000). Construction of plant bacterial artificial chromosome (BAC) libraries: An illustrated guide. *Journal of Agricultural Genomics*, *5*. http://wheat.pw.usda.gov/jag/.

Peterson, D. G., Wessler, S. R., & Paterson, A. H. (2002). Efficient capture of unique sequences from eukaryotic genomes. *Trends in Genetics*, *18*, 547–550.

Posner, E., & Skutil, J. (1968). The great neglect: The fate of Mendel's classic paper between 1865 and 1900. *Medical History*, *12*, 122–136.

Richard, G. F., Kerrest, A., & Dujon, B. (2008). Comparative genomics and molecular dynamics of DNA repeats in eukaryotes. *Microbiology and Molecular Biology Reviews*, *72*, 686–727.

Rubin, G. M., & Lewis, E. B. (2000). A brief history of Drosophila's contributions to genome research. *Science*, *287*, 2216–2218.

Sanger, F., Nicklen, S., & Coulson, A. R. (1977). DNA sequencing with chain-terminating inhibitors. *Proceedings of the National Academy of Sciences of the United States of America, 74*, 5463–5467.

Satou, Y., Mineta, K., Ogasawara, M., Sasakura, Y., Shoguchi, E., Ueno, K., et al. (2008). Improved genome assembly and evidence-based global gene model set for the chordate *Ciona intestinalis*: New insight into intron and operon populations. *Genome Biology, 9*, R152.

Sax, K. (1938). Chromosome aberrations induced by X-rays. *Genetics, 23*, 494–516.

Sax, K. (1940). An analysis of X-ray induced chromosomal aberrations in *Tradescantia. Genetics, 25*, 68.

Schlotterer, C. (2004). The evolution of molecular markers—Just a matter of fashion? *Nature Reviews Genetics, 5*, 63–69.

Sherman, J. D., & Stack, S. M. (1995). Two-dimensional spreads of synaptonemal complexes from solanaceous plants. VI. High-resolution recombination nodule map for tomato (*Lycopersicon esculentum*). *Genetics, 141*, 683–708.

Shizuya, H., Birren, B., Kim, U.-J., Mancino, V., Slepak, T., Tachiiri, Y., et al. (1992). Cloning and stable maintenance of 300-kilobase-pair fragments of human DNA in *Escherichia coli* using an F-factor-based vector. *Proceedings of the National Academy of Sciences of the United States of America, 89*, 8794–8797.

Singh, R. J. (2003). *Plant cytogenetics*. Boca Raton, FL: CRC Press.

Stack, S. M. (1991). Staining plant cells with silver. II. Chromosome cores. *Genome, 34*, 900–908.

Stack, S. M., Royer, S. M., Shearer, L. A., Chang, S. B., Giovannoni, J. J., Westfall, D. H., et al. (2009). Role of fluorescence *in situ* hybridization in sequencing the tomato genome. *Cytogenetics and Genome Research, 124*, 339–350.

Stack, S. M., Shearer, L., Szinay, D., van de Belt, J., & de Jong, J. H. (2013). Cytogenetics of the tomato genome sequence. In *Abstract presented at the Solanaceae Workshop, Plant & Animal Genome XXI. San Diego, CA.*

Stadler, L. J. (1928). Mutations in barley induced by X-rays and radium. *Science, 68*, 186–187.

Steuernagel, B., Taudien, S., Gundlach, H., Seidel, M., Ariyadasa, R., Schulte, D., et al. (2009). *De novo* 454 sequencing of barcoded BAC pools for comprehensive gene survey and genome analysis in the complex genome of barley. *BMC Genomics, 10*, 547.

Sturtevant, A. H. (1913). The linear arrangement of six sex-linked factors in *Drosophila*, as shown by their mode of association. *Journal of Experimental Zoology, 14*, 43–59.

Sturtevant, A. H. (1965). *A history of genetics*. New York, NY: Harper and Row.

Sutton, W. S. (1902). On the morphology of the chromosome group in *Brachystola magna. Biological Bulletin, 4*, 24–39.

Sutton, W. S. (1903). The chromosomes in heredity. *Biological Bulletin, 4*, 231–251.

Swift, H. (1950). The constancy of desoxyribose nucleic acid in plant nuclei. *Proceedings of the National Academy of Sciences of the United States of America, 36*, 643–654.

The Arabidopsis Genome Initiative (2000). Analysis of the genome sequence of the flowering plant *Arabidopsis thaliana. Nature, 408*, 796–815.

Thomas, C. A. (1971). The genetic organization of chromosomes. *Annual Review of Genetics, 5*, 237–256.

Tomato Genome Consortium (2012). The tomato genome sequence provides insights into fleshy fruit evolution. *Nature, 485*, 635–641.

Vendrely, R., & Vendrely, C. (1948). La teneur du noyau cellulaire en acide désoxyribonucléique à travers les organes, les individus et les espèces animales. *Experientia, 15*, 434–436.

Watson, J. D., & Crick, F. H. C. (1953). Molecular structure of nucleic acids: A structure for deoxyribose nucleic acid. *Nature, 171*, 737–738.

standard of gene and genome analysis is nucleic acid sequencing, and the technology to generate sequence information has evolved very rapidly since the first plant genes were isolated, amplified, and subjected to molecular analysis. Among the first plant genes to be isolated, cloned, and sequenced were ribosomal RNA and tRNA genes from chloroplasts and mitochondria (Bawnik, Beckmann, Sarid, & Daniel, 1983; Bohnert, Gordon, & Crouse, 1980; Delihas, Andersen, Sprouse, & Dudock, 1981; Graf, Roux, Stutz, & Kossel, 1982), seed storage genes (Lycett et al., 1983; Schuler, Schmitt, & Beachy, 1982; Vodkin, 1983; Vodkin, Rhodes, & Goldberg, 1983; Wiborg, Hyldig-Nielsen, Jensen, Paludan, & Marcker, 1983), genes transferred by infecting pathogens (Bevan, Barnes, & Chilton, 1983; De Greve et al., 1982; Dhaese et al., 1983; Heidekamp, Dirkse, Hille, & van Ormondt, 1983; Klee et al., 1984; Leemans et al., 1981; Simpson et al., 1982), genes encoding major structural proteins (Shah, Hightower, & Meagher, 1983), and genes encoding abundant enzymes (Dunsmuir, Smith, & Bedbrook, 1983; Shinozaki, Yamada, Takahata, & Sugiura, 1983; Spielmann & Stutz, 1983) or enzymes in the biochemical pathway of visible pigments. Genes encoding the regulatory and enzymatic proteins responsible for visual markers, like the maize purple anthocyanin pigments, were isolated and subjected to DNA sequencing relatively early in the history of plant gene discovery (Chandler, Radicella, Robbins, Chen, & Turks, 1989; Lechelt et al., 1989; Ludwig, Habera, Dellaporta, & Wessler, 1989; Paz-Ares, Wienand, Peterson, & Saedler, 1986). These visible markers were important in the early studies of maize genetics and transposable elements by researchers like Barbara McClintock (McClintock, 1984).

DNA sequencing followed the development of cloning technology that allowed isolated pieces of plant (or animal) DNA to be cloned into plasmid or bacteriophage vectors and amplified in appropriate bacterial host cells like *Escherichia coli*. Hybridization with homologous sequences from nonplant species was often used to identify evolutionarily conserved genes from plants. Cloning and amplification were necessary steps to create a sufficient mass of the targeted genes for early DNA sequencing technologies. Sequencing of tens to hundreds of nucleotides was a very laborious manual process compared to currently available automated sequencing technology. Early sequencing technologies relied on either chemical degradation approaches (Maxam–Gilbert sequencing; Maxam & Gilbert, 1977, 1980, 1992) or enzymatic chain-termination methods (Sanger sequencing; Sanger, Nicklen, & Coulson, 1977, 1992). Both techniques required separation of DNA fragments with single nucleotide size differences in long polyacrylamide gels using high-voltage potential differences. The very early technologies used radioactively labelled nucleotides and exposure of X-ray films to create banding patterns that could be 'read' by biologists as nucleic acid sequences. The early sequencing technologies provided approaches to sequence a few hundred to a few thousand bases at a time in increments of tens to a few hundred bases per single sequencing reaction. Multiple sequencing reactions were run on a single polyacrylamide gel, typically overnight. Many of the early sequencing labs used significant amounts of ^{32}P-labelled chemicals that contaminated portions of the labs for months to even years.

The drive to determine the sequence of the human genome created a significant motivation and opportunity to enhance the efficiency of nucleic acid sequencing technology. Both the chemistry of DNA sequencing and the instruments for automating the process were significantly enhanced with high rates of funding from the National Institutes of Health (NIH) and National Science Foundation (NSF). Better DNA polymerase enzymes for chain-termination sequencing and fluorescent nucleotides, as well as polymerase chain reaction amplification technologies, set the stage for automating Sanger sequencing. Applied Biosystems Incorporated, along with other companies, designed and developed the first automated DNA sequencing instruments in collaboration with academic groups. These early instruments also used polyacrylamide slab gels poured manually between glass plates and laser-stimulated fluorescence with one or more labelled nucleotide(s) and sensitive cameras to automate 'reading' of the DNA sequence.

2. RANDOM SEQUENCING OF EXPRESSED GENES FOR DISCOVERY

Following the realization that random complementary DNA (cDNA) sequencing could be very useful for gene discovery in mammals (Adams et al., 1991; Adams, Kerlavage, Fields, & Venter, 1993), this 'expressed sequence tag' or EST sequencing approach was rapidly adopted for gene discovery across many species. Pharmaceutical companies supported the approach either internally or through service companies/organizations like Human Genome Sciences (HGS) and Incyte Pharmaceuticals. Hundreds of thousands of random cDNAs were sequenced from a broad variety of animal and human tissues, and this led to similar EST sequencing efforts for plant gene discovery. Large agribusiness corporations contracted HGS and Incyte to generate EST sequences from various plant tissues and from different stages of development. A similar approach was undertaken in the public sector supported by NSF. The National Center for Biotechnology Information (NCBI) established a public database for ESTs and a software repository for analysis tools (Boguski, Lowe, & Tolstoshev, 1993). The targeted plant species were initially a small handful of model systems like Arabidopsis (Asamizu, Nakamura, Sato, & Tabata, 2000b; Cooke et al., 1996; Hofte et al., 1993; Neuteboom et al., 1999; Newman et al., 1994; Rounsley et al., 1996) and important crop plants like maize, rice (Wu et al., 2002; Yamamoto & Sasaki, 1997), and tomato (Yamamoto et al., 2005). In addition to Arabidopsis and the major crops, gene discovery by EST sequencing has been applied to a wide variety of different plant species due to its low cost and its ability to rapidly identify transcribed genes and primary metabolism genes found in all tissues. EST sequencing of plant genes has also been used to rapidly identify polymorphisms and allelic variants in a number of crop varieties and model system accessions. For example, EST datasets have been created for *Brassica napus* (Park et al., 1993), eucalyptus (Tagu & Martin, 1995), lotus (Asamizu, Nakamura, Sato, & Tabata, 2000a, 2005; Endo et al., 2000; Poulsen & Podenphant, 2002), castor beans (Van De Loo, Turner, & Somerville, 1995), Chinese cabbage (Lim et al., 1996), *Brassica campestris* (Kwak, Kim, Hong, & Nam, 1997), *Medicago truncatula* (Covitz, Smith, & Long, 1998), poplar (Sterky et al., 1998), apple (Sung et al., 1998), green alga (Asamizu, Nakamura, Sato, Fukuzawa, & Tabata, 1999), moss (Machuka et al., 1999), petunia (Yu et al., 1999), and dozens of other plant species as a means to discover new genes or identify genes expressed in specific tissues. These early EST collections were generated with Sanger chain-termination sequencing technology with newly available automated DNA sequencers. ESTs were often used in hybridization reactions to create linkage points (sequence-tagged sites) between the physical map of the genome and bacterial artificial chromosome (BAC) clones or contigs (Wu et al., 2002) and are still used to validate computational gene predictions and the extent of genome coverage.

3. EVOLUTION OF DNA SEQUENCING BEYOND SANGER METHODOLOGY

Gene sequencing and early approaches to sequencing expressed genes began in individual research labs and targeted relatively small regions of the genome or a small number of expressed genes. As sequencing technologies became more automated, it also became more applicable to larger and larger genomic regions or even entire genomes. The high expense of the equipment and the expertise to run the instruments and analyse sequences forced the larger projects into a relatively small number of genome centres. The Genome Institute at Washington University, DOE's Joint Genome Institute in Walnut Creek, the Broad Institute at MIT, the Sanger Institute in the United Kingdom, Beijing Genomics and Informatics (BGI) group in China, and the Genome Sciences Center at RIKEN in Japan are all examples of large genome centres established to sequence, assemble, and analyse genomes with rapidly evolving technology. Despite the very large investments made in capital equipment, genome sequencing is a difficult process and no eukaryotic genomes are truly 'complete'. Repetitive DNA in centromeres or located throughout the genome, by various natural events, creates a technical challenge for both sequencing and accurate assembly. Long repeats or long stretches of short repeats are often unstable when cloned for amplification in bacteria, are difficult to accurately align computationally, and can be problematic in the sequencing reactions. A high-quality genome sequence is one with a minimal number of missing regions, but it remains unclear what is actually still missing in some high-quality plant genomes. It is also much more expensive to generate a high-quality genome sequence relative to a draft sequence, and 'finishing' the last few percent of a genome can account for the majority of the overall cost. For this reason, most plant genomes sequenced to date are 'draft' sequences rather than 'complete' genomes.

Due to massive investments in enhancing DNA sequencing technologies, with the eventual goal to use sequencing as a medical diagnostic procedure, the costs of generating a genome sequence have been driven down several orders of magnitude. They are continuing to decrease in cost. The finished base cost 15 years ago was approximately one dollar but now is considerably less than one cent. Some estimates state that the cost of sequencing has decreased five orders of magnitude. The decreasing cost of genome sequencing is dramatically changing the landscape as well as the logistics of genome sequencing. It is much more common now to find high-throughput DNA sequencing instruments in institutional core facilities available as a service or even in single investigator labs. The decreased cost has also had an impact on the targets for sequencing. More plant varieties and more phylogenetically diverse plants are being sequenced. Sequencing technologies are rapidly advancing towards single-molecule approaches that eliminate amplification and towards massively parallel technologies that create a large amount of data in a single instrument run.

by Ecker (1998) and Martienssen and McCombie (2001). Sequencing entire genomes has the advantage of being able to discover genes that are expressed only under specific conditions that are not commonly recreated in the lab or production field. Another advantage of genome sequencing is providing information on the conserved sequences that do not encode proteins.

Arabidopsis was sequenced chromosome by chromosome by first constructing a BAC library from large fragments (~80–100 kbp) of the entire genome and then mapping individual BACs to one of the five Arabidopsis chromosomes. Thousands of BACs were 'fingerprinted' with restriction enzymes (cleaved to reveal the sizes of the fragments on an agarose gel) and assembled into an overlapping set of BACs representing coverage of the BAC library on each specific chromosome. This 'map-based' approach allows researchers to choose minimally overlapping BAC clones for DNA sequencing and allows very accurate sequencing with high chromosome coverage and the minimum number of gaps. A potential artefact of this map-based approach is that some BACs may contain repetitive DNA that is unstable in the bacterial amplification step or segments of DNA that are toxic to bacteria and therefore become underrepresented in the original library.

A 1.9 Mb region of Arabidopsis chromosome 4 was published before the genome was complete since this represented the largest contiguous plant genome sequence delineated and analysed to date (Bevan et al., 1998). Several similar papers described mapping- and sequencing-specific megabase regions of the Arabidopsis genome (Asamizu et al., 1998; Kaneko et al., 1999, 1998; Kotani, Hosouchi, & Tsuruoka, 1999; Kotani, Nakamura, et al., 1997; Kotani, Sato, et al., 1997; Nakamura et al., 1998; Quigley, Dao, Cottet, & Mache, 1996; Sato et al., 1998; Tremousaygue et al., 1997; Tutois et al., 1999; Williams, Clark, & Meyerowitz, 1999). It was clearly unique to generate and analyse megabase regions of any genome. Chromosomes 2 and 4 (of five Arabidopsis chromosomes) were published (Lin et al., 1999; Mayer, Lemcke, Schuller, Rudd, & Zaccaria, 2000; Mayer et al., 1999) prior to completion of the entire genome in December 2000 (Arabidopsis Genome, 2000).

Although the Arabidopsis genome was sequenced with this high-accuracy 'gold-standard' approach, it still has a few small coverage gaps and problem areas of the genome that are difficult or impossible to sequence with the current technology. This is true of any plant genome sequenced to date and will likely remain true well into the future. Nevertheless, the Arabidopsis genome represents the highest-quality and most complete plant genome sequenced, and the real challenge is the function of the sequence, its encoded proteins, and many other functional elements. See Chapter 4 for more information on the functional analysis of the Arabidopsis genome and the excellent review by Martienssen and McCombie (2001) published the year after completion of the genome sequence.

Since the publication of the Arabidopsis genome, the number of computationally predicted protein-coding genes has varied between approximately 25,000 and 30,000 genes. The human genome and possibly the *Caenorhabditis elegans* genome are probably the only equivalent genomes in terms of sequence quality and completion. Note that the Arabidopsis genome project (both the mapping and sequencing phases) took approximately 10 years at a total estimated cost of $100 million. The genome coverage is very high, making the Arabidopsis genome the gold standard of plant genomes. This is comparable to the cost of sequencing similar size genomes such as *C. elegans* or *Drosophila melanogaster*. Shotgun sequencing of genomes (random fragment approaches) had been proposed as a strategy to lower the cost and time required to sequence entire genomes (Venter et al., 1998; Venter, Smith, & Hood, 1996), but counterarguments were made against this somewhat risky approach (Green, 1997). Clone-by-clone approaches were considered to have a high probability of success and achieve the highest-quality outcome so were originally supported as the more reliable, low-risk approach. However, now, most genome projects are executed by the whole genome shotgun (WGS) approach.

In 1997, a meeting organized by Ron Phillips and Mike Freeling was held and an Interagency Working Group, targeted at sequencing the first cereals, was created. At this meeting, Takuji Sasaki from Japan announced that the Japanese government was initiating a 7-year programme to sequence the rice genome (reviewed by Briggs & Helentjaris, 1997). Whole genome sequencing of plants was beginning to be taken seriously, even though the cost remained relatively high with the available technology. It was frequently estimated to cost one dollar per finished base pair at that time, placing the rice genome project at over $500M using a clone-by-clone map-based approach (see further details later).

The rice genome became the next major target for whole genome sequencing and the first crop plant to be sequenced. Not only is rice a very important cereal (60–70% of staple crop production comes from rice, maize, and wheat) but also is an excellent model for cereal genomics; its relatively small genome and available genetic toolkit made it of interest to public and corporate groups. Both Monsanto and Novartis (later to become Syngenta) had privately funded efforts in cereal gene discovery targeted at sequencing the rice genome. The public effort to sequence the rice genome was an international effort led by researchers in Japan and was focused on generating a high-accuracy complete genome sequence very similar in scope and goals to the Arabidopsis genome project. Japonica rice was chosen as the variety to sequence since this international project was led by researchers in Japan where japonica varieties predominate. Indica rice represents about 80% of global rice production, and a second public effort initiated by the BGI generated a draft sequence of indica (see in the succeeding text).

The Rockefeller Foundation funded the construction of the japonica BAC library at the Clemson University Genomics Institute, an effort led by Rod Wing and collaborators (Chen et al., 2002). Novartis funded the fingerprinting and BAC-end sequencing of 60,000 clones that were used by the International Rice Genome Sequencing Project. Fingerprinting the BACs was accomplished using ethidium bromide staining of DNA fragments separated by agarose gel electrophoresis followed by image analysis and computational assembly using the software Contig FPC (Soderlund, Longden, & Mott, 1997). Monsanto established a contract with Leroy Hood's lab at the University of Washington and funded an effort to sequence 3391 japonica rice BAC clones via the clone-by-clone approach. This effort used unmapped BACs, and due to overlap of the individual clones, it represented approximately two-thirds (259 Mb) of the complete rice genome (420 Mb) (Barry, 2001). Both Novartis and the BGI used a whole genome random fragment approach (referred to as a WGS) strategy to generate the primary sequence data for assembly and mapping to the BAC contigs described earlier (Goff et al., 2002; Yu et al., 2002). By this time, automated sequencers had progressed to support chain-termination Sanger sequencing with fluorescent fragments separated inside 96- or 384-capillary arrays. The sequencing reactions were carried out in commercially available PCR thermocyclers and loaded into the automated sequencing instruments available from Applied Biosystems or Amersham in 96- or 384-well thermocycler plates. Plates were typically created by robotic picking of bacterial colonies carrying subclones of randomly sheared BACs prior to generating the amplified sequencing reactions. The automation and barcoding of sample plates helped cut down on human errors typical of large-scale projects.

The International Rice Genome Sequencing Project (IRGSP) was scheduled to complete the map-based sequencing of rice by 2008, but the project was accelerated and completed in 2005 (International Rice Genome Sequencing Project, 2005). Monsanto and Syngenta sequences were used by some partners in the IRGSP to deepen the coverage of specific regions of the japonica genome. The BGI created deeper sequencing depth, to both their published indica and japonica random fragment data, and reassembled both to create higher-quality genome sequences of the indica genome sequence and the Syngenta japonica genome sequence (Yu et al., 2005). Together, the various rice projects established its genome sequence as one of the highest-quality reference genomes available today. This first crop genome sequence generated a great deal of news (Bennetzen, 2002; Cantrell & Reeves, 2002; Goff, 2002; Leach, McCouch, Slezak, Sasaki, & Wessler, 2002; Normile & Pennisi, 2002; Ronald & Leung, 2002) and brought the BGI to the forefront of plant genome sequencing (Normile, 2002).

The challenge posed by the whole genome random fragment or WGS approach to sequencing was the assembly of the resulting fragments into contiguous segments representing a portion of the genome (contiguous sequences or 'contigs') and mapping of those contigs to the specific locations in the chromosome they represent. Random fragment approaches were routine for individual BACs that are typically 100 kb in size and had been applied to many prokaryotic genomes up to a few megabases (Mb) in size. The BGI and Syngenta rice genome projects were the first attempts to use a random fragment approach on a genome in the hundreds of megabases (420–450 Mb for rice). Assembly of the Syngenta raw sequence data was accomplished by binning overlapping random fragment sequences (typically a few hundred to one thousand bases long) and then assembling them in small groups since assembly of the entire set of sequences together was considered computationally impractical. The initial binning of sequences was a very computational intensive step, particularly for the time period these data were generated. The software and computational systems available today are much better at handling this scale of assembly, even with much greater depth of coverage and smaller error-prone raw sequences.

Since the publication of the rice genome, several rice varieties representing the spectrum of genetic diversity of rice have been resequenced in the Oryza Map Alignment Project (Jacquemin, Bhatia, Singh, & Wing, 2013). In addition, 50 accessions of rice, including ten genetically diverse ancestral species, have been resequenced (Xu et al., 2012). The BGI has announced that it has resequenced a few thousand varieties of rice and is targeting the resequencing of >100k varieties to low depth coverage.

sequences rather than repetitive). These pooled BAC clones were subjected to random fragment sequencing and assembly using Sanger sequencing. In addition, some clone-by-clone sequencing was performed. The genome coverage was estimated at two-thirds of the entire genome, while the gene coverage was estimated at >91%. In a related resource development effort, a high-density linkage map of Lotus was created from 94 F6 individuals of a cross between two varieties (Wang, Sato, Tabata, & Kawasaki, 2008). Together, the reference genome and the linkage maps provide a resource for phenotype–genotype association studies, although at relatively low coverage compared to other genome projects.

The *Sorghum bicolor* genome was published in 2009 (Paterson et al., 2009). It was the largest plant genome sequenced to this point (~730 Mb) and was sequenced to a depth of 8.5-fold using standard Sanger sequencing technology in a WGS approach. Small insert libraries in plasmids (2–3 kb and 5–8 kb), a fosmid library, and two BAC libraries (13.5-fold genome coverage) were used to create sequence covering 97.3% of the genome. Approximately 90% of the genome was aligned to 127 scaffolds and contained 1476 fingerprinted contigs. This is very good coverage for a random fragment whole genome sequencing approach, especially considering the large size of the genome.

The cucumber genome (*Cucumis sativus* L.) was sequenced with low-coverage Sanger sequences (3.85-fold coverage) and greater than 68-fold coverage with Illumina short reads (42–53 base average lengths) (Huang et al., 2009). More than 350,000 ESTs created using Roche 454 technology, along with other ESTs, were used to determine the 96.8% genome coverage, although the genome sequence covered an estimated 30% less than the predicted genome size. This novel approach using a mixture of older and newer sequencing technologies is one route towards reducing the high cost of plant genome sequencing. Many of the plant genome projects published after this point used combinations of different sequencing technologies to accelerate the project and drive down the costs. Mixing sequencing technology platforms is becoming much more common as new computational approaches for assembly and analysis become available. These new technologies and the drivers behind their introduction will now be described.

6. EXAMPLES OF THE MANY RECENTLY PUBLISHED PLANT GENOME PROJECTS

An increasing number of plant genomes have been published since 2009. Some of these projects were started and even finished with older approaches like WGS Sanger sequencing and construction and anchoring of the assembled sequence using BAC fingerprinting, contiging, and end sequencing. However, many of the newly published genomes use a mixture of platforms or even novel, less expensive technologies to create the published genome products. These will be briefly reviewed here in the order in which they were published.

The genome of one of the most important crops, and a critical model cereal plant, maize, was published in 2009 (Feuillet & Eversole, 2009). This was one of the largest genomes sequenced to date to create a high-quality reference genome. It will probably remain one of the few in this class, although the sequence diversity within maize and the presence–absence differences between varieties make it difficult to claim that any single specific maize variety is *the* representative reference genome. The Maize Genome Sequencing Consortium was launched over a decade ago with the specific aims to generate a complete maize genome sequence integrated with the physical and the genetic maps (Coe et al., 2002; Cone et al., 2002; Davis et al., 1999; Fu et al., 2006; Lee et al., 2002; Sharopova et al., 2002; Wei et al., 2007, 2009). Even an optical map of the maize genome has been created as a part of this project (Zhou et al., 2009). See optical mapping technology description in the succeeding text. In addition, a haplotype map of variation has been generated using 27 genetically diverse lines and next-generation sequencing of a reduced but representative genome space (Gore et al., 2009). The maize physical map consists of approximately 17k BACs in 435 contigs covering 93% of the genome (version #1) (Wei et al., 2009). Approximately 94% of this physical map is ordered and oriented. This is very high coverage considering the repetitive nature of the maize genome. The maize genome was sequenced in a clone-by-clone high-accuracy approach using the physical map and minimally overlapping clones. BAC clones were sequenced via standard Sanger sequencing with four- to sixfold coverage (Schnable et al., 2009). Small insert plasmids and large insert fosmids were sequenced to generate approximately 380 Mb of finished unique sequence and 61,161 scaffolds covering 2.3 Gb. An estimated 250 Mb of genome sequence is missing from version 1 and most of that is missing also from the physical map, suggesting it to be repetitive sequence. Maize is likely to represent the most challenging plant genome sequenced to date, but larger genomes like wheat will face similar technical hurdles.

Resequencing additional maize varieties beyond the reference sequence (of the genotype B73) has been initiated and will be done in an accelerated time frame as sequencing costs continue to decrease and both sequencing and assembly technologies improve (Chia et al., 2012; Ganal et al., 2011). Next-generation sequencing is being used to identify polymorphisms and determine genotypes for molecular breeding. In fact, next-generation sequencing has become inexpensive enough to stimulate the development of 'genotype-by-sequencing' technology (Elshire et al., 2011). Coupled with association studies, these tools and datasets will provide an accelerated route for molecular breeding and crop variety improvement.

The castor bean (*Ricinus communis*) genome was published in 2010 (Chan et al., 2010). A genetic map of the castor bean did not exist, and at the time, limited genomic information was available. The J. Craig Venter Institute and the Institute for Genome Science generated 2.1 million Sanger sequence reads from plasmid and fosmid libraries representing approximately 4.6-fold coverage of the estimated 329 Mb genome. As expected from this relatively low fold coverage, the sequence assemblies clustered into 25,800 scaffolds (3500 larger than 2 kb). This genome is of interest because it encodes the ribosome-inactivating ricin toxin subunits.

In 2010, the genomes of domesticated soybean *Glycine max* (Schmutz et al., 2010), undomesticated soybean (*Glycine soja* Sieb. and Zucc) (Kim et al., 2010), the model grass *Brachypodium distachyon*, and the domesticated apple (*Malus* x *domestica* Borkh.) were published. Domesticated soybean (*G. max*) and Brachypodium were both sequenced by the standard Sanger approach with >15 million reads assembled into 3363 scaffolds for soybean (950 Mb, 85% of the genome) and 9.43-fold coverage of the 272 Mb genome for Brachypodium (International Brachypodium Initiative, 2010). The raw sequence read data for the *G. soja* project were generated with short reads from Illumina GA II technology (35 and 76 bp reads) and with Roche 454 technology. The assembled *G. max* genome was used as a reference genome to assemble short reads from *G. soja*.

The apple genome (Velasco et al., 2010) sequencing project incorporated a mixture of Sanger sequencing to form the larger sequence read scaffolds and 454 sequencing to fill in gaps and build coverage depth. The apple genome has an estimated genome size of 742.3 Mb. Sequence coverage was 16.9-fold total (26% by Sanger long reads and 74% by Roche 454 technology). BACs and fosmids were used for large insert sequencing templates and plasmids with 2–12 kb inserts were used for the 454 sequencing templates. A genetically diverse collection of soybean varieties have also been resequenced to identify genetic diversity and help build useful tools for molecular breeding applications (Lam et al., 2010). These genome projects show that the introduction of next-generation sequencing technology, to increase depth of coverage on a standard Sanger assembly, created a greater and less expensive depth of coverage.

The *Medicago* Genome Initiative was established over 10 years ago to sequence this model legume (Bell et al., 2001). The genome was sequenced in a collaborative project between researchers in the United States, France, the United Kingdom, the Netherlands, Germany, Belgium, and Nigeria (Young et al., 2011). This project used a variety of different technologies at different centres. Seed BACs for initial sequencing were identified by hybridization to expressed sequence tags, and the initial sequencing was done using Sanger chain-termination approaches on 3–6 kb plasmid vectors. Subsequent sequencing was done using pyrosequencing approaches with evolving technologies provided by 454, first 105 bp reads and later 210 bp reads (GSFLX and Titanium technologies from 454 Life Sciences, respectively). They also used short Illumina reads (36, 54, and 76 nucleotides) to generate depth of coverage across the whole genome using a shotgun strategy on 0.375 and 5 kb mate-paired libraries. This strategy created approximately 27 Gb of sequence for depth of coverage. To align contig sequences to the physical map of *Medicago*, optical mapping was used (Anantharaman, Mishra, & Schwartz, 1999; Aston, Mishra, & Schwartz, 1999; Samad, Huff, Cai, & Schwartz, 1995; Schwartz & Samad, 1997). Approximately one million single DNA restriction maps were generated from large chromosomal fragments to create 26 contigs between 958 kb and 38.6 Mb that cover 377.97 Mb of the genome. This project demonstrated the use of mixed sequencing technologies as well as a unique mapping technology.

In addition to sequencing the genome to provide a reference sequence, the collaborators are also using next-generation sequencing to identify diversity in 384 genetically diverse inbred lines (see http://www.medicagohapmap.org/?genome). This increasingly common approach will facilitate the identification of allelic variants, single nucleotide polymorphisms (SNPs), insertions/deletions, and copy number variants. Thirty of the *Medicago* diversity panel lines have been sequenced at 20 × coverage or more, while the remaining lines are being sequenced to a depth of at least 5 ×. These sequences should provide an excellent set of materials for introducing genetic diversity into breeding materials.

Theobroma cacao (the source of chocolate) was sequenced by a global collaborating group and published in 2011 (Argout et al., 2011). Members of this collaboration had previously published on the transcriptome of Theobroma (Argout et al., 2008). The team generated 26 Gb from over 26 million Roche 454 and ~400 million Illumina technology raw sequencing reads. They also generated paired BAC-end Sanger sequences of 44k clones to help

genome with 96.6 Gb of raw data (over 100-fold coverage depth of a genome estimated at ~844 Mb). Sanger-generated BAC-end sequences were used in a standard mapping approach and >181k Sanger EST sequences were used to validate gene predictions and estimate gene coverage. The genome was estimated to be 93.9% covered and the gene space 97.1% covered. This combined technology yields a fairly high-quality genome sequence at a moderate cost, but the decreasing cost of even newer sequencing technologies will displace the use of Sanger and Roche 454 technologies even further.

The *Brassica rapa* (Chinese cabbage) genome was sequenced by an international consortium (Wang et al., 2011). A WGS approach generated 36 Gb of short Illumina reads at 72.36 genome coverage depth and was compared to a conventional clone-by-clone BAC Sanger sequence of a single chromosome done by other researchers (Coe et al., 2002). The authors reported that the BAC-based Sanger sequence generated more gaps (1035) than the WGS assembly (858), although these are close in number. The assemblies matched well, and not surprisingly, the Sanger longer-read approach was superior at sequence coverage in areas of repeats. This project used 44, 75, and 101 bp raw read lengths from the Illumina GA II platform. They also integrated ~200k BAC-end sequences to help create the physical map.

Next-generation sequencing technology was also used to create a draft sequence of the extremophile *Thellungiella parvula* (Dassanayake et al., 2011) and the rubber tree *Hevea brasiliensis* (Rahman et al., 2013). Both projects used Roche 454 and Illumina raw reads to generate relatively low genome coverage (50- and 43-fold, respectively), and both projects were completed and published by small teams. The rubber tree sequencing also used the very short-read SOLiD technology and the overall coverage was relatively low with a large number of gaps. These two projects represent minimalist genome sequencing projects that will need additional efforts to complete the sequence and/or make the available data useful for breeders rather than research molecular biologists.

A very small collaborative team published the genome of *Cannabis sativa* in 2011 (van Bakel et al., 2011). They used Illumina 100 bp reads and 44 bp paired-end raw reads from six libraries (200–660 bp in length) to create 92 Gb of raw sequence data (110-fold coverage of the estimated 820 Mb genome). They also generated 18.8 Gb of raw RNA-Seq data from six tissues and used low-coverage Roche 454 mate pairs to confirm predicted genes. The resulting genome was 136,290 scaffolds, indicating this project is a low-coverage gene discovery effort, but relatively inexpensive. Hemp was also scan sequenced to determine the difference between cannabinoid-producing and nonproducing varieties (van Bakel et al., 2011).

A draft genome sequence of the African staple crop cassava was published in early 2012 (Prochnik et al., 2012). This project was done by a relatively small team and was initiated as the Global Cassava Partnership in 2003. Approximately 800k Sanger sequencing reads and >70k Roche 454 reads were used to create >29-fold coverage of the genome. The assembled sequence represents approximately 69% of the genome and 96% of the protein-coding space. Existing simple sequence repeats and SNPs between a genetically diverse cross were used to generate the physical map. The cassava genome is estimated to be 770 and 532.5 Mb is covered with 12,977 scaffolds. Half the genome is covered by only 487 of these scaffolds. Although this project was published in 2012, it was initiated and planned in a relatively early phase of next-generation sequencing and relied mainly on expensive and time-consuming Sanger technology.

A draft sequence of another African staple, the pigeon pea (*Cajanus cajan*) genome, was published in early 2012 (Varshney et al., 2012). Next-generation sequencing (Illumina GA II and HiSeq, 2000) technology was used to create 163.4-fold coverage or 130.7 Gb of raw data (50–100 bp in length) that covered an estimated 72.7% of the genome and 97% of the transcriptome (from 21,434 assembled transcripts). The *de novo* assembled sequence was anchored to the genomic map using 88,860 BAC-end sequences (Sanger sequencing technology on 41,302 pairs of BAC ends) and aligned to the genetic map with 833 markers. The coverage represents 578 Mb and is present in 6534 scaffolds. The approaches used represent a common strategy to generate high coverage of the genome and the transcriptome at a reasonable cost and are useful for breeding since it is anchored to the physical and genetic maps. A likely improvement will be to eliminate the Sanger sequencing used to create the physical map with either long single-molecule sequencing (e.g. Pacific Biosciences' technology) or optical mapping.

Foxtail millet and a related wild species (*Setaria italica* and *Setaria viridis*) were sequenced by an international group and published in mid-2012 (Bennetzen et al., 2012). This crop is a major cereal staple in developing nations. An inbred variety of *S. italica* (Yugu1) was used to create the reference Setaria genome, and the map was created by a cross of two divergent lines. Sanger sequencing created >63k ESTs and Roche 454 technology created an additional 1.2 million ESTs. 784 million RNA-Seq reads were generated on the Illumina GA II platform, so several different sequencing technologies were used in this project. The raw genome sequence data were generated by Sanger technology (5.7 million reads resulting in 8.29-fold genome coverage). The sequence contig length was just over 400 Mb and the scaffold length approximately 406 Mb, representing 80% of the estimated 510 Mb genome. Due to the use of Sanger long-read technology, the genome is represented in only 6791 sequence contigs assembled into 336 scaffolds. This is very good coverage with a high degree of linkage, but probably one of the last traditional Sanger-based genome sequencing projects to be published. To generate the genetic map, 992 SNPs between the two diverse species were used.

Interestingly, a second genome project was published for *S. italica* and was undertaken by the BGI and collaborators in China and Denmark (Zhang, Liu, et al., 2012). This project used Illumina GA II and HiSeq 2000 technology in a WGS approach with 127-fold depth. Several different library sizes ranging from 170 bp to 40 kb were used, and they also sequenced a second variety to generate SNPs to create the genetic map. Optical mapping was used to generate the physical map and 16,903 sequence contigs were anchored to 258 scaffolds. Coverage of genes was estimated at 96%. This project was likely considerably faster and significantly less expensive to execute than the Setaria project described earlier, but the coverage and accuracy appear to be less. The trade-off in expense and resources required is probably well worth the compromise in coverage and quality given that the end product of both projects is likely to create roughly the same value to the end user.

The salt-tolerant plant *Thellungiella salsuginea* has recently been sequenced by the BGI and their collaborators (Wu et al., 2012). They generated 34.6 Gb of raw data on the Illumina GA II platform resulting in 134-fold coverage of the estimated 260 Mb genome. The assembly covered 233.7 Mb in 2682 scaffolds (90% of the genome). Since they did not have a genetic or physical map, they used comparative chromosomal painting to assign sequence contigs to the genome using Arabidopsis as a reference genome. 515 scaffolds were anchored onto 7 chromosomes to attain 80% genome coverage. The approaches used in this genome project represent a minimal effort to acquire a genome sequence and map but do require a closely related reference with a defined genome.

The genome of *Cucumis melo* L. (melon) was sequenced by a collaborative group mainly located in Spain (Garcia-Mas et al., 2012). A double haploid was used to avoid problems with heterozygosity. Approximately 30k sequence contigs were assembled from ~22 million 454 reads and 53k

Sanger BAC-end sequences. The 450 Mb genome was covered with a 13.52-fold depth of raw data and 90% of the genome was reported to be in 78 of 1594 scaffolds. As in other genome projects done by WGS approaches, the sequence coverage had gaps where transposable elements were predicted.

Flax (*Linum usitatissimum*) was sequenced and assembled *de novo* using entirely next-generation (Illumina GA II) sequencing on various-sized libraries (Wang, Hobson, et al., 2012). 34.98 Gb of raw data was generated and filtered for quality to create 69.4-fold genome coverage. Only 25.88 Gb of the raw sequence was used for assembly of the estimated 373 Mb genome. 286,852 ESTs were used to estimate coverage at 93–95%. Previous efforts had created a BAC-based physical map using Sanger sequencing (Fu et al., 2006). An estimated 81% of the genome was covered by sequence assemblies and 85% covered by scaffolds. The authors conclude that next-generation sequencing can be used to efficiently create a genome *de novo*, but the long-range accuracy of the assembly still needs further improvement. Many of the subsequent genome projects using next-generation sequencing, as the main source of raw data, created greater depth of coverage to achieve higher levels of assembly and accuracy. The trend appears to be towards coverage depths in excess of 100-fold and occasionally significantly higher than this.

A draft diploid cotton genome (*Gossypium raimondii*) was published by a relatively small collaborating group from China, Denmark, and the United States (Wang, Wang, et al., 2012). The majority of the raw sequence data was generated using the Illumina paired-end sequencing on several libraries ranging in size from 170 bp to 40 kb. The raw data represented 103.6-fold coverage of the genome estimated to be 775.2 Mb. In total, 41,307 assembled sequence contigs were joined in 4715 scaffolds and 567.2 Mb was anchored to the physical map. The cotton D-genome physical map was published previously using a large BAC library (>92k clones), BAC high information content fingerprinting and anchoring to the genetic map (Davis et al., 1999). For the D-genome, 8064 BACs were used for paired-end sequencing by Sanger technology and contigs were aligned to the Arabidopsis and grape genomes. A total of 3973 contigs were aligned to the physical map. The genome sequencing effort achieved 93.4% coverage of 58,061 ESTs in the assembled sequence data, resulting in moderate coverage of the genome.

Another cotton genome project focused on higher-contiguity of *G. raimondii* and compared it to draft sequences for several additional

The genome of the fruit tree *Prunus mume* (Chinese plum or Japanese apricot) was sequenced and assembled by a combination of next-generation sequencing and optical mapping (Zhang, Chen, et al., 2012). Illumina GA II technology was used to create >100-fold genome coverage using short insert libraries ranging from 180 to 800 bp and large insert libraries from 2 to 40 kb. The genome size of 280 Mb was estimated using K-mer statistics, and the genome coverage by 1449 scaffolds was estimated to be 84.6%. Optical mapping with two independent restriction enzyme digests (BamHI and NdeI) and 300-fold depth of coverage created the physical map. The physical map was aligned and oriented using 779 SNPs identified by RAD on a segregating F1 hybrid. One of the primary lessons of this project is the advantage of a highly homozygous line for next-generation sequencing. The authors were aware from previous projects that heterozygosity in the genome sequenced will create problems with assembly, so they searched for a highly inbred and homozygous individual, test sequenced it to confirm high homozygosity, and then proceeded with the sequencing.

The watermelon (*Citrullus lanatus*) genome project added a very useful effort to create a draft genome sequence and 20 resequenced varieties to discover SNPs useful for molecular breeding efforts (Guo et al., 2013). The collaborative team from BGI, Boyce Thompson Institute, UC Davis, USDA, and Copenhagen generated 108.6-fold coverage of the estimated 425 Mb genome using the Illumina sequencing platform. They generated approximately 1 million ESTs and anchored 330 Mb of sequence to the genome (93.5% of the assembled sequence). The resequenced varieties help clarify the population structure of breeding materials for use in future enhanced breeding strategies. The publication includes a very useful table comparing the *de novo* assembly of watermelon with the assembly of eight other plant genome projects that used only Illumina, a combination of Illumina and other next-generation technologies, or a combination of Sanger and other next-generation technologies. Table 3.1 shows the trends in genome projects towards less expensive approaches and the team size used to complete the various plant genome projects. The use of new sequencing technologies does not need to be at the expense of genome assembly or quality.

The peach genome (*Prunus persica*) is another example of a fruit tree genome project executed by an international consortium from the United States, Italy, France, and several other countries (The International Peach Genome Initiative et al., 2013). The project was executed by a standard Sanger technology approach with 8.5-fold coverage and 454 resulting

Table 3.1 A list of published plant genomes, the year of publication, the number of team members on the publication (where known), and the various technologies used to generate the genome sequence

Year	Species	References	Technology used	Author #
2000	*Arabidopsis thaliana*	Arabidopsis Genome (2000)	S, SB, CC	146
2002	Rice (*Oryza sativa* L. ssp. japonica)	Goff et al. (2002)	S, SB, WGS	55
2002	Rice (*Oryza sativa* L. ssp. indica)	Yu et al. (2002)	S, CC	100
2005	Rice (*Oryza sativa* L. ssp. japonica)	International Rice Genome Sequencing Project (2005)	S, SB, CC	249
2006	Black cottonwood (*Populus trichocarpa*)	Tuskan et al. (2006)	S, SB, WGS	110
2007	Grapevine (*Vitis vinifera*)	Jaillon et al. (2007)	S, SB, WGS	57
2007	*Chlamydomonas reinhardtii*	Merchant et al. (2007)	S, SB, WGS	117
2008	Papaya (*Carica papaya* Linnaeus).	Ming et al. (2008)	S, SB, WGS	85
2008	*Lotus japonicus*	Sato et al. (2008)	S, SB, PBCC	29
2008	Moss (*Physcomitrella*)	Rensing et al. (2008)	S, WGS	71
2009	Cucumber (*Cucumis sativus* L.)	Huang et al. (2009)	S, R, W, I, GS	96
2009	Maize (*Zea mays*)	Schnable et al. (2009)	S, CC, OM	157
2009	*Sorghum bicolor*	Paterson et al. (2009)	S, SB, WGS	45
2010	Apple (Malus x domestica Borkh.)	Velasco et al. (2010)	S, SB, R, WGS	86
2010	*Brachypodium distachyon*	International Brachypodium Initiative (2010)	S, WGS	174
2010	Castor bean (*Ricinus communis*)	Chan et al. (2010)	S, WGS	18

Continued

Table 3.1 A list of published plant genomes, the year of publication, the number of team members on the publication (where known), and the various technologies used to generate the genome sequence—cont'd

Year	Species	References	Technology used	Author #
2010	Soybean (*Glycine soja* Sieb. and Zucc.)	Kim et al. (2010)	S, WGS	29
2010	Soybean (*Glycine max*)	Schmutz et al. (2010)	I, R, WGS	45
2011	*Arabidopsis lyrata*	Hu et al. (2011)	S, SB, WGS	30
2011	*Brassica rapa*	Wang et al. (2011)	SB, I, WGS	111
2011	*Theobroma cacao*	Argout et al. (2011)	SB, R, I, WGS	61
2011	*Cannabis sativa*	van Bakel et al. (2011)	R, I, WGS	7
2011	Date palm (*Phoenix dactylifera*)	Al-Dous et al. (2011)	I, WGS	17
2011	*Medicago truncatula*	Young et al. (2011)	S, R, I, WGS, OM	124
2011	Potato (*Solanum tuberosum*)	Potato Genome Sequencing Consortium et al. (2011)	SB, I, WGS	98
2011	*Selaginella moellendorffii*	Banks et al. (2011)	S, WGS	103
2011	Woodland strawberry (*Fragaria vesca*)	Shulaev et al. (2011)	I, L, WGS	72
2011	*Thellungiella parvula*	Dassanayake et al. (2011)	R, I, WGS	11
2012	Banana (*Musa acuminata*)	D'Hont et al. (2012)	S, SB, R, I, WGS	64
2012	Barley (*Hordeum vulgare* L.)	International Barley Genome Sequencing Consortium et al. (2012)	SB, R, I, CC	13
2012	Cassava (*Manihot esculenta*)	Prochnik et al. (2012)	S, SB, R, WGS	12

Table 3.1 A list of published plant genomes, the year of publication, the number of team members on the publication (where known), and the various technologies used to generate the genome sequence—cont'd

Year	Species	References	Technology used	Author #
2012	*Prunus mume*	Zhang, Liu, et al. (2012)	I, WGS, OM	32
2012	Diploid Cotton (*Gossypium raimondii*)	Wang, Wang, et al. (2012)	BS, I, WGS	29
2012	Polyploid Cotton (*Gossypium hirsutum*)	Paterson et al. (2012)	S, BS, R I, WGS	74
2012	Flax (*Linum usitatissimum*)	Wang, Hobson, et al. (2012)	BS, I, WGS	23
2012	Foxtail millet (*Setaria italica*)	Zhang, Chen, et al. (2012)	S, R, WGS	37
2012	*Setaria italica* and *Setaria viridis*	Bennetzen et al. (2012)	I, WGS, OM	34
2012	Melon (*Cucumis melo* L.)	Garcia-Mas et al. (2012)	SB, R, WGS	35
2012	Pigeonpea (*Cajanus cajan*)	Varshney et al. (2012)	SB, I, WGS	31
2012	Tomato (*Solanum lycopersicum*)	Tomato Genome (2012)	S, SB, R, I, CC, PBCC	317
2013	Chickpea (*Cicer arietinum*)	Varshney et al. (2013)	I, WGS	49
2013	Moso bamboo (*Phyllostachys heterocycla*)	Peng et al. (2013)	SB, I, WGS	40
2013	Peach (*Prunus persica*)	The International Peach Genome Initiative et al. (2013)	S, SB, WGS	54
2013	Pear (*Pyrus bretschneideri* Rehd.)	Wu et al. (2013)	BI, WGS, CC	60
2013	Rubber tree (*Hevea brasiliensis*)	Rahman et al. (2013)	R, I, L, WGS	29

Continued

for assembly and analysis approaches that are tuned to the inherit errors or shortcomings (e.g. limited read length or high error rate but longer read length) of each sequencing platform. Combining data from different platforms will be a regular need and an evolving one given the short technology cycle. Resequencing genomes of varieties will be a more common approach towards discovery of allelic diversity and genotype by sequencing will become a gold-standard approach for high-quality mapping and molecular breeding. As clearly seen in the last few years, more plant genome sequencing projects will be done on a larger scale (see Fig. 3.1), to rapidly empower molecular breeding applications globally. An effort to sequence 'orphan crops' for developing countries has already been initiated (see news item at http://www.merid.org/Content/News_Services/Food_Security_and_AgBiotech_News/Articles/2011/Sep/26/a.aspx). Expensive mapping techniques like BAC fingerprinting and end sequencing with Sanger chain-termination approaches will be supplemented or even eventually replaced with optical mapping of restriction enzyme fragments and marker discovery using polymorphisms discovered through targeted sequencing of a small fraction of the genome. Confirmation of draft sequence coverage and validation of gene identification will be achieved by highly efficient transcriptome sequencing. Variety resequencing in crops like switchgrass is being used to discover SNP markers for breeding (Ersoz et al., 2012). In fact, SNP discovery efforts have been applied to many different crops using next-generation sequencing (Bachlava et al., 2012; Blanca et al., 2012; Durstewitz et al., 2010; Ersoz et al., 2012; Gaur et al., 2012; Geraldes et al., 2011; Han et al., 2011; Hendre, Kamalakannan, & Varghese, 2012; Hyten, Cannon, et al., 2010; Hyten, Song, et al., 2010; Oliver et al., 2013; Sharpe et al., 2013; Shirasawa et al., 2010; Sonah et al., 2013; Sun et al., 2013; Trebbi et al., 2011; Trick et al., 2012; Vandepitte et al., 2013; Wu et al., 2010). Likewise, next-generation sequencing of varieties has been used for trait mapping via bulk segregant analysis adaptation (Hu et al., 2012; Salunkhe et al., 2011; Yuan, Sagawa, Di Stilio, & Bradshaw, 2013). The rapid pace of plant genome sequencing will make this review out-of-date in a relatively short time, but online resources will be maintained to keep researchers current on the published and available plant genomes. Many more genomes are available now that have not been published, and these can be found at websites like the JGI's Phytozome (http://www.phytozome.net/), the Genomes Online Database (Pagani et al., 2012) (GOLD, http://www.genomesonline.org/cgi-bin/GOLD/index.cgi), and the CoGepedia link from the comparative genomics web service CoGe maintained by Eric Lyons and

James Schnable (http://genomevolution.org/wiki/index.php/Sequenced_plant_genomes).

In summary, DNA sequencing of plants is only now beginning to cover the rich diversity of species and varieties of interest to plant biologists, crop breeders, and evolutionary ecologists. Efforts in the coming decades promise to deliver massive datasets that will challenge our analyses and interpretive skills as well as create opportunities for cross-disciplinary efforts to both understand the complexity of biology and create rational approaches towards true synthetic genes, pathways, and ultimately organisms.

ACKNOWLEDGEMENT

Stephen A. Goff is funded by the National Science Foundation under The iPlant Collaborative (NSF Grant #DBI-0735191).

REFERENCES

Adams, M. D., Kelley, J. M., Gocayne, J. D., Dubnick, M., Polymeropoulos, M. H., Xiao, H., et al. (1991). Complementary DNA sequencing: Expressed sequence tags and human genome project. *Science*, *252*(5013), 1651–1656.

Adams, M. D., Kerlavage, A. R., Fields, C., & Venter, J. C. (1993). 3,400 new expressed sequence tags identify diversity of transcripts in human brain. *Nature Genetics*, *4*(3), 256–267. http://dx.doi.org/10.1038/ng0793-256.

Al-Dous, E. K., George, B., Al-Mahmoud, M. E., Al-Jaber, M. Y., Wang, H., Salameh, Y. M., et al. (2011). De novo genome sequencing and comparative genomics of date palm (Phoenix dactylifera). *Nature Biotechnology*, *29*(6), 521–527. http://dx.doi.org/10.1038/nbt.1860.

Anantharaman, T., Mishra, B., & Schwartz, D. (1999). Genomics via optical mapping. III: Contiging genomic DNA. *Proceedings International Conference on Intelligent Systems for Molecular Biology*, 18–27.

Arabidopsis Genome, I. (2000). Analysis of the genome sequence of the flowering plant Arabidopsis thaliana. *Nature*, *408*(6814), 796–815. http://dx.doi.org/10.1038/35048692.

Argout, X., Fouet, O., Wincker, P., Gramacho, K., Legavre, T., Sabau, X., et al. (2008). Towards the understanding of the cocoa transcriptome: Production and analysis of an exhaustive dataset of ESTs of Theobroma cacao L. generated from various tissues and under various conditions. *BMC Genomics*, *9*, 512. http://dx.doi.org/10.1186/1471-2164-9-512.

Argout, X., Salse, J., Aury, J. M., Guiltinan, M. J., Droc, G., Gouzy, J., et al. (2011). The genome of Theobroma cacao. *Nature Genetics*, *43*(2), 101–108. http://dx.doi.org/10.1038/ng.736.

Asamizu, E., Nakamura, Y., Sato, S., Fukuzawa, H., & Tabata, S. (1999). A large scale structural analysis of cDNAs in a unicellular green alga, Chlamydomonas reinhardtii. I. Generation of 3433 non-redundant expressed sequence tags. *DNA Research*, *6*(6), 369–373.

Asamizu, E., Nakamura, Y., Sato, S., & Tabata, S. (2000a). Generation of 7137 non-redundant expressed sequence tags from a legume, Lotus japonicus. *DNA Research*, 7(2), 127–130.

Asamizu, E., Nakamura, Y., Sato, S., & Tabata, S. (2000b). A large scale analysis of cDNA in Arabidopsis thaliana: Generation of 12,028 non-redundant expressed sequence tags from normalized and size-selected cDNA libraries. *DNA Research*, 7(3), 175–180.

Asamizu, E., Nakamura, Y., Sato, S., & Tabata, S. (2005). Comparison of the transcript profiles from the root and the nodulating root of the model legume Lotus japonicus by serial analysis of gene expression. *Molecular Plant-Microbe Interactions*, *18*(5), 487–498. http://dx.doi.org/10.1094/MPMI-18-0487.

Asamizu, E., Sato, S., Kaneko, T., Nakamura, Y., Kotani, H., Miyajima, N., et al. (1998). Structural analysis of Arabidopsis thaliana chromosome 5. VIII. Sequence features of the regions of 1,081,958 bp covered by seventeen physically assigned P1 and TAC clones. *DNA Research*, *5*(6), 379–391.

Aston, C., Mishra, B., & Schwartz, D. C. (1999). Optical mapping and its potential for large-scale sequencing projects. *Trends in Biotechnology*, *17*(7), 297–302.

Bachlava, E., Taylor, C. A., Tang, S., Bowers, J. E., Mandel, J. R., Burke, J. M., et al. (2012). SNP discovery and development of a high-density genotyping array for sunflower. *PLoS One*, 7(1), e29814. http://dx.doi.org/10.1371/journal.pone.0029814.

Banks, J. A., Nishiyama, T., Hasebe, M., Bowman, J. L., Gribskov, M., dePamphilis, C., et al. (2011). The Selaginella genome identifies genetic changes associated with the evolution of vascular plants. *Science*, *332*(6032), 960–963. http://dx.doi.org/10.1126/science.1203810.

Barry, G. F. (2001). The use of the Monsanto draft rice genome sequence in research. *Plant Physiology*, *125*(3), 1164–1165.

Bawnik, N., Beckmann, J. S., Sarid, S., & Daniel, V. (1983). Isolation and nucleotide sequence of a plant tRNA gene: Petunia asparagine tRNA. *Nucleic Acids Research*, *11*(4), 1117–1122.

Bell, C. J., Dixon, R. A., Farmer, A. D., Flores, R., Inman, J., Gonzales, R. A., et al. (2001). The Medicago Genome Initiative: A model legume database. *Nucleic Acids Research*, *29*(1), 114–117.

Bennett, M. D. (1998). Plant genome values: How much do we know? *Proceedings of the National Academy of Sciences of the United States of America*, *95*(5), 2011–2016.

Bennetzen, J. (2002). The rice genome. Opening the door to comparative plant biology. *Science*, *296*(5565), 60–63. http://dx.doi.org/10.1126/science.1071402.

Bennetzen, J. L., Schmutz, J., Wang, H., Percifield, R., Hawkins, J., Pontaroli, A. C., et al. (2012). Reference genome sequence of the model plant Setaria. *Nature Biotechnology*, *30*(6), 555–561. http://dx.doi.org/10.1038/nbt.2196.

Bevan, M. (1997). Objective: The complete sequence of a plant genome. *Plant Cell*, *9*(4), 476–478. http://dx.doi.org/10.1105/tpc.9.4.476.

Bevan, M., Bancroft, I., Bent, E., Love, K., Goodman, H., Dean, C., et al. (1998). Analysis of 1.9 Mb of contiguous sequence from chromosome 4 of Arabidopsis thaliana. *Nature*, *391*(6666), 485–488. http://dx.doi.org/10.1038/35140.

Bevan, M., Barnes, W. M., & Chilton, M. D. (1983). Structure and transcription of the nopaline synthase gene region of T-DNA. *Nucleic Acids Research*, *11*(2), 369–385.

Blanca, J., Esteras, C., Ziarsolo, P., Perez, D., Ferna Ndez-Pedrosa, V., Collado, C., et al. (2012). Transcriptome sequencing for SNP discovery across Cucumis melo. *BMC Genomics*, *13*, 280. http://dx.doi.org/10.1186/1471-2164-13-280.

Boguski, M. S., Lowe, T. M., & Tolstoshev, C. M. (1993). dbEST—Database for "expressed sequence tags" *Nature Genetics*, *4*(4), 332–333. http://dx.doi.org/10.1038/ng0893-332.

Bohnert, H. J., Gordon, K. H., & Crouse, E. J. (1980). Homologies among ribosomal RNA and messenger RNA genes in chloroplasts, mitochondria and E. coli. *Molecular and General Genetics*, *179*(3), 539–545.

Briggs, S. P., & Helentjaris, T. (1997). Plant genomics moves into the limelight. *Genome Research*, 7(9), 856–857.

Brunner, A. M., Busov, V. B., & Strauss, S. H. (2004). Poplar genome sequence: Functional genomics in an ecologically dominant plant species. *Trends in Plant Science*, *9*(1), 49–56. http://dx.doi.org/10.1016/j.tplants.2003.11.006.

Cantrell, R. P., & Reeves, T. G. (2002). The rice genome. The cereal of the world's poor takes center stage. *Science*, *296*(5565), 53. http://dx.doi.org/10.1126/science.1070721.

Chan, A. P., Crabtree, J., Zhao, Q., Lorenzi, H., Orvis, J., Puiu, D., et al. (2010). Draft genome sequence of the oilseed species Ricinus communis. *Nature Biotechnology*, *28*(9), 951–956. http://dx.doi.org/10.1038/nbt.1674.

Chandler, V. L., Radicella, J. P., Robbins, T. P., Chen, J., & Turks, D. (1989). Two regulatory genes of the maize anthocyanin pathway are homologous: Isolation of B utilizing R genomic sequences. *Plant Cell*, *1*(12), 1175–1183. http://dx.doi.org/10.1105/tpc.1.12.1175.

Chen, M., Presting, G., Barbazuk, W. B., Goicoechea, J. L., Blackmon, B., Fang, G., et al. (2002). An integrated physical and genetic map of the rice genome. *Plant Cell*, *14*(3), 537–545.

Chia, J. M., Song, C., Bradbury, P. J., Costich, D., de Leon, N., Doebley, J., et al. (2012). Maize HapMap2 identifies extant variation from a genome in flux. *Nature Genetics*, *44*(7), 803–807. http://dx.doi.org/10.1038/ng.2313.

Coe, E., Cone, K., McMullen, M., Chen, S. S., Davis, G., Gardiner, J., et al. (2002). Access to the maize genome: An integrated physical and genetic map. *Plant Physiology*, *128*(1), 9–12.

Cone, K. C., McMullen, M. D., Bi, I. V., Davis, G. L., Yim, Y. S., Gardiner, J. M., et al. (2002). Genetic, physical, and informatics resources for maize. On the road to an integrated map. *Plant Physiology*, *130*(4), 1598–1605. http://dx.doi.org/10.1104/pp. 012245.

Cooke, R., Raynal, M., Laudie, M., Grellet, F., Delseny, M., Morris, P. C., et al. (1996). Further progress towards a catalogue of all Arabidopsis genes: Analysis of a set of 5000 non-redundant ESTs. *The Plant Journal*, *9*(1), 101–124.

Covitz, P. A., Smith, L. S., & Long, S. R. (1998). Expressed sequence tags from a root-hair-enriched medicago truncatula cDNA library. *Plant Physiology*, *117*(4), 1325–1332.

Dassanayake, M., Oh, D. H., Haas, J. S., Hernandez, A., Hong, H., Ali, S., et al. (2011). The genome of the extremophile crucifer Thellungiella parvula. *Nature Genetics*, *43*(9), 913–918. http://dx.doi.org/10.1038/ng.889.

Davis, G. L., McMullen, M. D., Baysdorfer, C., Musket, T., Grant, D., Staebell, M., et al. (1999). A maize map standard with sequenced core markers, grass genome reference points and 932 expressed sequence tagged sites (ESTs) in a 1736-locus map. *Genetics*, *152*(3), 1137–1172.

De Greve, H., Dhaese, P., Seurinck, J., Lemmers, M., Van Montagu, M., & Schell, J. (1982). Nucleotide sequence and transcript map of the Agrobacterium tumefaciens Ti plasmid-encoded octopine synthase gene. *Journal of Molecular and Applied Genetics*, *1*(6), 499–511.

Delihas, N., Andersen, J., Sprouse, H. M., & Dudock, B. (1981). The nucleotide sequence of the chloroplast 5S ribosomal RNA from spinach. *Nucleic Acids Research*, *9*(12), 2801–2805.

Dhaese, P., De Greve, H., Gielen, J., Seurinck, L., Van Montagu, M., & Schell, J. (1983). Identification of sequences involved in the polyadenylation of higher plant nuclear transcripts using Agrobacterium T-DNA genes as models. *The EMBO Journal*, *2*(3), 419–426.

D'Hont, A., Denoeud, F., Aury, J. M., Baurens, F. C., Carreel, F., Garsmeur, O., et al. (2012). The banana (Musa acuminata) genome and the evolution of monocotyledonous plants. *Nature*, *488*(7410), 213–217. http://dx.doi.org/10.1038/nature11241.

Dunsmuir, P., Smith, S. M., & Bedbrook, J. (1983). The major chlorophyll a/b binding protein of petunia is composed of several polypeptides encoded by a number of distinct nuclear genes. *Journal of Molecular and Applied Genetics*, *2*(3), 285–300.

Durstewitz, G., Polley, A., Plieske, J., Luerssen, H., Graner, E. M., Wieseke, R., et al. (2010). SNP discovery by amplicon sequencing and multiplex SNP genotyping in the allopolyploid species Brassica napus. *Genome*, *53*(11), 948–956. http://dx.doi.org/10.1139/G10-079.

Sieb. and Zucc.) genome. *Proceedings of the National Academy of Sciences of the United States of America, 107*(51), 22032–22037. http://dx.doi.org/10.1073/pnas.1009526107.

Klee, H., Montoya, A., Horodyski, F., Lichtenstein, C., Garfinkel, D., Fuller, S., et al. (1984). Nucleotide sequence of the tms genes of the pTiA6NC octopine Ti plasmid: Two gene products involved in plant tumorigenesis. *Proceedings of the National Academy of Sciences of the United States of America, 81*(6), 1728–1732.

Kotani, H., Hosouchi, T., & Tsuruoka, H. (1999). Structural analysis and complete physical map of Arabidopsis thaliana chromosome 5 including centromeric and telomeric regions. *DNA Research, 6*(6), 381–386.

Kotani, H., Nakamura, Y., Sato, S., Kaneko, T., Asamizu, E., Miyajima, N., et al. (1997). Structural analysis of Arabidopsis thaliana chromosome 5. II. Sequence features of the regions of 1,044,062 bp covered by thirteen physically assigned P1 clones. *DNA Research, 4*(4), 291–300.

Kotani, H., Sato, S., Fukami, M., Hosouchi, T., Nakazaki, N., Okumura, S., et al. (1997). A fine physical map of Arabidopsis thaliana chromosome 5: Construction of a sequence-ready contig map. *DNA Research, 4*(6), 371–378.

Kwak, J. M., Kim, S. A., Hong, S. W., & Nam, H. G. (1997). Evaluation of 515 expressed sequence tags obtained from guard cells of Brassica campestris. *Planta, 202*(1), 9–17.

Lam, H. M., Xu, X., Liu, X., Chen, W., Yang, G., Wong, F. L., et al. (2010). Resequencing of 31 wild and cultivated soybean genomes identifies patterns of genetic diversity and selection. *Nature Genetics, 42*(12), 1053–1059. http://dx.doi.org/10.1038/ng.715.

Leach, J., McCouch, S., Slezak, T., Sasaki, T., & Wessler, S. (2002). Why finishing the rice genome matters. *Science, 296*(5565), 45.

Lechelt, C., Peterson, T., Laird, A., Chen, J., Dellaporta, S. L., Dennis, E., et al. (1989). Isolation and molecular analysis of the maize P locus. *Molecular and General Genetics, 219*(1–2), 225–234.

Lee, M., Sharopova, N., Beavis, W. D., Grant, D., Katt, M., Blair, D., et al. (2002). Expanding the genetic map of maize with the intermated B73 x Mo17 (IBM) population. *Plant Molecular Biology, 48*(5–6), 453–461.

Leemans, J., Shaw, C., Deblaere, R., De Greve, H., Hernalsteens, J. P., Maes, M., et al. (1981). Site-specific mutagenesis of Agrobacterium Ti plasmids and transfer of genes to plant cells. *Journal of Molecular and Applied Genetics, 1*(2), 149–164.

Lim, C. O., Kim, H. Y., Kim, M. G., Lee, S. I., Chung, W. S., Park, S. H., et al. (1996). Expressed sequence tags of Chinese cabbage flower bud cDNA. *Plant Physiology, 111*(2), 577–588.

Lin, X., Kaul, S., Rounsley, S., Shea, T. P., Benito, M. I., Town, C. D., et al. (1999). Sequence and analysis of chromosome 2 of the plant Arabidopsis thaliana. *Nature, 402*(6763), 761–768. http://dx.doi.org/10.1038/45471.

Ling, H. Q., Zhao, S., Liu, D., Wang, J., Sun, H., Zhang, C., et al. (2013). Draft genome of the wheat A-genome progenitor Triticum urartu. *Nature, 496*(7443), 87–90. http://dx.doi.org/10.1038/nature11997.

Ludwig, S. R., Habera, L. F., Dellaporta, S. L., & Wessler, S. R. (1989). Lc, a member of the maize R gene family responsible for tissue-specific anthocyanin production, encodes a protein similar to transcriptional activators and contains the myc-homology region. *Proceedings of the National Academy of Sciences of the United States of America, 86*(18), 7092–7096.

Lycett, G. W., Delauney, A. J., Gatehouse, J. A., Gilroy, J., Croy, R. R., & Boulter, D. (1983). The vicilin gene family of pea (Pisum sativum L.): A complete cDNA coding sequence for preprovicilin. *Nucleic Acids Research, 11*(8), 2367–2380.

Machuka, J., Bashiardes, S., Ruben, E., Spooner, K., Cuming, A., Knight, C., et al. (1999). Sequence analysis of expressed sequence tags from an ABA-treated cDNA library identifies stress response genes in the moss Physcomitrella patens. *Plant and Cell Physiology, 40*(4), 378–387.

Mardis, E. R. (2008). Next-generation DNA sequencing methods. *Annual Review of Genomics and Human Genetics, 9*, 387–402. http://dx.doi.org/10.1146/annurev.genom.9.081307.164359.

Mardis, E. R. (2009). New strategies and emerging technologies for massively parallel sequencing: Applications in medical research. *Genome Medicine, 1*(4), 40. http://dx.doi.org/10.1186/gm40.

Martienssen, R., & McCombie, W. R. (2001). The first plant genome. *Cell, 105*(5), 571–574.

Maxam, A. M., & Gilbert, W. (1977). A new method for sequencing DNA. *Proceedings of the National Academy of Sciences of the United States of America, 74*(2), 560–564.

Maxam, A. M., & Gilbert, W. (1980). Sequencing end-labeled DNA with base-specific chemical cleavages. *Methods in Enzymology, 65*(1), 499–560.

Maxam, A. M., & Gilbert, W. (1992). A new method for sequencing DNA. 1977. *Biotechnology, 24*, 99–103.

Mayer, K. F., Lemcke, K., Schuller, C. M., Rudd, S., & Zaccaria, P. (2000). Arabidopsis genome analysis as exemplified by analysis of chromosome 4. *Briefings in Bioinformatics, 1*(4), 389–397.

Mayer, K., Schuller, C., Wambutt, R., Murphy, G., Volckaert, G., Pohl, T., et al. (1999). Sequence and analysis of chromosome 4 of the plant Arabidopsis thaliana. *Nature, 402*(6763), 769–777. http://dx.doi.org/10.1038/47134.

McClintock, B. (1984). The significance of responses of the genome to challenge. *Science, 226*(4676), 792–801.

Merchant, S. S., Prochnik, S. E., Vallon, O., Harris, E. H., Karpowicz, S. J., Witman, G. B., et al. (2007). The Chlamydomonas genome reveals the evolution of key animal and plant functions. *Science, 318*(5848), 245–250. http://dx.doi.org/10.1126/science.1143609.

Ming, R., Hou, S., Feng, Y., Yu, Q., Dionne-Laporte, A., Saw, J. H., et al. (2008). The draft genome of the transgenic tropical fruit tree papaya (Carica papaya Linnaeus). *Nature, 452*(7190), 991–996. http://dx.doi.org/10.1038/nature06856.

Nakamura, Y., Sato, S., Asamizu, E., Kaneko, T., Kotani, H., Miyajima, N., et al. (1998). Structural analysis of Arabidopsis thaliana chromosome 5. VII. Sequence features of the regions of 1,013,767 bp covered by sixteen physically assigned P1 and TAC clones. *DNA Research, 5*(5), 297–308.

Neuteboom, L. W., Ng, J. M., Kuyper, M., Clijdesdale, O. R., Hooykaas, P. J., & van der Zaal, B. J. (1999). Isolation and characterization of cDNA clones corresponding with mRNAs that accumulate during auxin-induced lateral root formation. *Plant Molecular Biology, 39*(2), 273–287.

Newman, T., de Bruijn, F. J., Green, P., Keegstra, K., Kende, H., McIntosh, L., et al. (1994). Genes galore: A summary of methods for accessing results from large-scale partial sequencing of anonymous Arabidopsis cDNA clones. *Plant Physiology, 106*(4), 1241–1255.

Normile, D. (2002). The rice genome. Beijing Genomics Institute: From standing start to sequencing superpower. *Science, 296*(5565), 36–39. http://dx.doi.org/10.1126/science.296.5565.36.

Normile, D., & Pennisi, E. (2002). The rice genome. Rice: Boiled down to bare essentials. *Science, 296*(5565), 32–36. http://dx.doi.org/10.1126/science.296.5565.32.

Oliver, R. E., Tinker, N. A., Lazo, G. R., Chao, S., Jellen, E. N., Carson, M. L., et al. (2013). SNP discovery and chromosome anchoring provide the first physically-anchored hexaploid oat map and reveal synteny with model species. *PLoS One, 8*(3), e58068. http://dx.doi.org/10.1371/journal.pone.0058068.

Pagani, I., Liolios, K., Jansson, J., Chen, I. M., Smirnova, T., Nosrat, B., et al. (2012). The Genomes OnLine Database (GOLD) v.4: Status of genomic and metagenomic projects and their associated metadata. *Nucleic Acids Research*, *40*(Database issue), D571–D579. http://dx.doi.org/10.1093/nar/gkr1100.

Park, Y. S., Kwak, J. M., Kwon, O. Y., Kim, Y. S., Lee, D. S., Cho, M. J., et al. (1993). Generation of expressed sequence tags of random root cDNA clones of Brassica napus by single-run partial sequencing. *Plant Physiology*, *103*(2), 359–370.

Paterson, A. H., Bowers, J. E., Bruggmann, R., Dubchak, I., Grimwood, J., Gundlach, H., et al. (2009). The Sorghum bicolor genome and the diversification of grasses. *Nature*, *457*(7229), 551–556. http://dx.doi.org/10.1038/nature07723.

Paterson, A. H., Wendel, J. F., Gundlach, H., Guo, H., Jenkins, J., Jin, D., et al. (2012). Repeated polyploidization of Gossypium genomes and the evolution of spinnable cotton fibres. *Nature*, *492*(7429), 423–427. http://dx.doi.org/10.1038/nature11798.

Paz-Ares, J., Wienand, U., Peterson, P. A., & Saedler, H. (1986). Molecular cloning of the c locus of Zea mays: A locus regulating the anthocyanin pathway. *The EMBO Journal*, *5*(5), 829–833.

Peng, Z., Lu, Y., Li, L., Zhao, Q., Feng, Q., Gao, Z., et al. (2013). The draft genome of the fast-growing non-timber forest species moso bamboo (Phyllostachys heterocycla). *Nature Genetics*, *45*(4), 456–461. http://dx.doi.org/10.1038/ng.2569.

Potato Genome Sequencing Consortium, Xu, X., Pan, S., Cheng, S., Zhang, B., Mu, D., et al. (2011). Genome sequence and analysis of the tuber crop potato. *Nature*, *475*(7355), 189–195. http://dx.doi.org/10.1038/nature10158.

Poulsen, C., & Podenphant, L. (2002). Expressed sequence tags from roots and nodule primordia of Lotus japonicus infected with Mesorhizobium loti. *Molecular Plant-Microbe Interactions*, *15*(4), 376–379. http://dx.doi.org/10.1094/MPMI.2002.15.4.376.

Prochnik, S., Marri, P. R., Desany, B., Rabinowicz, P. D., Kodira, C., Mohiuddin, M., et al. (2012). The cassava genome: Current progress, future directions. *Tropical Plant Biology*, *5*(1), 88–94. http://dx.doi.org/10.1007/s12042-011-9088-z.

Quigley, F., Dao, P., Cottet, A., & Mache, R. (1996). Sequence analysis of an 81 kb contig from Arabidopsis thaliana chromosome III. *Nucleic Acids Research*, *24*(21), 4313–4318.

Rahman, A. Y., Usharraj, A. O., Misra, B. B., Thottathil, G. P., Jayasekaran, K., Feng, Y., et al. (2013). Draft genome sequence of the rubber tree Hevea brasiliensis. *BMC Genomics*, *14*, 75. http://dx.doi.org/10.1186/1471-2164-14-75.

Rensing, S. A., Lang, D., Zimmer, A. D., Terry, A., Salamov, A., Shapiro, H., et al. (2008). The Physcomitrella genome reveals evolutionary insights into the conquest of land by plants. *Science*, *319*(5859), 64–69. http://dx.doi.org/10.1126/science.1150646.

Ronald, P., & Leung, H. (2002). The rice genome. The most precious things are not jade and pearls. *Science*, *296*(5565), 58–59. http://dx.doi.org/10.1126/science.1071337.

Rounsley, S. D., Glodek, A., Sutton, G., Adams, M. D., Somerville, C. R., Venter, J. C., et al. (1996). The construction of Arabidopsis expressed sequence tag assemblies. A new resource to facilitate gene identification. *Plant Physiology*, *112*(3), 1177–1183.

Salunkhe, A. S., Poornima, R., Prince, K. S., Kanagaraj, P., Sheeba, J. A., Amudha, K., et al. (2011). Fine mapping QTL for drought resistance traits in rice (Oryza sativa L.) using bulk segregant analysis. *Molecular Biotechnology*, *49*(1), 90–95. http://dx.doi.org/10.1007/s12033-011-9382-x.

Samad, A., Huff, E. F., Cai, W., & Schwartz, D. C. (1995). Optical mapping: A novel, single-molecule approach to genomic analysis. *Genome Research*, *5*(1), 1–4.

Sanger, F., Nicklen, S., & Coulson, A. R. (1977). DNA sequencing with chain-terminating inhibitors. *Proceedings of the National Academy of Sciences of the United States of America*, *74*(12), 5463–5467.

Sanger, F., Nicklen, S., & Coulson, A. R. (1992). DNA sequencing with chain-terminating inhibitors. 1977. *Biotechnology*, *24*, 104–108.

Sato, S., Kaneko, T., Kotani, H., Nakamura, Y., Asamizu, E., Miyajima, N., et al. (1998). Structural analysis of Arabidopsis thaliana chromosome 5. IV. Sequence features of the regions of 1,456,315 bp covered by nineteen physically assigned P1 and TAC clones. *DNA Research*, *5*(1), 41–54.

Sato, S., Nakamura, Y., Kaneko, T., Asamizu, E., Kato, T., Nakao, M., et al. (2008). Genome structure of the legume, Lotus japonicus. *DNA Research*, *15*(4), 227–239. http://dx.doi.org/10.1093/dnares/dsn008.

Schmutz, J., Cannon, S. B., Schlueter, J., Ma, J., Mitros, T., Nelson, W., et al. (2010). Genome sequence of the palaeopolyploid soybean. *Nature*, *463*(7278), 178–183. http://dx.doi.org/10.1038/nature08670.

Schnable, P. S., Ware, D., Fulton, R. S., Stein, J. C., Wei, F., Pasternak, S., et al. (2009). The B73 maize genome: Complexity, diversity, and dynamics. *Science*, *326*(5956), 1112–1115. http://dx.doi.org/10.1126/science.1178534.

Schuler, M. A., Schmitt, E. S., & Beachy, R. N. (1982). Closely related families of genes code for the alpha and alpha' subunits of the soybean 7S storage protein complex. *Nucleic Acids Research*, *10*(24), 8225–8244.

Schwartz, D. C., & Samad, A. (1997). Optical mapping approaches to molecular genomics. *Current Opinion in Biotechnology*, *8*(1), 70–74.

Shah, D. M., Hightower, R. C., & Meagher, R. B. (1983). Genes encoding actin in higher plants: Intron positions are highly conserved but the coding sequences are not. *Journal of Molecular and Applied Genetics*, *2*(1), 111–126.

Sharopova, N., McMullen, M. D., Schultz, L., Schroeder, S., Sanchez-Villeda, H., Gardiner, J., et al. (2002). Development and mapping of SSR markers for maize. *Plant Molecular Biology*, *48*(5–6), 463–481.

Sharpe, A. G., Ramsay, L., Sanderson, L. A., Fedoruk, M. J., Clarke, W. E., Li, R., et al. (2013). Ancient orphan crop joins modern era: Gene-based SNP discovery and mapping in lentil. *BMC Genomics*, *14*, 192. http://dx.doi.org/10.1186/1471-2164-14-192.

Shinozaki, K., Yamada, C., Takahata, N., & Sugiura, M. (1983). Molecular cloning and sequence analysis of the cyanobacterial gene for the large subunit of ribulose-1,5-bisphosphate carboxylase/oxygenase. *Proceedings of the National Academy of Sciences of the United States of America*, *80*(13), 4050–4054.

Shirasawa, K., Isobe, S., Hirakawa, H., Asamizu, E., Fukuoka, H., Just, D., et al. (2010). SNP discovery and linkage map construction in cultivated tomato. *DNA Research*, *17*(6), 381–391. http://dx.doi.org/10.1093/dnares/dsq024.

Shulaev, V., Sargent, D. J., Crowhurst, R. N., Mockler, T. C., Folkerts, O., Delcher, A. L., et al. (2011). The genome of woodland strawberry (Fragaria vesca). *Nature Genetics*, *43*(2), 109–116. http://dx.doi.org/10.1038/ng.740.

Simpson, R. B., O'Hara, P. J., Kwok, W., Montoya, A. L., Lichtenstein, C., Gordon, M. P., et al. (1982). DNA from the A6S/2 crown gall tumor contains scrambled Ti-plasmid sequences near its junctions with plant DNA. *Cell*, *29*(3), 1005–1014.

Soderlund, C., Longden, I., & Mott, R. (1997). FPC: A system for building contigs from restriction fingerprinted clones. *Computer Applications in the Biosciences*, *13*(5), 523–535.

Sonah, H., Bastien, M., Iquira, E., Tardivel, A., Legare, G., Boyle, B., et al. (2013). An improved genotyping by sequencing (GBS) approach offering increased versatility and efficiency of SNP discovery and genotyping. *PLoS One*, *8*(1), e54603. http://dx.doi.org/10.1371/journal.pone.0054603.

Spielmann, A., & Stutz, E. (1983). Nucleotide sequence of soybean chloroplast DNA regions which contain the psb A and trn H genes and cover the ends of the large single copy region and one end of the inverted repeats. *Nucleic Acids Research*, *11*(20), 7157–7167.

Sterky, F., Regan, S., Karlsson, J., Hertzberg, M., Rohde, A., Holmberg, A., et al. (1998). Gene discovery in the wood-forming tissues of poplar: Analysis of 5, 692 expressed sequence tags. *Proceedings of the National Academy of Sciences of the United States of America*, *95*(22), 13330–13335.

Yamamoto, K., & Sasaki, T. (1997). Large-scale EST sequencing in rice. *Plant Molecular Biology*, *35*(1–2), 135–144.

Yamamoto, N., Tsugane, T., Watanabe, M., Yano, K., Maeda, F., Kuwata, C., et al. (2005). Expressed sequence tags from the laboratory-grown miniature tomato (Lycopersicon esculentum) cultivar Micro-Tom and mining for single nucleotide polymorphisms and insertions/deletions in tomato cultivars. *Gene*, *356*, 127–134. http://dx.doi.org/10.1016/j.gene.2005.04.026.

Yang, R., Jarvis, D. E., Chen, H., Beilstein, M. A., Grimwood, J., Jenkins, J., et al. (2013). The reference genome of the halophytic plant Eutrema salsugineum. *Frontiers in Plant Science*, *4*, 46. http://dx.doi.org/10.3389/fpls.2013.00046.

Young, N. D., Debelle, F., Oldroyd, G. E., Geurts, R., Cannon, S. B., Udvardi, M. K., et al. (2011). The Medicago genome provides insight into the evolution of rhizobial symbioses. *Nature*, *480*(7378), 520–524. http://dx.doi.org/10.1038/nature10625.

Yu, J., Hu, S., Wang, J., Wong, G. K., Li, S., Liu, B., et al. (2002). A draft sequence of the rice genome (Oryza sativa L. ssp. indica). *Science*, *296*(5565), 79–92. http://dx.doi.org/10.1126/science.1068037.

Yu, H. J., Moon, M. S., Lee, H. S., Mun, J. H., Kwon, Y. M., & Kim, S. G. (1999). Analysis of cDNAs expressed during first cell division of petunia petal protoplast cultures using expressed sequence tags. *Molecules and Cells*, *9*(3), 258–264.

Yu, J., Wang, J., Lin, W., Li, S., Li, H., Zhou, J., et al. (2005). The Genomes of Oryza sativa: A history of duplications. *PLoS Biology*, *3*(2), e38. http://dx.doi.org/10.1371/journal.pbio.0030038.

Yuan, Y. W., Sagawa, J. M., Di Stilio, V. S., & Bradshaw, H. D., Jr. (2013). Bulk segregant analysis of an induced floral mutant identifies a MIXTA-like R2R3 MYB controlling nectar guide formation in Mimulus lewisii. *Genetics*, *194*(2), 523–528. http://dx.doi.org/10.1534/genetics.113.151225.

Zhang, Q., Chen, W., Sun, L., Zhao, F., Huang, B., Yang, W., et al. (2012). The genome of Prunus mume. *Nature Communications*, *3*, 1318. http://dx.doi.org/10.1038/ncomms2290.

Zhang, G., Liu, X., Quan, Z., Cheng, S., Xu, X., Pan, S., et al. (2012). Genome sequence of foxtail millet (Setaria italica) provides insights into grass evolution and biofuel potential. *Nature Biotechnology*, *30*(6), 549–554. http://dx.doi.org/10.1038/nbt.2195.

Zhou, S., Wei, F., Nguyen, J., Bechner, M., Potamousis, K., Goldstein, S., et al. (2009). A single molecule scaffold for the maize genome. *PLoS Genetics*, *5*(11), e1000711. http://dx.doi.org/10.1371/journal.pgen.1000711.

Arabidopsis is amenable to most known tissue culture techniques and is transformable by a number of methods (Lloyd et al., 1986) including non-tissue culture methods that make it practical to do T-DNA insertion mutagenesis screens (Bechtold & Pelletier, 1998; Clough & Bent, 1998; Feldmann & Marks, 1987). There are a wide variety of land races with many different morphological and physiological characteristics. Many of the biological resources from seeds to cDNAs are available through the Arabidopsis Biological Resource Center and The European Arabidopsis Stock Centre (Nottingham Arabidopsis Seed Center—NASC). Finally, it is a member of an agronomically important group of plants, the brassica or mustard family. However, the one feature of Arabidopsis that cannot be overstated is its small genome size as demonstrated by publications from the Meyerowitz group in 1984. These publications brought this species to the attention of many molecular biologists around the world and the size of the Arabidopsis community exploded over the next 5 years.

The initiative to sequence the *Arabidopsis* genome was proposed in 1989 by the Biological, Behavioral, and Social Sciences Directorate (BBS) of the National Science Foundation (NSF) with considerable input from academic and industrial scientists. Although not directly stated, the agency wanted to spend $100 million to develop a genome project equivalent to the National Institute of Health's human genome project. A series of meetings and workshops, with scientists from the United States, Europe, Japan and Australia, was held to plan a framework for developing the resources necessary to sequence the genome. As Arabidopsis was the first plant genome, and one of the earliest eukaryotes to be sequenced, there were many strategies to be worked out and efficiencies to be gained. Fortunately, as with the worm and fly research communities, the Arabidopsis community was very collaborative. A plan to coordinate Arabidopsis genome research was described in a 1990 publication 'A Long-Range Plan for the Multinational Coordinated *A. thaliana* Genome Research Project' (NSF 90–80). Given the state of sequencing technology at that time, it was estimated that the genome could be sequenced by the year 2000. As such, the Arabidopsis research community began to establish the biological resources needed for sequencing the genome. In 1996, the Arabidopsis Genome Initiative (AGI) was formed 'to facilitate cooperation among international sequencing projects' so that the genome could be sequenced by the year 2004, except for the difficult-to-sequence repetitive regions such as the nucleolar organizing regions (NORs) and centromeres. With improvements in sequencing technologies and competition between the Arabidopsis sequencing groups and

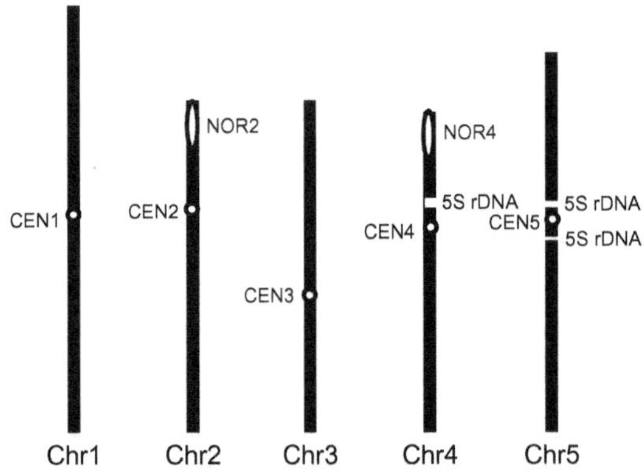

Figure 4.1 The five chromosomes of *Arabidopsis thaliana*. The centromeres (CEN) are shown along with the two nucleolar organizing regions (NORs) and the 5S rDNA regions. *Adapted from Haas et al. (2005).*

1995), YAC (Camilleri et al., 1998) and TAC (transformation-competent artificial chromosome) libraries (Liu et al., 1999) and a small number of cosmid clones. A physical map, integrated with the genetic map, was used to anchor clones and contigs. The map was built using a combination of fingerprint analysis of BACs (Marra et al., 1999), PCR of sequenced tagged sites (Sato et al., 1998) and hybridization (Bent, Johnson, & Bancroft, 1998; Mozo et al., 1999). Different approaches were used for different chromosomes. End sequencing of 47,788 BACs aided in extending and integrating the developing contigs. In this way, 10 contigs were developed from just 1569 clones (as described earlier). PCR was also used to develop a minimal tiling path. Clones were double-strand-sequenced with fewer than 1 in 10,000 to 1 in 100,000 errors. Any ambiguities were resolved and corrected.

One of the goals of the Arabidopsis sequencing effort as proposed in 1990 was the 'creation of cDNA and EST libraries representing different tissues and cell types'. By 2000, there were 150,000 ESTs in GenBank and full-length cDNA projects were under way. These early ESTs were generated by a number of researchers (1152, Hofte et al., 1993; 1447, Newman et al., 1994; 4998, Cooke et al., 1996; 30,000, Delseny, Cooke, Raynal, & Grellet, 1997; 10,500, White et al., 2000). ESTs would prove critical in identifying genes in long stretches of genomic DNA. Likewise, full-length cDNAs were important for correct annotation of genes and for identifying alternative splicing as well as alternative transcription start sites.

Sequencing began with a 1.9 Mb contiguous sequence on chromosome 4 (Bevan et al., 1998). This project relied on sequencing ordered cosmids. The analysis of this sequenced segment revealed that the Arabidopsis genome was gene rich, with a gene on average every 4.8 kb, and that 54% of predicted genes had similarity to known genes. This sequence also revealed several classes of genes that had not previously been observed in plants. From the analyses of this 1.9 Mb, and the 13 Mb that was available from other ongoing genomic sequencing projects with the five chromosomes, it was estimated that Arabidopsis contained 21,000 protein-coding genes. The complete sequences of chromosomes 2 and 4 were published a year later (Lin et al., 1999; Mayer et al., 1999). Publications describing the sequence of chromosomes 1, 3 and 5 appeared the following year (Salanoubat et al., 2000; Tabata et al., 2000; Theologis et al., 2000). We will describe the highlights from sequencing each of these chromosomes as of their publication date. It should be noted that as more ESTs and full-length cDNAs were sequenced and utilized to train software to find genes in Arabidopsis, and as other plant genomes were sequenced and annotated and used to compare against the Arabidopsis sequence, the number of annotated genes increased (see later).

2.1. Chromosome 2

Chromosome 2 sequencing was initiated from BACs that had been anchored to a physical map (Table 4.1). In total, 257 BAC and P1 clones were sequenced to produce 24 Mb of finished sequence (Lin et al., 1999). The upper arm of chromosome 2 from the NOR to the centromeric region was 3.6 Mb (Table 4.1 and Fig. 4.1), while the lower arm of this

Table 4.1 The Arabidopsis genome sequencing effort (2000)

Chromosome	**1**	**2**	**3**	**4**	**5**
Sequencing groups[a]	SPP	TIGR	CNS and KI	EU/AGP and CSHSC	CSHSC and KI
Total length (Mb)	29.1	19.6	23.2	17.5	26
Length of arms (top and bottom)	14.4 and 14.6 Mb	3.6 and 16 Mb	13.5 and 9.6 Mb	3.0 and 14.5 Mb	11.2 and 14.8 Mb
Number of protein-coding genes	6543 (7078)[b]	4036 (4245)	5220 (5437)	3825 (4124)	5874 (6318)
Gene density	4.0	4.9	4.5	4.6	4.4

[a]SPP, Stanford University/University of Pennsylvania/Plant Gene Expression Laboratory Consortium; TIGR, The Institute for Genome Research; CNS, Centre National de Séquençage; KI, Kazusa DNA Research Institute; EU/AGP, EU Arabidopsis Genome Project; CSHSC, Cold Spring Harbor Sequencing Consortium.
[b]Numbers in parentheses represent gene estimates for TAIR 10—27,202 total protein-coding genes.

acrocentric chromosome was represented by 16 Mb of sequence. The total length of the two arms, excluding the NOR and centromeres, was 45% longer than the original estimate. As a similar pattern was being observed in the other sequenced chromosomes, the genome size was substantially larger than the 70–100 Mb originally estimated (Meyerowitz & Pruitt, 1985). The NOR, consisting of tandemly repeated ribosomal RNA genes, was estimated to be 3.6 Mb in length. The ~180 bp repeat block in the centromere was estimated to be approximately 820–830 kb (Lin et al., 1999).

The authors identified protein-coding genes at average intervals of 4.4 kb, slightly denser than what was observed in a small portion of chromosome 4 (Bevan et al., 1998). Of 4057 genes identified on chromosome 2, 51.5% could be assigned to a functional category by homology to known genes, whereas 48.5% had no predicted function (21.4% had an unknown function and another 27.1% encoded hypothetical proteins). Interestingly, 60% of the genes encoding predicted proteins (2542) had a significant match with another protein in the available genomic DNA, with a majority of the matches being to another protein encoded on chromosome 2. In fact, 593 of these matches were found in 239 tandem duplications ranging in size from two to nine genes. It was also observed that duplicated genes were found within segmental chromosome duplications, consistent with an earlier suggestion based on parallel organization of duplicated DNA markers in a genetic map. For example, a 170 gene segment of chromosome 2 was found to be duplicated in chromosome 1. Of the 170 genes, 57 were found as gene pairs in the duplicated region (Lin et al., 1999). Several other large segments of chromosome 2 were found as duplications in other chromosomes. Four hundred pseudogenes, located near the centromere, were also identified on chromosome 2. An unexpected finding was that 270 kb of sequence in the centromeric region was nearly identical to that of the Arabidopsis mitochondrial genome. This finding, in addition to the presence of 135 putative chloroplast genes found in chromosome 2, attests to the frequent lateral transfer of genes from organelles to the nucleus. Finally, Lin et al. (1999) identified 562 transposons and retroelements in the pericentromeric regions where gene density (kb/gene) was found to be sparse.

2.2. Chromosome 4

This acrocentric chromosome was sequenced using primarily BAC clones optimized to contain minimal overlap (Mayer et al., 1999). In total, 131 BAC, 4 P1 and 56 cosmid clones along with 10 PCR products were sequenced to generate 17.4 Mb of nonredundant sequence in three contigs. These contigs are represented by 2.6 Mb of DNA on the top arm, 14.5 Mb on the longer bottom arm (Table 4.1) and a third shorter contig in the centromeric heterochromatin. The sequenced region of chromosome 4 encodes 3744 genes, four snRNAs and 81 transfer RNAs. The number of genes annotated for chromosome 4 would increase by more than 300 genes by TAIR 10 (see later). A similar increase was observed for each of the other chromosomes. For chromosome 4, 34% of the predicted genes matched a cDNA or EST, very similar to the 33.5% match for chromosome 2 (Lin et al., 1999). In terms of organellar gene transfer during evolution, 18% of the genes on chromosome 4 have a potential N-terminal chloroplast and mitochondrial transit peptide. Gene density was found to be one gene per 4.6 kb, similar to what was observed for a 1.9 Mb piece of this chromosome previously (Bevan et al., 1998) and for chromosome 2 (Lin et al., 1999). Both gene and segmental duplications were identified in the sequence. As expected, the gene density in heterochromatin is 1/10th of that found in the distal euchromatic regions of the chromosome. The centromeric region of chromosome 4 was mapped cytogenetically to a region of 4 Mb in length, much larger than the 830 kb estimated for chromosome 2, and consisted of '200 kb of 5S rDNA and 1 Mb of pAL1-rich sequence flanked by dispersed retroelements and other repeats' (Mayer, Lemcke, Schuller, Rudd, & Zaccaria, 2000).

2.3. Chromosome 1

Chromosome 1 is metacentric and the longest of the five Arabidopsis chromosomes. To sequence the chromosome, Theologis et al. (2000) used one YAC clone and 369 BAC clones. The sequence of the top arm was 14.2 Mb, whereas the bottom arm was 14.6 Mb in length, with three sequencing gaps left to be resolved (Theologis et al., 2000). The authors identified 6848 protein-coding genes, at a density of one gene per 4.1 kb, 236 tRNAs and 12 sn RNAs (Table 4.1). The percent of genes lacking introns (18%) is similar to that observed for chromosome 2 (23%). As with the previously published chromosomes, there are gene families ($n=312$) that contain clustered duplications. A much higher percent (50%) of the genes on chromosome 1 match a cDNA or EST than those on chromosomes 2 and 4 (34%), likely a reflection of the increasing number of ESTs in the database in the year since chromosomes 2 and 4 were published (e.g. Galaud et al., 1999; White et al., 2000). Chromosome 1 contains 4 Mb-sized segmental

The genome also contains >4000 transposable elements of various types and accounts for 10% of the genome, 20% of the intergenic DNA.

2.7. The Arabidopsis gene set post-2000 and its comparison to those of other biota

A determination by Hosouchi et al. (2002) indicates that the centromeres are longer than originally estimated, with the exception of chromosome 4. The lengths are now estimated to be 9, 4, 4, 5.3 and 4.35 Mb for chromosome 1–5, respectively, almost four times the original estimates. These longer estimated centromere lengths would make the Arabidopsis genome 146 Mb. In addition, Hall, Kettler and Preuss (2003) showed that the mean length of the repeats in the centromere was 178 bp rather than 180 bp with 72% being 178 bp, 18% at 177 bp and 8% at 179 bp. There are polymorphisms in the centromeric regions between different ecotypes of Arabidopsis (Hall et al., 2003). Centromeric regions are gene-poor relative to the remainder of the genome, but at least 47 expressed genes have been identified in the centromeric regions (Yamada et al., 2003).

ESTs and full-length cDNAs, and more recently RNA-Seq data, have been and continue to be instrumental in annotating the Arabidopsis genome both for direct comparison and for serving as benchmarks for developing software for gene annotation. As of 2013, there were almost 2 million Arabidopsis EST sequences in NCBI. The first EST collections were small (1152, Hofte et al., 1993; 1447, Newman et al., 1994; 4998, Cooke et al., 1996; 10,500, White et al., 2000), but the collection grew quickly (e.g. 155 k, Seki et al., 2002; 200 k, Alexandrov et al., 2006). Using a collection of 5000 sequenced full-length cDNAs, Haas et al. (2002) corrected the annotation of 35% of the genes annotated in 2000 and showed that 5% of the cDNAs represented newly discovered genes. Yamada et al. (2003) used these and other collections of full-length cDNAs to improve the annotation of the genome. They predicted 25,540 protein-coding genes among the 26,828 genes predicted. The other predicted genes included tRNAs and other RNA genes, pseudogenes and transposons. By 2005, the number of protein-coding genes had increased to 26,207, in addition to 3786 transposons or pseudogenes (Bevan & Walsh, 2005). With each new TAIR annotation release, the number has increased.

The final TAIR genome annotation release (TAIR10; Table 4.2) contains 27,202 nuclear protein-coding genes, 4827 pseudogenes and transposable element genes and 1359 nc RNAs (689 tRNAs, 15 rRNAs, 90 snRNAs or small nucleolar RNAs, 177 miRNAs and 394 other RNAs). Gene density is 4.35 kb/gene with an average of 5.89 exons/gene, average exon length of

Table 4.2 Chromosome statistics from TAIR 10

Chromosome	Protein-coding genes	Pre-tRNAs	Other RNAs	Pseudogenes	Transposons	Totals
1	7078	240	191	241	683	8433
2	4245	96	129	217	826	5513
3	5437	93	120	202	878	6730
4	4124	79	101	121	711	5410
5	6318	123	118	143	805	7507
Totals	27,202	631	659	924	3903	

296 nt and average intron length of 165 nt. Arabidopsis appears to contain significantly fewer nuclear protein-coding genes than any other sequenced plant species (Table 4.3), except for *S. bicolor* and *C. papaya* where the current annotation resulted in fewer than 28,000 genes. As with the increasing number of genes in Arabidopsis, it is likely that the number of genes in the agronomically important species will also increase concomitantly with experimentation such as RNA-Seq, especially keeping in mind that these genomes have generally been less intensively scrutinized than Arabidopsis. TAIR10 also reported 88 and 122 protein-coding genes in the chloroplast and mitochondria genome, respectively, which are sometimes added in with the total number of protein-coding genes. In addition, there are 37 and 21 pre-tRNAs, along with eight and three rRNA genes, in the chloroplast and mitochondria genomes, respectively.

The annotation of the Arabidopsis genome, and other plant genomes, will continue to improve as additional types of experiments are analysed. For example, when considering alternative splicing, the percent of intron-containing genes in the Arabidopsis genome with multiple splice variants has increased from 1.2% in 2003 to >61% by 2013 (Loraine, McCormick, Estrada, Patel, & Qin, 2013; Syed, Kalyna, Marquez, Barta, & Brown, 2012). In TAIR10, the number of genes identified with splice variants increased to 5885 (18%). Filichkin et al. (2010) using RNA-Seq data report that 42% of intron-containing genes in Arabidopsis are alternatively spliced. The novel splice sites for some of these were confirmed *in vivo*. More recently, Marquez, Brown, Simpson, Barta and Kalyna (2012) had shown that the percent of intron-containing genes with alternative splice variants in Arabidopsis is 61%. The most frequent type of alternative splicing in plants is intron retention (Alexandrov et al., 2006; Haas et al., 2005; Marquez et al., 2012). Of the splice variants, 70% were in

Table 4.3 Number of genes in selected plant genomes

Species	Length of sequence	No. of protein-coding genes	References
A. thaliana	119 Mb	27,202	The Arabidopsis Genome Initiative (2000)
A. lyrata	206.7 Mb	32,670	Hu et al. (2011)
B. rapa	283.8 Mb	41,174	Wang et al. (2011)
B73 maize	2.3 Gb	32,540	Schnable et al. (2009)
C. papaya	372 Mb	~28,000	Ming et al. (2008)
G. max	950 Mb	46,430	Schmutz et al. (2010)
G. raimondii	775.2 Mb	40,976	Wang et al. (2012)
M. truncatula	375 Mb	48,066	Young et al. (2012)
P. trichocarpa	410 Mb	45,555	Tuskan et al. (2006)
S. bicolor	679.9 Mb	27,640	Paterson et al. (2009)
S. tuberosum	727 Mb	39,031	Potato Genome Sequencing Consortium (2011)
S. lycopersicum	760 Mb	34,727	Tomato Genome Consortium (2012)
V. vinifera	487 Mb	30,434	Jaillon et al. (2007)

Not all genomes are completely sequenced.

the untranslated regions and resulted in identical proteins. Finally, Loraine et al. (2013), using RNA-Seq of Arabidopsis pollen, and seedlings for comparison, detected 14 regions in the genome not previously annotated as expressed; 12 were confirmed by polymerase chain reaction. They also identified 1908 new splicing events. Some caution needs to be taken in interpreting these results as RNA is in a very dynamic state when extracted from any given cell type. As such, unspliced introns could result from errors in pre-mRNA splicing or other RNA processing events.

2.7.1 Functional annotation

Similar to flies and worms, a large number of mutants have been identified and mapped in Arabidopsis (Lloyd & Meinke, 2012). In fact, mutant phenotypes have been identified for ~10% of the genes. Lloyd and Meinke (2012) mapped loss- or change-of-function mutant phenotypes to 2400 loci (8.7% of annotated genes). Of these mutant phenotypes, 30% of the underlying genes are essential for early development and survival, 36% are responsible for morphology, 12% are responsible for cellular or biochemical pathways, and 22% were classified as conditional. In addition, they have identified a list of 401 genes that exhibit a mutant phenotype only when in combination with a mutation in a paralogous gene. In combination, there is phenotypic data for 10% of the Arabidopsis genes. Alonso and Ecker (2006) had generated and made available a collection of mapped T-DNA insertions in Arabidopsis. These lines are made available as homozygous knockouts by the biological resource centres and many thousands of gene knockouts have been screened for a variety of phenotypes. Screening genes in protein families has shown that most genes do not show an alteration in phenotype when individually disrupted (e.g. six sucrose synthase genes (Bieniawska et al., 2007), four members of the five genes in the *CRINKLY4* gene family (Cao, Li, Suh, Guo, & Becraft, 2005), four members of the UDP-glucuronic acid gene family (Kanter et al., 2005), three members of the MAPKKK gene family (Krysan, Jester, Gottwald, & Sussman, 2002), 23 of the 33 GRAS family members tested (Lee et al., 2008), and 54 of 55 subtilisin-like serine proteases tested (Rautengarten et al., 2005)).

2.7.2 Arabidopsis gene families

Gene family size variation is an important mechanism in plants that allows them to adapt to a changing environment. The Arabidopsis genes can be grouped into an estimated 9723 families (Guo, 2013), far fewer than had been identified by the AGI in 2000. Of these, 5980 are singletons, 1689 contain two members, and 2054 contain three or more members. These numbers are very similar to those observed for rice, poplar, sorghum and *A. lyrata* (Guo, 2013).

With so many plant genomes sequenced now, it is relatively easy to identify genes in one species that lack a homologous sequence in any other species (orphan genes). This is an interesting class of genes as their orphan status would suggest that they have arisen more recently in evolution. In fact, in Drosophila, Chen, Zhang and Long (2010) had shown that orphan genes can quickly become essential by playing a role in development. Of the 9723 gene families in Arabidopsis, 1328 (13.7%) are orphan gene families compared to rice where this percentage is much higher, 58.8% (Guo, 2013). The percentage of orphan gene families in the other angiosperms examined is 24.6% for *A. lyrata*, 36% for *S. bicolor* and 44.5% for poplar. In the five angiosperm species examined, orphan gene families occur mostly as singletons (96% for *A. thaliana* to 82% for poplar). For a more primitive

2.7.3 Discovering natural variation in Arabidopsis

To discover much of the natural variation in Arabidopsis, in 2008, researchers proposed generating DNA sequences from 1001 inbred strains of *A. thaliana*. These land races have adapted to various natural environments across the world and contain many types of polymorphisms that can be discovered by sequencing and comparison to the reference sequence described earlier. As of April 2013, 840 ecotypes had been sequenced, with the data for most of these released, and 31 were being sequenced. This project provides genotyping data that enable researchers to identify sequences that are responsible for phenotypic differences discovered from testing the various land races under a broad spectrum of conditions (Cao et al., 2011; Schneeberger et al., 2011).

3. EVOLUTIONARY HISTORY

Gene duplication events that are important for evolution can occur via four mechanisms: genome duplication, segmental duplication, tandem duplication and transposition events. All angiosperm species examined have undergone one or more whole-genome duplication events (Paterson, Freeling, Tang, & Wang, 2010). For Arabidopsis, there is evidence of three genome duplications/tripications (Blanc, Hokamp, & Wolfe, 2003; Bowers, Chapman, Rong, & Paterson, 2003). While there was some early controversy about the timing of these events due to limited genomic data available for comparison (Bowers et al., 2003), the sequencing of other related genomes has clarified that the triplication, referred to as γ, was earliest and shared by most if not all dicots (Jaillon et al., 2007; Tang, Bowers, et al., 2008; Tang, Wang, et al., 2008). The two duplications (β and more recent α) happened more recently than the divergence of Arabidopsis from the Brassicales (Ming et al., 2008), but prior to the divergence of Arabidopsis from Brassica 14.5–20.4 million years ago (Bowers et al., 2003). Tandem and segmental duplication events contributed much to the expansion of gene families (Cannon, Mitra, Baumgarten, Young, & May, 2004), which have also contributed to functional divergence. A number of gene families in Arabidopsis have been analysed against the same gene families in other plant species to gain insight into the evolution of various sets of genes. Studies of a few protein families are highlighted later.

3.1. Comparison of protein families

Of the more than 1000 protein kinases in Arabidopsis, the receptor-like kinase (RLK)/Pelle family is the largest gene family with more than 600 members (Lehti-Shiu, Zou, Hanada, & Shiu, 2009; Shiu et al., 2004). Rice not only has nearly twice as many RLK/Pelle genes with 1100 members but also has double the total number of genes (32,000–55,000; Goff et al., 2002; Yu et al., 2005). RLKs play important roles in plant growth, development and defence responses. The expansion of this family coincided with the establishment of land plants. Phylogenetic analysis of RLKs from Arabidopsis and rice suggests that the common ancestor of these taxa had >440 RLKs. For Arabidopsis, the expansions are attributed to both tandem and large-scale duplications, but for rice, tandem duplication seems to be the major mechanism. Further, the RLKs involved in development have not expanded since the Arabidopsis–rice split, but those involved in defence/disease have expanded, suggesting that the defence genes are under strong selective pressure (Shiu et al., 2004).

The number of TFs in Arabidopsis is even larger than the number of protein kinases. Guo et al. (2005) identified 1789 different TFs that fall into 49 families. One of the larger families of TFs to be studied phylogenetically across Arabidopsis and rice is the ERF family of genes. The Arabidopsis and rice genomes contain 122 and 139 ERF genes, respectively (Nakano, Suzuki, Fujimura, & Shinshi, 2006). Phylogenetic analysis shows that these genes can be divided into 12 and 15 groups in Arabidopsis and rice, respectively. Eleven of the groups are present in both species showing that much of the diversification occurred before the monocot–dicot split. The location of the 122 Arabidopsis genes in the genome shows that 90 of them are in previously identified duplicated segmental regions that resulted from a polyploidy that occurred around 24–40 million years ago, close to the emergence of the crucifer family (Nakano et al., 2006) and most probably the α duplication noted earlier. Approximately 75% of ERF genes, which lie within recently duplicated segmental chromosomes, have a clear paralog in these regions. Ten pairs of genes were due to tandem duplications. This finding is consistent with a previous report demonstrating that duplicated genes involved in signal transduction and transcription are preferentially retained (Blanc & Wolfe, 2004).

Another family of plant-specific TFs that have been studied across Arabidopsis and rice is the SBP-domain proteins (Squamosa promoter binding protein; 76 amino acids), which bind specifically to related motifs in the *Antirrhinum majus* SQUA promoter and the orthologous Arabidopsis AP1 promoter (Yang, Wang, Hu, Xu, & Xu, 2008). While Arabidopsis contains 17 SBP-domain genes, rice contains 19. Phylogenetic analysis indicates that these genes existed before the monocot–dicot split and that they expanded in number after the split. Phylogenetic analysis divides the genes into nine subgroups based on the motifs and their order in the protein. Analysis of nucleotide substitution rates revealed that the SBP domain has gone through purifying selection, whereas some regions outside the SBP domain have gone through positive or relaxed purifying selection (Yang, Wang, et al., 2008).

The CPP-like TF genes encode proteins with two similar Cys-rich domains termed CXC domains and are distributed widely in plants and animals (Yang, Gu, et al., 2008). Members of this gene family play a role in the control of cell division and development of reproductive tissues. Eight CPP-like genes were found in Arabidopsis and 11 in rice. Phylogenetic analysis of the CPP-like gene family results in two subfamilies (A and B) with both containing Arabidopsis and rice genes, suggesting that this gene family was also formed before the dicot–monocot split. Most interestingly, subfamily A could be divided into three distinct orthologous groups with A1 containing only dicot members and A2 containing only monocot members indicating that the subgroups likely diverged after the monocot–dicot split. Finally, the third group of CPP-like genes contained both monocot and dicot genes but no Arabidopsis genes indicating that Arabidopsis lost this set of genes during evolution. To ascertain whether gene expansion occurred via segmental or tandem duplication, the positions of the genes in Arabidopsis and rice were examined. In Arabidopsis, CPP-like genes are located on all four chromosomes except 1. One pair of genes, showing a close evolutionary relationship, was observed to be located in a tandem repeat indicating that the pair arose from a tandem duplication event. By analysing the position and the sequences of other members of this family, the authors concluded that several arose from segmental duplications. In rice, the 11 CPP-like genes were distributed on nine of the 12 chromosomes. One pair was located in a tandem repeat, but they did not show a close phylogenetic relationship so this pair was not likely to have arisen from a tandem duplication. Other members of the family had highly conserved genes around the flanking genes and were reported to be the result of segmental duplication events. Finally, two CPP-like genes in Arabidopsis and two in rice were hypothesized to have come into existence after the monocot–dicot split as suggested by their position on the phylogenetic tree (Yang, Gu, et al., 2008). There are additional similar phylogenetic analyses of gene families in Arabidopsis and rice that show similar patterns of duplications (e.g. the GRAS transcriptional regulator family, Tian et al., 2004; trehalose-6-phosphate synthase gene family, Yang et al., 2012).

Cyclophilins are ubiquitous proteins found in all organisms ranging from bacteria to mammals and act as molecular chaperones in various molecular and biochemical pathways. Cyclophilins have peptidylpropyl isomerase activity that facilitates efficient protein folding and are therefore needed in every cell type. Arabidopsis contains 35 cyclophilin genes, whereas rice contains 28 (Trivedi et al., 2012). Phylogenetic analysis of the cyclophilins in Arabidopsis and rice showed that the proteins were highly variable but more closely related to each other than to the cyclophilins from yeast. Sequence divergence among the cyclophilins in Arabidopsis and rice suggests that the species experienced different environments and therefore different selection pressures over the course of their evolution. Phylogenetic analysis suggests that the homologues have not arisen from tandem duplication events.

4. CONCLUSIONS

The sequence of the Arabidopsis genome has accelerated our understanding of specific genes as well as gene families more than we could have predicted when NSF proposed funding the sequencing project in 1989. We could not have envisioned the scientific advancements that would be made in related areas that make the complete genome more valuable such as full-length cDNAs (e.g. Haas et al., 2002), the use of T-DNA insertions in reverse genetics (Alonso & Ecker, 2006), microarray technology and next generation sequencing. In the early part of the sequencing project, academic scientists were concerned about the limited funding for research programmes and that the investment in sequencing a complete genome at that time would be too high. In fact, it took seven more years before the AGI was formed to complete the sequencing of the genome. In retrospect, having the sequence in hand years earlier would have been highly advantageous for advancing all of plant sciences. Arabidopsis has served as one of the most important (if not the most important) model plant species and has been, and continues to be, utilized to lead the way in many areas of plant biology. Now that the genome is completed, it is clear that we still have a lot to learn about, for example, (1) what constitutes a gene, (2) novel classes of regulation (epigenetics and nc RNAs and the role of alternative splicing and alternative transcription start sites and others yet to be discovered), (3) the evolution of genes and gene families and (4) how genes can be used to improve crops in either a breeding or a transgenic approach.

Guo, Y.-L. (2013). Gene family evolution in green plants with emphasis on the origination and evolution of *Arabidopsis thaliana* genes. *The Plant Journal*, *73*, 941–951.

Guo, A., Kun He, K., Liu, D., Bai, S., Gu, X., Wei, L., et al. (2005). DATF: A database of *Arabidopsis* transcription factors. *Bioinformatics*, *21*, 2568–2569.

Haas, B. J., Volfovsky, N., Town, C. D., Troukhan, M., Alexandrov, N., & Feldmann, K. A. (2002). Full-length messenger RNA sequences greatly improve genome annotation. *Genome Biology*, *3*, 1–12.

Haas, B. J., Wortman, J. R., Ronning, C. M., Hannick, L. I., Smith, R. K., Jr., Maiti, R., et al. (2005). Complete reannotation of the Arabidopsis genome: Methods, tools, protocols and the final release. *BMC Biology*, *3*, 7.

Hall, S. E., Kettler, G., & Preuss, D. (2003). Centromere satellites from Arabidopsis populations: Maintenance of conserved and variable domains. *Genome Research*, *13*, 195–205.

Hofte, H., Desprez, T., Amselem, J., Chiapello, H., Rouzé, P., Caboche, M., et al. (1993). An inventory of 1152 expressed sequence tags obtained by partial sequencing of cDNAs from Arabidopsis thaliana. *The Plant Journal*, *4*, 1051–1061.

Hosouchi, T., Kumekawa, N., Tsuruoka, H., & Kotani, H. (2002). Physical map-based sizes of the centromeric regions of *Arabidopsis thaliana* chromosomes 1, 2, and 3. *DNA Research*, *9*, 117–121.

Hu, T. T., Pattyn, P., Bakker, E. G., Cao, J., Cheng, J.-F., Clark, R. M., et al. (2011). The Arabidopsis lyrata genome sequence and the basis of rapid genome size change. *Nature Genetics*, *43*, 476–483.

Jaillon, O., Aury, J.-M., Noel, B., Policriti, A., Clepet, C., Casagrande, A., et al. (2007). The grapevine genome sequence suggests ancestral hexaploidization in major angiosperm phyla. *Nature*, *449*, 463–467.

Jang, J. Y., Kim, D. G., Kim, Y. O., Kim, J. S., & Kang, H. (2004). An expression analysis of a gene family encoding plasma membrane aquaporins in response to abiotic stresses in *Arabidopsis thaliana*. *Plant Molecular Biology*, *54*, 713–725.

Kanter, U., Usadel, B., Guerineau, F., Li, Y., Pauly, M., & Tenhaken, R. (2005). The inositol oxygenase gene family of Arabidopsis is involved in the biosynthesis of nucleotide sugar precursors for cell-wall matrix polysaccharides. *Planta*, *221*, 243–254.

Krysan, P. J., Jester, P. J., Gottwald, J. R., & Sussman, M. R. (2002). An Arabidopsis mitogen-activated protein kinase kinase kinase gene family encodes essential positive regulators of cytokinesis. *Plant Cell*, *14*, 1109–1120.

Lee, M.-H., Kim, B., Song, S.-K., Heo, J.-O., Yu, N.-I., Lee, S. A., et al. (2008). Large-scale analysis of the GRAS gene family in *Arabidopsis thaliana*. *Plant Molecular Biology*, *67*, 659–670.

Lehti-Shiu, M. D., Zou, C., Hanada, K., & Shiu, S.-H. (2009). Evolutionary history and stress regulation of plant receptor-like kinase/pelle genes. *Plant Physiology*, *150*, 12–26.

Leutwiler, L. S., Hough-Evans, B. R., & Meyerowitz, E. M. (1984). The DNA of Arabidopsis. *Molecular and General Genetics*, *194*, 15–23.

Lin, X., Kaul, S., Rounsley, S., Shea, T. P., Benito, M.-I., Town, C. D., et al. (1999). Sequence and analysis of chromosome 2 of the plant *Arabidopsis thaliana*. *Nature*, *402*, 761–768.

Ling, H. (2008). Sequence analysis of GDSL lipase gene family in *Arabidopsis thaliana*. *Pakistan Journal of Biological Sciences*, *11*, 763–767.

Liu, Y.-G., Mitsukawa, N., Vazquez-Tello, A., & Whittier, R. F. (1995). Generation of a high-quality P1 library of Arabidopsis suitable for chromosome walking. *The Plant Journal*, *7*, 351–358.

Liu, Y. G., Shirano, Y., Fukaki, H., Yanai, Y., Tasaka, M., Tabata, S., et al. (1999). Complementation of plant mutants with large genomic DNA fragments by a transformation-competent artificial chromosome vector accelerates positional cloning. *Proceedings of the National Academy of Sciences of the United States of America*, *96*, 6535–6540.

Lloyd, A. M., Barnason, A. R., Rogers, S. G., Byrne, M. C., Fraley, R. T., & Horsch, R. B. (1986). Transformation of Arabidopsis thaliana with *Agrobacterium tumefaciens*. *Science, 234*, 464–466.

Lloyd, J., & Meinke, D. (2012). A comprehensive dataset of genes with a loss-of-function mutant phenotype in Arabidopsis. *Plant Physiology, 158*, 1115–1129.

Loraine, A. E., McCormick, S., Estrada, A., Patel, K., & Qin, P. (2013). RNA-Seq of Arabidopsis pollen uncovers novel transcription and alternative splicing. *Plant Physiology, 162*, 1092–1109.

Louvet, R., Cavel, E., Gutierrez, L., Guenin, S., Roger, D., Gillet, F., et al. (2006). Comprehensive expression profiling of the pectin methylesterase gene family during silique development in *Arabidopsis thaliana*. *Planta, 224*, 782–791.

Marquez, Y., Brown, J. W. S., Simpson, C., Barta, A., & Kalyna, M. (2012). Transcriptome survey reveals increased complexity of the alternative splicing landscape in Arabidopsis. *Genome Research, 22*, 1184–1195.

Marra, M., Kucaba, T., Sekhon, M., Hillier, L., Martienssen, R., Chinwalla, A., et al. (1999). A map for sequence analysis of the *Arabidopsis thaliana* genome. *Nature Genetics, 22*, 265–270.

Mayer, K. F. X., Lemcke, K., Schuller, C. N., Rudd, S., & Zaccaria, P. (2000). Arabidopsis genome analysis as exemplified by analysis of chromosome 4. *Briefings in Bioinformatics, 1*, 389–397.

Mayer, K., Schuller, C., Wambutt, R., Murphy, G., Volckaert, G., Pohl, T., et al. (1999). Sequence and analysis of chromosome 4 of the plant *Arabidopsis thaliana*. *Nature, 208*, 769–777.

Meyerowitz, E. M. (1994). Structure and organization of the *Arabidopsis thaliana* nuclear genome. In E. M. Meyerowitz & C. R. Somerville (Eds.), *Arabidopsis* (pp. 21–36). New York, NY: Cold Spring Harbor Laboratory Press.

Meyerowitz, E. M., & Pruitt, R. E. (1985). *Arabidopsis thaliana* and plant molecular genetics. *Science, 229*, 1214–1218.

Ming, R., Hou, S., Feng, Y., Yu, Q., Dionne-Laporte, A., Saw, J. H., et al. (2008). The draft genome of the transgenic tropical fruit tree papaya (*Carica papaya Linnaeus*). *Nature, 452*, 991–996.

Miyashima, S., Honda, M., Hashimoto, K., Tatematsu, K., Hashimoto, T., Sato-Nara, K., et al. (2013). A comprehensive expression analysis of the Arabidopsis MICRORNA165/6 gene family during embryogenesis reveals a conserved role in meristem specification and a non-cell-autonomous function. *Plant and Cell Physiology, 54*, 375–384.

Mozo, T., Dewar, K., Dunn, P., Ecker, J. R., Fischer, S., Kloska, S., et al. (1999). A complete BAC-based physical map of the *Arabidopsis thaliana* genome. *Nature Genetics, 22*, 271–275.

Mozo, T., Fischer, S., Shizuya, H., & Altmann, T. (1998). Construction and characterization of the IGF Arabidopsis BAC library. *Molecular and General Genetics, 258*, 562–570.

Nakano, T., Suzuki, K., Fujimura, T., & Shinshi, H. (2006). Genome-wide analysis of the ERF gene family in Arabidopsis and rice. *Plant Physiology, 140*, 411–432.

Newman, T., de Bruijn, F. J., Green, P., Keegstra, K., Kende, H., McIntosh, L., et al. (1994). Genes galore: A summary of methods for accessing results from large-scale partial sequencing of anonymous Arabidopsis cDNA clones. *Plant Physiology, 106*, 1241–1255.

Okada, T., Endo, M., Singh, M. B., & Bhalla, P. L. (2005). Analysis of the histone H3 gene family in Arabidopsis and identification of the male-gamete-specific variant AtMGH3. *The Plant Journal, 44*, 557–568.

Okushima, Y., Overvoorde, P. J., Arima, K., Alonso, J. M., Chan, A., Chang, C., et al. (2005). Functional genomic analysis of the AUXIN RESPONSE FACTOR gene family members in Arabidopsis thaliana: Unique and overlapping functions of ARF7 and ARF19. *Plant Cell, 17*, 444–463.

Paterson, A. H., Bowers, J. E., Bruggmann, R., Dubchak, I., Grimwood, J., Gundlach, H., et al. (2009). The *Sorghum bicolor* genome and the diversification of grasses. *Nature, 457*, 551–556.

Paterson, A. H., Chapman, B. A., Kissinger, J. C., Bowers, J. E., Feltus, F. A., & Estill, J. C. (2006). Many gene and domain families have convergent fates following independent whole-genome duplication events in Arabidopsis, Oryza, Saccharomyces and Tetraodon. *Trends in Genetics, 22*, 597–602.

Paterson, A. H., Freeling, M., Tang, H., & Wang, X. (2010). Insights from the comparison of plant genome sequences. *Annual Review of Plant Biology, 61*, 349–372.

Potato Genome Sequencing Consortium (2011). Genome sequence and analysis of the tuber crop potato. *Nature, 475*, 189–195.

Pruitt, R. E., & Meyerowitz, E. M. (1986). Characterization of the genome of Arabidopsis thaliana. *Journal of Molecular Biology, 187*, 169–183.

Rautengarten, C., Steinhauser, D., Büssis, D., Stintzi, A., Schaller, A., Kopka, J., et al. (2005). Inferring hypotheses on functional relationships of genes: Analysis of the *Arabidopsis thaliana* subtilase gene family. *PLoS Computational Biology, 1*, e40.

Salanoubat, M., Lemcke, K., Rieger, M., Ansorge, W., Unseld, M., Fartmann, B., et al. (2000). Sequence and analysis of chromosome 3 of the plant *Arabidopsis thaliana. Nature, 408*, 820–822.

Sato, S., Kotani, H., Hayahi, R., Liu, Y.-G., Shibata, D., & Tabata, S. (1998). A physical map of *Arabidopsis thaliana* chromosome 3 represented by two contigs of CIC YAC, P1, TAC, and BAC clones. *DNA Research, 5*, 163–168.

Schmutz, J., Cannon, S. B., Schlueter, J., Ma, J., Mitros, T., Nelson, W., et al. (2010). Genome sequence of the palaeopolyploid soybean. *Nature, 463*, 178–183.

Schnable, P. S., Ware, D., Fulton, R. S., Stein, J. C., Wei, F., Pasternak, S., et al. (2009). The B73 maize genome: Complexity, diversity, and dynamics. *Science, 326*, 1112–1115.

Schneeberger, K., Ossowski, S., Ott, F., Klein, J. D., Wang, X., & Lanz, C. (2011). Reference-guided assembly of four diverse *Arabidopsis thaliana* genomes. *Proceedings of the National Academy of Sciences of the United States of America, 108*, 10249–10254.

Seki, M., Narusaka, M., Kamiya, A., Ishida, J., Satour, M., Sakurai, T., et al. (2002). Functional annotation of a full-length Arabidopsis cDNA collection. *Science, 296*, 141–145.

Shiu, S. H., Karlowski, W. M., Pan, R., Tzeng, Y. H., Mayer, K. F., & Li, W. H. (2004). Comparative analysis of the receptor-like kinase family in Arabidopsis and rice. *Plant Cell, 16*, 1220–1234.

Smalle, J., & Vierstra, R. D. (2004). The ubiquitin 26S proteasome proteolytic pathway. *Annual Review of Plant Biology, 55*, 555–590.

Sparrow, A. H., & Miksche, J. P. (1961). Correlation of nuclear volume and DNA content with higher plant tolerance to chromic radiation. *Science, 134*, 282–283.

Sparrow, A. H., Price, H. J., & Underbrink, A. G. (1972). A survey of DNA content per cell and per chromosome of prokaryotic and eukaryotic organisms: Some evolutionary considerations. *Brookhaven Symposia in Biology, 23*, 451–494.

Sung, D. Y., Vierling, E., & Guy, C. L. (2001). Comprehensive expression profile analysis of the Arabidopsis Hsp70 gene family. *Plant Physiology, 126*, 789–800.

Syed, N. H., Kalyna, M., Marquez, Y., Barta, A., & Brown, J. W. (2012). Alternative splicing in plants—Coming of age. *Trends in Plant Science, 17*, 616–623.

Tabata, S., Kaneko, T., Nakamura, Y., Kotani, H., Kato, T., Asamizu, E., et al. (2000). Sequence and analysis of chromosome 5 of the plant *Arabidopsis thaliana. Nature, 408*, 823–826.

Tang, H., Bowers, J. E., Wang, X., Ming, R., Alam, M., & Paterson, A. H. (2008). Synteny and collinearity in plant genomes. *Science, 320*, 486–488.

Tang, H., Wang, X., Bowers, J. E., Ming, R., Alam, M., & Paterson, A. H. (2008). Unraveling ancient hexaploidy through multiply-aligned angiosperm gene maps. *Genome Research, 18*, 1944–1954.

The Arabidopsis Genome Initiative (2000). Analysis of the genome sequence of the flowering plant *Arabidopsis thaliana*. *Nature*, *408*, 796–815.

The C. elegans Sequencing Consortium (1998). Genome Sequence of the Nematode *C. elegans*: A platform for investigating biology. *Science*, *282*, 2012–2018.

Theologis, A., Ecker, J., Palm, C. J., Federspiel, M. A., Kaul, S., White, O., et al. (2000). Sequence and analysis of chromosome 1 of the plant *Arabidopsis thaliana*. *Nature*, *408*, 816–820.

Tian, C., Wan, P., Sun, S., Li, J., & Chen, M. (2004). Genome-wide analysis of the GRAS gene family in rice and Arabidopsis. *Plant Molecular Biology*, *54*, 519–532.

Tognolli, M., Penel, C., Greppin, H., & Simon, P. (2002). Analysis and expression of the class III peroxidase large gene family in Arabidopsis thaliana. *Gene*, *288*, 129–138.

Tomato Genome Consortium (2012). The tomato genome sequence provides insights into fleshy fruit evolution. *Nature*, *485*, 635–641.

Trivedi, D. K., Yadav, S., Vaid, N., & Tuteja, N. (2012). Genome wide analysis of Cyclophilin gene family from rice and Arabidopsis and its comparison with yeast. *Plant Signaling & Behavior*, 7, 1653–1666.

Turlapati, P. V., Kim, K. W., Davin, L. B., & Lewis, N. G. (2011). The laccase multigene family in Arabidopsis thaliana: Towards addressing the mystery of their gene function(s). *Planta*, *233*, 439–470.

Tuskan, G. A., Difazio, S., Jansson, S., Bohlmann, J., Grigoriev, I., Hellsten, U., et al. (2006). The genome of black cottonwood, Populus trichocarpa (Torr. & Gray). *Science*, *313*, 1596–1604.

Umate, P., Tuteja, R., & Tuteja, N. (2010). Genome-wide analysis of helicase gene family from rice and Arabidopsis: A comparison with yeast and human. *Plant Molecular Biology*, *73*, 449–465.

Van Bel, M., Proost, S., Wischnitzki, E., Movahedi, S., Scheerlinck, C., Van de Peer, U., et al. (2012). Dissecting plant genomes with the PLAZA comparative genomics platform. *Plant Physiology*, *158*, 590–600.

Wang, K., Wang, Z., Li, F., Ye, W., Wang, J., Song, G., et al. (2012). The draft genome of a diploid cotton *Gossypium raimondii*. *Nature Genetics*, *44*, 1098–1103.

Wang, X., Wang, H., Wang, J., Sun, R., Wu, J., Liu, S., et al. (2011). The genome of the mesopolyploid crop species *Brassica rapa*. *Nature Genetics*, *43*, 1035–1039.

White, J. A., Todd, J., Newman, T., Focks, N., Girke, T., de Ilarduya, O. M., et al. (2000). A new set of Arabidopsis expressed sequence tags from developing seeds. The metabolic pathway from carbohydrates to seed oil. *Plant Physiology*, *124*, 1582–1594.

Yamada, K., Lim, J., Dale, J. M., Chen, H., Shinn, P., Palm, C. J., et al. (2003). Empirical analysis of transcriptional activity in the Arabidopsis genome. *Science*, *302*, 842–846.

Yang, Z., Gu, S., Wang, X., Li, W., Tang, Z., & Xu, C. (2008). Molecular evolution of the CPP-like gene family in plants: Insights from comparative genomics of Arabidopsis and rice. *Journal of Molecular Evolution*, *67*, 66–277.

Yang, H. L., Liu, Y. J., Wang, C. L., & Zeng, Q. Y. (2012). Molecular evolution of trehalose-6-phosphate synthase (TPS) gene family in Populus, Arabidopsis and rice. *PLoS One*, 7, e42438.

Yang, Z., Wang, X., Hu, Z., Xu, H., & Xu, C. (2008). Comparative study of SBP-box gene family in Arabidopsis and rice. *Gene*, *407*, 1–11.

Young, N. D., Debellé, F., Oldroyd, G. E. D., Geurts, R., Cannon, S. B., Udvardi, M. K., et al. (2012). The Medicago genome provides insight into the evolution of rhizobial symbioses. *Nature*, *480*, 520–524.

Yu, J., Wang, J., Lin, W., Li, S., Li, H., Zhou, J., et al. (2005). The genomes of Oryza sativa: A history of duplications. *PLoS Biology*, *3*, e38.

Zhang, X., Zong, J., Liu, J., Yin, J., & Zhang, D. (2010). Genome-wide analysis of WOX gene family in rice, sorghum, maize, Arabidopsis and poplar. *Journal of Integrative Plant Biology*, *52*, 1016–1026.

1. SEQUENCING STRATEGIES AND OUTCOME

Many plants have been introduced as staples in the human diet during the course of evolution. However, although there are 250,000–300,000 known edible plant species on earth, only about 150–200 species are used by humans (http://www.fao.org). The level of cultivation differs from one plant species to another; some are cultivated in a large scale to meet global demands whereas others are cultivated in a more limited scale for local consumption. The most widely cultivated plants are the main sources of energy and are often rich in protein, carbohydrate and/or fats. These include the major cereal crops, which are essentially available for cultivation in one or several growing seasons per year, easy to grow and harvest, suitable for long-term storage and adaptable to various cooking preferences.

Human evolution has been closely associated with the continuous quest for plants as sources of food, forage and fuel indispensable for sustaining life. In particular, one of the driving forces that have brought about the expansion of agriculture in the modern world is the intensification of the processes to extract more resources from the environment resulting in cultivation of the major sources of food to massive proportions. Currently, the major cereal crops such as rice, wheat and corn have a total world production of 0.48, 0.7 and 0.9 billion tons (USDA PSD online database), respectively, becoming the leading sources of food for about 7 billion people.

Among the major cereal crops, rice is grown in about 148 million hectares or 3% of the world's agricultural lands, ranging from flooded fields to dry lands under temperate, subtropical and tropical climates. Deliberate human interventions in the last 10,000 years that include selection and breeding resulted in tremendous improvements in rice productivity, tolerance to biotic and abiotic stresses and adaptability to a wide range of soil and climate. Two types of Oryza species are currently domesticated, *Oryza sativa* cultivated mainly in China and most parts of Asia and *Oryza glaberrima* cultivated in limited regions in Africa. The cultivation of O. *sativa* originated in Southeast Asia and is therefore referred to as Asian rice. On the other hand, O. *glaberrima* is limited to West Africa, hence the reference to African rice. As the main species currently cultivated worldwide, O. *sativa* is further categorized into two subspecies, O. *sativa* ssp. *japonica* and O. *sativa* ssp. *indica*, based on ecogeographic adaptation, morphological features of flowers and seeds, eating texture, habitation and crossing ability among others. Furthermore, recent developments in phylogenetic analysis using sequence polymorphism revealed that *O. sativa* is divided into five groups based on genetic structure, namely, *indica*, *aus*, aromatic, temperate *japonica* and tropical *japonica* (Garris, Tai, Coburn, Kresovich, & McCouch, 2005).

Rice is the major source of food for about half of the world population, mainly living in Asia, Africa, and Latin America. Rapid increase in population in the last 50–60 years has necessitated the breeding of high-yielding rice varieties to provide a stable food supply for mankind. Two agronomic traits, namely, semidwarfism (Khush, 1990) and hybrid vigour (Yuan, 1994), have played major roles in the quest for rice varieties with high yield. Both relied on rice plants with mutations. Semidwarfism is caused by mutation of the gene GA20 oxidase involved in gibberellin biosynthesis (Ashikari et al., 2002). On the other hand, although hybrid vigour is a well-known phenomenon in genetics, the exact molecular mechanism by which it confers increased yield has not yet been fully elucidated (Schnable & Springer, 2013). Finding a male sterile mutant of wild rice has enabled the production of hybrid rice with about a 1.3-fold increase in yield. These breeding strategies have had significant impacts on the world food supply in the last three to four decades. However, rice production must increase by more than 50% over the next three to four decades to keep up with the continuing increase in the world population. This time, the challenge is even more enormous as new rice varieties with higher yield will need to be grown under impending constraints brought about by environmental degradation, rapid depletion of arable lands and water resources and global warming. To accomplish these goals, a genetic blueprint of the rice plant, the genome sequence, has become indispensable (Sasaki & Burr, 2000).

As in other species, rice genome analysis started with the development of polymorphic DNA markers such as RAPD, SSR, AFLP and RFLP followed by extensive mapping of these markers to provide early insight into the genetic structure of rice (Harushima et al., 1998; Kurata et al., 1994; McCouch et al., 1988). A mapping population was generated by crossing a *japonica* 'Nipponbare' and an *indica* 'Kasalath' to obtain a high frequency of polymorphism. Analysis and use of rice expressed sequence tags (ESTs) greatly facilitated the development of genetic markers with primer sequences for PCR (Sasaki et al., 1994). The genetically mapped ESTs were converted to 6591 PCR-based markers and used for reliable ordering of rice DNA fragments cloned in YAC, BAC and PAC vectors along the 12 chromosomes (Wu et al., 2002). Additionally, the rice genome was reconstructed after digesting rice BAC clones by restriction enzymes and assembling the clones into contigs (Soderlund, Humphray, Dunham,

& French, 2000). This method does not necessarily require any sequence information although alignment of the assembled BAC contigs requires genetic markers and BAC end sequences to assure correct reconstruction of the genome.

In September 1997, the international rice scientific community adopted a collaborative effort to sequence the rice genome. A consortium consisting of research organizations from 10 countries or regions organized the International Rice Genome Sequencing Project (IRGSP) with the ultimate goal of decoding the genome sequence of the *japonica* rice variety, 'Nipponbare' (Sasaki & Burr, 2000). The basic goal and standard agreement in IRGSP was to obtain a precisely map-based, accurate genome sequence with less than one base-pair (bp) error in 10,000 bp. This quality was achieved through a combination of high-quality shotgun sequence reads, sevenfold redundancy and sequencing of all bases on both strands using two chemistries. In the case of rice, a high level of accuracy is indispensable so that the resulting genome sequence could be used as the fundamental reference for SNP discovery among the thousands of rice varieties worldwide, map-based cloning of agronomically important genes and comparative analysis among plant taxa based on syntenic relationships. In 2004, the IRGSP finished the genome sequencing of *O. sativa* ssp. *japonica* 'Nipponbare', representing the first monocot and first cereal crop to be completely sequenced. The total sequence length of 370.7 Mb corresponds to 95% of the 389 Mb rice genome, including virtually all of the euchromatin and complete centromeres of chromosomes 4 and 8 (Nagaki et al., 2004; Wu et al., 2004; Zhang et al., 2004). Later, the structure and gene expression pattern of the centromere region of chromosome 5 was also elucidated (Mizuno et al., 2011). Additionally, BAC/PAC clones carrying telomere-specific repeat sequences, CCCTAAA, were identified at 15 of the 24 rice telomeres, an indication that the physical map of rice was also largely covered. This map-based genome sequence information facilitated a comprehensive characterization of rice genome structure including the gene content, tandem gene duplication, segmental duplication, types of transposable elements and organellar insertions in the nuclear genome (IRGSP, 2005).

The efforts of the IRGSP were complemented with the establishment of the Rice Annotation Project (RAP), an international collaborative organization launched in 2005 to undertake a comprehensive characterization of the genome sequence via evidence-based and reliable annotation. The primary concept of RAP is to map the full-length cDNAs onto the reference genome sequence (Sakai et al., 2013; The Rice Annotation Project, 2007;

The Rice Full-Length cDNA Consortium, 2003). An annotation database, named Rice Annotation Project Database (RAP-DB, http://rapdb.dna.affrc.go.jp/), has been developed to provide a comprehensive set of gene annotations for the entire genome sequence. Since the completion of the IRGSP pseudomolecules in 2004, the rice genome assembly has been updated twice. The first update was made in 2008 with the addition of seven new telomeres or telomere-associated sequences, revision of the chromosome 5 centromere sequence and gap-filling on chromosome 11. In the second update in 2011, sequence errors were comprehensively corrected using next-generation sequencing data as a joint collaboration of the RAP and the Michigan State University Rice Genome Annotation Project resulting in a genome assembly almost 10× more accurate than the first published genome (Kawahara et al., 2013). This latest genome assembly is now referred to as Os-Nipponbare-Reference-IRGSP-1.0 or IRGSP-1.0 for short. Although there are still two main rice annotation databases, RAP-DB (http://rapdb.dna.affrc.go.jp/) and MSU Rice Genome Annotation Project DB (http://rice.plantbiology.msu.edu/), the genome sequence is now unified and both annotations can now be easily compared to each other. In the case of RAP-DB, a total of 37,869 loci including 35,679 protein-coding and 2190 non-protein-coding loci were identified (Sakai et al., 2013). In particular, the RAP-DB has incorporated literature-based manually curated data, commonly used gene names, gene symbols and RNA-seq transcriptome data. The Illumina sequence reads of a leading Japanese *japonica* 'Koshihikari' and a Chinese *indica* 'Guangluai-4' have been incorporated in a browser for short-read assemblies to show alignments, SNPs and gene functional annotations. These additional features further enhance the utility of RAP-DB for structural and functional characterization of the genome such as analysis of disrupted genes, comparative analysis of syntenic relationship among Oryza or Poaceae species and other purposes (Bolot et al., 2009; Goicoechea et al., 2010).

So far, the map-based high-quality 'Nipponbare' rice genome sequence has effectively accelerated basic research on rice biology and applied research in agriculture. However, rice is a complex plant system and many cultivars have been bred among *indica*, *aus*, aromatic, temperate *japonica* and tropical *japonica* types. Therefore, a high-quality map-based sequence of an *indica* genome is highly desirable. The whole-genome sequences for two cultivated subspecies of rice should more effectively enable both rice researchers and breeders to directly identify many biological processes with direct roles in agricultural productivity and will offer great opportunities for comparative

gene number and content as shown in hexaploid wheat and grape (Brenchley et al., 2012; Velasco et al., 2007).

Comparative analyses have also revealed conservation of gene structures and nucleotide compositions around the genic regions between rice and Arabidopsis. The intron positions in genic regions and nucleotide compositions around the transcription start sites were also conserved (Roy & Penny, 2007; Tanaka, Koyanagi, & Itoh, 2009). In comparative genomics, these conserved characteristics represent one of the universal trends in plant evolution.

As expected, rice and Arabidopsis differ in genome structure at several levels. For example, the difference in genome size was more reflected in intron length (423 bp in rice and 168 bp in Arabidopsis) than exon length (The Rice Annotation Project, 2007). Moreover, homology search showed thousands of lineage-specific genes and functional diversification has been reported even in orthologs. For example, genes associated with regulation of flowering time are highly conserved among plants. However, the gene functions are highly diversified. *CONSTANS* (*CO*) was isolated as a gene promoting flowering time in long-day (LD) conditions in Arabidopsis (Putterill, Robson, Lee, Simon, & Coupland, 1995). On the other hand, *Hd1* gene in rice is a *CO* ortholog and promotes flowering in short-day (SD) but delays it in LD conditions (Yano et al., 2000). The functional diversification of these CO orthologs has occurred after speciation in each lineage or in one lineage, even though the orthologs function in the same biological process. The different function does not reflect dichotomy between monocots and dicots, because barley and wheat are LD plants with *CO*-like genes that affect flowering time in parallel with the Arabidopsis *CO* gene (Griffiths, Dunford, Coupland, & Laurie, 2003).

Recently, alternative variants and novel transcribed regions have been detected based on transcriptome data generated by RNA-seq (Lu et al., 2010; Marquez, Brown, Simpson, Barta, & Kalyna, 2012). In addition to protein-coding RNAs, functional RNA genes such as microRNA (miRNA), small interfering RNAs (siRNAs) and long intergenic noncoding RNAs (lincRNAs) have been widely identified. The miRBase (http://www.mirbase.org) contains 708 and 338 mature sequences for rice and Arabidopsis (release 19) with 547 and 271 nonredundant sequences, respectively, and 19 identical miRNAs in the two species. Among them, 133 and 95 nonredundant sequences in rice and Arabidopsis, respectively, were homologous miRNAs (with less than three mismatches), suggesting that similar to protein-coding genes, mature miRNAs were highly conserved between monocots and dicots. The lncRNAs have been studied extensively in Arabidopsis, and 6480 lincRNAs were predicted in reverse strands of coding regions, introns and intergenic regions (Liu et al., 2012). As is true of gene diversification, regulation of gene expression by these RNA genes and other factors such as methylation and posttranscript modification has generated species-specific characteristics.

Comparative analyses of the genomes of rice and Arabidopsis provide evidence on similarities and differences that are characteristic of the two major groups of higher plant species. As the first two species to be sequenced and with the most complete high-quality genome sequence, these two plants serve as reference genomes for a wide range of monocot and dicot species. Moreover, the highest quality genome sequences and annotation data also contribute to the many genome analyses in plants. One of the most advantageous analyses using genome-wide annotation data is to elucidate existence of a particular gene, especially finding lost genes and mutated genes. Even though expression and function of genes can be evaluated by experiments, lack of evidence of expression does not prove the absence of a gene since many genes are expressed only briefly and/or at only very low levels. Genome sequence enables us to definitively determine the presence or absence of genes and gene orientation and gene colinearity. However, many draft genome sequences determined by whole-genome shotgun sequencing and NGS assembly can only be assembled into many contigs and scaffolds, with insufficient positional information for physical mapping. Then, unlike annotation data of rice and Arabidopsis, the predicted gene sets may show erroneous results. In addition, TAIR in Arabidopsis and RAP in rice continue to maintain the annotation data into which the latest experimental data are incorporated. Even though new methods to concatenate contigs have been developed, the quality of assembly might not reach the quality of rice and Arabidopsis genomes, unless read lengths are much longer and the number of experimental validations is increased. In this context, rice and Arabidopsis may remain as the most reliable platforms for comprehensive comparative genome analysis even beyond the post-genome sequencing era.

3. EVOLUTIONARY HISTORY (ESPECIALLY GENOME DUPLICATION)

One interesting feature of the genome evolution of genus *Oryza* is the variation in genome size among species. Among the major genome groups, *Oryza australiensis* has the largest diploid genome, which is more than twice as large as that of *O. sativa*, whereas *O. glaberrima* is almost one-third that of *O. australiensis* (Ammiraju et al., 2006). One of the evolutionary processes driving the variation in genome size is the proliferation of transposable elements. Genome survey analysis of 12 *Oryza* species revealed that LTR retrotransposons, particularly the two families of Ty3–*gypsy* elements, accounted for the genome size variations (Zuccolo et al., 2007). Piegu et al. (2006) revealed that 60% of the genome of *O. australiensis* was composed of three types of LTR retrotransposons. Bursts of LTR retrotransposons appear to have occurred after the radiation of *O. australiensis* estimated at 8.5 MYA, resulting in doubling of genome size (Piegu et al., 2006). The two subspecies of rice, *japonica* and *indica*, gained an estimated 2% and 6% increase in genome sizes, respectively, after their divergence from a common ancestor shared with *O. glaberrima*, which primarily resulted from the amplification of LTR retrotransposons (Ma & Bennetzen, 2004). However, the rice genome appeared to have experienced extensive DNA losses during evolution, with an estimated two-thirds of the LTR-retrotransposon sequences eliminated through unequal homologous recombination and illegitimate recombination within the last 8 million years (Ma, Devos, & Bennetzen, 2004). These results suggested that retrotransposon insertion and deletion are the primary factors governing genome size variation among the species (Bennetzen, Ma, & Devos, 2005).

Another process leading to the genome size expansion is polyploidization, in which a genome is doubled either through the hybridization of two species or by the formation of unreduced gametes. Although the extant cultivated rice (*O. sativa*) is a diploid, nine wild species are recognized as allotetraploid (Ge, Sang, Lu, & Hong, 1999, Vaughan, Morishima, & Kadowaki, 2003). The *Oryza officinalis* complex contains two allotetraploid genome types, BBCC and CCDD, including six wild species (Ge et al., 1999; Vaughan et al., 2003). The remaining three species include two HHJJ tetraploids and one species with ambiguous (HHKK) genome type (Vaughan et al., 2003). Phylogenetic studies of a limited number of genes suggested that the most recent polyploidization event occurred 0.3–0.6 MYA, resulting in the formation of the BBCC tetraploids (Lu et al., 2009; Wang et al., 2009). CCDD tetraploids were derived 0.9–1.6 MYA (Lu et al., 2009; Wang et al., 2009). HHJJ tetraploids were derived earlier than the BBCC and CCDD formation (Ge et al., 1999). Despite the recent formation of the tetraploids, each genome in each species has experienced different and dynamic evolutionary processes. Multicolour fluorescent *in situ* hybridization (FISH) showed intergenomic translocations in some BBCC tetraploids (Wang et al., 2009). Lu et al. (2009) revealed that 5% of the duplicated genes derived from allopolyploidy events were pseudogenized in BBCC and CCDD tetraploids, whereas one-third of the duplicated genes were pseudogenized in HHJJ and HHKK tetraploids. These analyses suggested that allopolyploidy events occurred recurrently in the genus *Oryza* and that genome evolution in the allotetraploids is an ongoing process. Although no genome sequence of allotetraploid *Oryza* is available so far, the International *Oryza* Map Alignment Project (OMAP) was organized in 2007 to generate reference genome sequences of eight AA genome species and a representative species of nine genome types including four allotetraploids. Complete genome sequences of the allotetraploids and their progenitors will enhance understanding of the mechanisms by which the genomes of allotetraploids have been shaped, changed and maintained through evolution.

It is well established that most if not all flowering plant species underwent whole-genome duplication (WGD) events during their evolution (Van de Peer, Fawcett, Proost, Sterck, & Vandepoele, 2009) (see also Chapters 8 and 9). We can observe remnants of WGD events even in the compact genome of rice. Genome-wide intragenomic comparison of gene order revealed that almost half or more of the rice genes were associated with the genome duplication events, indicating that ancient polyploidy occurred in the ancestral genome (Paterson, Bowers, & Chapman, 2004; Wang, Shi, Hao, Ge, & Luo, 2005). There are nine major duplicated blocks maintained in the rice genome (Paterson et al., 2004), which is different from the extensive rearrangements of the *Arabidopsis thaliana* genome (Blanc, Barakat, Guyot, Cooke, & Delseny, 2000; Bowers, Chapman, Rong, & Paterson, 2003). Despite the maintenance of the large duplicated blocks, size of the individual blocks varied from less than a few Mbp to over 10 Mbp (Wang et al., 2005), suggesting that complex intra- and interchromosome rearrangements occurred also in the rice genome.

Distribution of the sequence divergence between syntenic duplicated genes estimated as the number of synonymous substitutions (Ks) revealed that gene pairs in almost all duplicated blocks showed comparable Ks values, suggesting that the gene duplications resulted from one WGD event (denoted as ρ) inferred to have occurred about 70 MYA (Paterson et al., 2004; Wang et al., 2005). One exception is the duplication block involving terminal segments of chromosomes 11 and 12 in which gene pairs showed extremely low Ks values, suggesting that large-scale segmental duplication occurred very recently, 5–7 MYA (The Rice Chromosomes 11 and 12 Sequencing Consortia, 2005; Wang et al., 2005; Yu et al., 2005). However, the shared terminal segments were observed also in other cereal genomes such as sorghum (Paterson et al., 2009), wheat (Singh et al.,

Chaw, S. M., Chang, C. C., Chen, H. L., & Li, W. H. (2004). Dating the monocot-dicot divergence and the origin of core eudicots using whole chloroplast genomes. *Journal of Molecular Evolution, 58*, 424–441.

Devos, K. M., Pittaway, T. S., Reynolds, A., & Gale, M. D. (2000). Comparative mapping reveals a complex relationship between the pearl millet genome and those of foxtail millet and rice. *Theoretical and Applied Genetics, 100*, 190–198.

D'Hont, A., Denoeud, F., Aury, J. M., Baurens, F. C., Carreel, F., Garsmeur, O., et al. (2012). The banana (*Musa acuminata*) genome and the evolution of monocotyledonous plants. *Nature, 488*, 213–217.

Garris, A. J., Tai, T. H., Coburn, J., Kresovich, S., & McCouch, S. (2005). Genetic structures and diversity in *Oryza sativa* L. *Genetics, 169*, 1631–1638.

Ge, S., Sang, T., Lu, B. R., & Hong, D. Y. (1999). Phylogeny of rice genomes with emphasis on origins of allotetraploid species. *Proceedings of the National Academy of Sciences of the United States of America, 96*, 14400–14405.

Goicoechea, J. L., Ammiraju, J. S. S., Marri, P. R., Chen, M., Jackson, S., Yu, Y., et al. (2010). The future of rice genomics: Sequencing the collective *Oryza* genome. *Rice, 3*, 89–97.

Griffiths, S., Dunford, R. P., Coupland, G., & Laurie, D. A. (2003). The evolution of *CONSTANS*-like gene families in barley, rice and Arabidopsis. *Plant Physiology, 131*, 1855–1867.

Harushima, Y., Yano, M., Shomura, A., Sato, M., Shimano, T., Kuboki, Y., et al. (1998). A high-density rice genetic linkage map with 2275 markers using a single F2 population. *Genetics, 148*, 479–494.

Huala, E., Dickerman, A. W., Garcia-Hernandez, M., Weems, D., Reiser, L., LaFond, F., et al. (2001). The Arabidopsis Information Resource (TAIR): A comprehensive database and web-based information retrieval, analysis, and visualization system for a model plant. *Nucleic Acids Research, 29*, 102–105.

Huang, X., Kurata, N., Wei, X., Wang, Z.-W., Wang, A., Zhao, Q., et al. (2012). A map of rice genome variation reveals the origin of cultivated rice. *Nature, 490*, 497–501.

Imanishi, T., Itoh, T., Suzuki, Y., O'Donovan, C., Fukuchi, S., Koyanagi, K. O., et al. (2004). Integrative annotation of 21,037 human genes validated by full-length cDNA clones. *PLoS Biology, 2*, e162.

International Rice Genome Sequencing Project (2005). The map-based sequence of the rice genome. *Nature, 436*, 793–800.

Kawahara, Y., de la Bastide, M., Hamilton, P. J., Kanamori, H., McCombie, R. W., Ouyang, S., et al. (2013). Improvement of the *Oryza sativa* Nipponbare reference genome using next generation sequence and optical map data. *Rice, 6*, 4.

Kellogg, E. A. (1998). Relationships of cereal crops and other grasses. *Proceedings of the National Academy of Sciences of the United States of America, 95*, 2005–2010.

Khush, G. S. (1990). Varietal needs for different environments and breeding strategies. In K. Muralidharan & E. A. Siddiq (Eds.), *New frontiers in rice research* (pp. 68–75). Hyderabad, India: Directorate of Rice Research.

Kurata, N., Nagamura, Y., Yamamoto, K., Harushima, Y., Sue, N., Wu, J., et al. (1994). A 300-kilobase-interval genetic map of rice including 883 expressed sequences. *Nature Genetics, 8*, 365–372.

Leroy, P., Guilhot, N., Sakai, H., Bernard, A., Choulet, F., Theil, S., et al. (2012). TriAnnot: A versatile and high performance pipeline for the automated annotation of plant genomes. *Frontiers in Plant Science, 3*, 1–14.

Liu, J., Jung, C., Xu, J., Wang, H., Deng, S., Bernad, L., et al. (2012). Genome-wide analysis uncovers regulation of long intergenic noncoding RNAs in *Arabidopsis*. *Plant Cell, 24*, 4333–4345.

Lu, F., Ammiraju, J. S., Sanyal, A., Zhang, S., Song, R., Chen, J., et al. (2009). Comparative sequence analysis of MONOCULM1-orthologous regions in 14 *Oryza* genomes. *Proceedings of the National Academy of Sciences of the United States of America, 106*, 2071–2076.

Lu, T., Lu, G., Fan, D., Zhy, C., Li, W., Zhao, Q., et al. (2010). Function annotation of the rice transcriptome at single-nucleotide resolution by RNA-seq. *Genome Research, 20*, 1238–1249.

Ma, J., & Bennetzen, J. L. (2004). Rapid recent growth and divergence of rice nuclear genomes. *Proceedings of the National Academy of Sciences of the United States of America, 101*, 12404–12410.

Ma, J., Devos, K. M., & Bennetzen, J. L. (2004). Analyses of LTR-retrotransposon structures reveal recent and rapid genomic DNA loss in rice. *Genome Research, 14*, 860–869.

Makino, T., & McLysaght, A. (2010). Ohnologs in the human genome are dosage balanced and frequently associated with disease. *Proceedings of the National Academy of Sciences of the United States of America, 107*, 9270–9274.

Marquez, Y., Brown, J. W., Simpson, C., Barta, A., & Kalyna, M. (2012). Transcriptome survey reveals increased complexity of the alternative splicing landscape in *Arabidopsis*. *Genome Research, 22*, 1184–1195.

McCouch, S. R., Kochert, G., Yu, Z. H., Wang, Z. Y., Khush, G. S., Coffman, W. R., et al. (1988). Molecular mapping of rice chromosomes. *Theoretical and Applied Genetics, 76*, 815–829.

Misra, S., Crosby, M. A., Mungall, C. J., Matthews, B. B., Campbell, K. S., Hradecky, P., et al. (2002). Annotation of the *Drosophila melanogaster* euchromatic genome: A systematic review. *Genome Biology, 3*, research0083.1–0083.22.

Mizuno, H., Kawahara, Y., Wu, J., Katayose, Y., Kanamori, H., Ikawa, H., et al. (2011). Asymmetric distribution of gene expression in the centromeric region of rice chromosome 5. *Frontiers in Plant Science, 2*, 16.

Nagaki, K., Cheng, Z., Ouyang, S., Talbert, P. B., Kim, M., Jones, K. M., et al. (2004). Sequencing of a rice centromere uncovers active gene. *Nature Genetics, 36*, 138–145.

Ohyanagi, H., Tanaka, T., Sakai, H., Shigenomoto, Y., Yamaguchi, K., Habara, T., et al. (2006). The rice annotation project database (RAP-DB): Hub for *Oryza sativa* ssp. *japonica* genome information. *Nucleic Acids Research, 34*, D741–D744.

Oono, Y., Kawahara, Y., Kanamori, H., Mizuno, H., Yamagata, H., Yamamoto, M., et al. (2011). mRNA-Seq reveals a comprehensive transcriptome profile of rice under phosphate stress. *Rice, 4*, 50–65.

Paterson, A. H., Bowers, J. E., Bruggmann, R., Dubchak, I., Grimwood, J., Gundlach, H., et al. (2009). The *Sorghum bicolor* genome and the diversification of grasses. *Nature, 457*, 551–556.

Paterson, A. H., Bowers, J. E., & Chapman, B. A. (2004). Ancient polyploidization predating divergence of the cereals, and its consequences for comparative genomics. *Proceedings of the National Academy of Sciences of the United States of America, 101*, 9903–9908.

Paterson, A. H., Chapman, B. A., Kissinger, J. C., Bowers, J. E., Feltus, F. A., & Estill, J. C. (2006). Many gene and domain families have convergent fates following independent whole-genome duplication events in *Arabidopsis*, *Oryza*, *Saccharomyces* and *Tetraodon*. *Trends in Genetics, 22*, 597–602.

Piegu, B., Guyot, R., Picault, N., Roulin, A., Sanyal, A., Kim, H., et al. (2006). Doubling genome size without polyploidization: Dynamics of retrotransposition-driven genomic expansions in *Oryza australiensis*, a wild relative of rice. *Genome Research, 16*, 1262–1269.

Putterill, J., Robson, F., Lee, K., Simon, R., & Coupland, G. (1995). The *CONSTANS* gene of Arabidopsis promotes flowering and encodes a protein showing similarities to zinc finger transcription factors. *Cell, 80*, 847–857.

Roy, S. W., & Penny, D. (2007). Patterns of intron loss and gain in plants: Intron loss-dominated evolution and genome-wide comparison of *O. sativa* and *A. thaliana*. *Molecular Biology and Evolution, 24*, 171–181.

Sakai, H., Lee, S. S., Tanaka, T., Numa, H., Kim, J., Kawahara, Y., et al. (2013). Rice Annotation Project Database (RAP-DB): An integrative and interactive database for rice genomics. *Plant Cell Physiology, 54*, e6.

Salmon, A., Ainouche, M. L., & Wendel, J. F. (2005). Genetic and epigenetic consequences of recent hybridization and polyploidy in Spartina (Poaceae). *Molecular Ecology, 14*, 1163–1175.

Sasaki, T., & Burr, B. (2000). International Rice Genome Sequencing Project: The effort to completely sequence the rice genome. *Current Opinion in Plant Biology, 3*, 138–141.

Sasaki, T., Song, J., Koga-Ban, Y., Matsui, E., Fang, F., Higo, H., et al. (1994). Toward cataloguing all rice genes: Large scale sequencing of randomly chosen rice cDNAs from a callus cDNA library. *Plant Journal, 6*, 615–624.

Schnable, P. S., & Springer, N. M. (2013). Progress toward understanding heterosis in crop plants. *Annual Review of Plant Biology, 64*, 71–88.

Singh, N. K., Dalal, V., Batra, K., Singh, B. K., Chitra, G., Singh, A., et al. (2007). Single-copy genes define a conserved order between rice and wheat for understanding differences caused by duplication, deletion, and transposition of genes. *Functional and Integrative Genomics, 7*, 17–35.

Soderlund, C., Humphray, S., Dunham, A., & French, L. (2000). Contigs built with fingerprints, markers, and FPC V4.7. *Genome Research, 10*, 1772–1787.

Tanaka, T., Koyanagi, K. O., & Itoh, T. (2009). Highly diversified molecular evolution of downstream transcription start sites in rice and Arabidopsis. *Plant Physiology, 149*, 1316–1324.

Tang, H., Bowers, J. E., Wang, X., & Paterson, A. H. (2010). Angiosperm genome comparisons reveal early polyploidy in the monocot lineage. *Proceedings of the National Academy of Sciences of the United States of America, 107*, 472–477.

The Rice Annotation Project (2007). Curated genome annotation *of Oryza sativa* ssp. *japonica* and comparative genome analysis with *Arabidopsis thaliana*. *Genome Research, 17*, 175–183.

The Rice Chromosomes 11 and 12 Sequencing Consortia (2005). The sequence of rice chromosomes 11 and 12, rich in disease resistance genes and recent gene duplications. *BMC Biology, 3*, 20.

The Rice Full-Length cDNA Consortium (2003). Collection, mapping, and annotation of over 28,000 cDNA clones from japonica rice. *Science, 301*, 376–379.

Van de Peer, Y., Fawcett, J. A., Proost, S., Sterck, L., & Vandepoele, K. (2009). The flowering world: A tale of duplications. *Trends in Plant Sciences, 14*, 680–688.

Vaughan, D. A., Morishima, H., & Kadowaki, K. (2003). Diversity in the *Oryza* genus. *Current Opinion in Plant Biology, 6*, 139–146.

Velasco, R., Zharkikh, A., Troggio, M., Cartwright, D. A., Cestaro, A., Pruss, D., et al. (2007). A high quality draft consensus sequence of the genome of a heterozygous grapevine variety. *PLoS One, 19*, e1326.

Wang, B., Ding, Z., Liu, W., Pan, J., Li, C., Ge, S., et al. (2009). Polyploid evolution in *Oryza officinalis* complex of the genus *Oryza*. *BMC Evolutionary Biology, 9*, 250.

Wang, X., Shi, X., Hao, B., Ge, S., & Luo, J. (2005). Duplication and DNA segmental loss in the rice genome: Implications for diploidization. *New Phytologist, 165*, 937–946.

Wang, X., Tang, H., Bowers, J. E., Feltus, F. A., & Paterson, A. H. (2007). Extensive concerted evolution of rice paralogs and the road to regaining independence. *Genetics, 177*, 1753–1763.

Wu, J., Maehara, T., Shimokawa, T., Yamamoto, S., Harada, C., Takazaki, Y., et al. (2002). A comprehensive rice transcript map containing 6,591 expressed sequence tag sites. *Plant Cell, 14*, 525–535.

Wu, J., Yamagata, H., Hayashi-Tsugane, M., Hijishita, S., Fujisawa, M., Shibata, M., et al. (2004). Composition and structure of the centromeric region of rice chromosome 8. *Plant Cell, 16*, 967–976.

Xiong, Y., Liu, T., Tian, C., Sun, S., Li, J., & Chen, M. (2005). Transcription factors in rice: A genome-wide comparative analysis between monocots and eudicots. *Plant Molecular Biology, 59*, 191–203.

Yano, M., Katayose, Y., Ashikari, M., Yamanouchi, U., Monna, L., Fuse, T., et al. (2000). *Hd1*, a major photoperiod sensitivity quantitative trait locus in rice, is closely related to the Arabidopsis flowering time gene *CONSTANS*. *Plant Cell, 12*, 2473–2483.

Yu, J., Wang, J., Lin, W., Li, S., Li, H., Zhou, J., et al. (2005). The genomes of *Oryza sativa*: A history of duplications. *PLoS Biology, 3*, e38.

Yuan, L. P. (1994). Increasing yield potential in rice by exploiting heterosis. In S. S. Virmani (Ed.), *Hybrid rice technology: New developments and future prospects* (pp. 1–6). Los Banos, Philippines: IRRI.

Yuan, Q., Ouyang, S., Liu, J., Suh, B., Cheung, F., Sultana, R., et al. (2003). The TIGR rice genome annotation resource: Annotating the rice genome and creating resources for plant biologists. *Nucleic Acids Research, 31*, 229–233.

Zhang, Y., Huang, Y., Zhang, L., Li, Y., Lu, T., Lu, Y., et al. (2004). Structural features of the rice chromosome 4 centromere. *Nucleic Acids Research, 32*, 2023–2030.

Zuccolo, A., Sebastian, A., Talag, J., Yu, Y., Kim, H., Collura, K., et al. (2007). Transposable element distribution, abundance and role in genome size variation in the genus *Oryza*. *BMC Evolutionary Biology, 7*, 152.

related regions via 'top-down' approaches recovers cryptic synteny by making use of transitive homeology, enabling deep comparisons of distantly related genomes despite extensive structural rearrangements. Paleohexaploidy (ancient genome triplication) seems to be a phenomenon particularly influential in eudicot plants, including one such event that occurred in the eudicot stem lineage, preceding the diversification of core eudicots. At the end of this chapter, we review recent research towards reconstructing the eudicot ancestral genome. Systematic genome comparisons promise better understanding and utilization of structural and functional correlations in eudicots and other groups.

1. INTRODUCTION

Having been a key factor in founding modern genetics, eudicot plants are also indispensable in sustaining the Earth and human society. Since the beginning of civilization, eudicot plants have provided us with food (such as many kinds of beans, nuts, leaves, and fruits), feed, oxygen, ornamentals, medication, and materials for many aspects of daily life. *Arabidopsis thaliana*, a member of the Brassicaceae family, was the first plant species to have its whole genome sequenced and remains the best-studied model organism in plant sciences (see Chapter 4). The eudicot clade contains about three quarters of extant angiosperm (flowering plant) species, distributed over the world's terrestrial ecosystems and even some aquatic ecosystems (such as mangrove swamps). The eudicot crown group ancestor likely lived in the early Cretaceous, rapidly radiating to produce all major extant lineages through several episodes of diversification. Because of their enormous diversity and widespread associations with humanity and the environment, eudicot plants have been popular genome sequencing targets, with the largest number of genomes sequenced or being sequenced among all plant clades. These growing data provide key resources to resolve previously equivocal phylogenetic relationships inside this diverse group, to deconvolute widespread paleopolyploidy events in eudicot lineages and thus to reconstruct their genome evolutionary history at both macro- and microscales, and to formulate hypotheses about the genetic bases of essential and unique phenotypes. The structure and sequence diversity of eudicots (and angiosperms generally; see Chapter 8) has also motivated development of novel computational algorithms to disentangle the reticulate mappings of homeology among the genomic regions. In this chapter, we review research progress in understanding the genome organization and evolution of eudicot plants.

2. PHYLOGENY AND EVOLUTION OF EUDICOT PLANTS

Eudicot plants constitute the largest monophyletic clade of angiosperms (flowering plants). They are conventionally defined by having two embryonic cotyledons (thus the name 'eudicotyledons') or, more narrowly, having tricolpate pollen grains. Angiosperms and eudicots are pervasive on the Earth with astonishing diversity and adaptation, from *Colobanthus quitensis* (Antarctic pearlwort) to *Lecythis ampla* (rainforest emergent), from cactus (desert specialist) to *Zostera* (marine eelgrass), from *Wolffia* (flower < 0.5 mm in diameter) to *Rafflesia* (flower one metre in diameter), and from the 'all-healing' *Panax ginseng* to the carnivorous *Nepenthes* (pitcher plant). Angiosperms consist of about 250,000 recorded species in about 450 families, of which about 198,000 species distributed among about 336 families are eudicots (Hedges & Kumar, 2009; Stevens, 2012; The Angiosperm Phylogeny Group, 2009).

Despite the tremendous diversity, extant angiosperms all evolved from a crown group common ancestor that diverged from other seed plants in the Jurassic period (Bell, Soltis, & Soltis, 2010; Soltis et al., 2009; Wikstrom, Savolainen, & Chase, 2001). Through several phases of morphological and functional diversification, by the late Cretaceous, angiosperms had dominated many habitats worldwide (Crane & Lidgard, 1989; Doyle & Donoghue, 1993 Friis, Pedersen, & Crane, 2006). The majority of extant plant taxa appeared so suddenly in the Earth's history that Darwin referred to their diversification as 'an abominable mystery'. More recent research showed that early angiosperms diversified extensively and very rapidly in the early Cretaceous (Crane, Friis, & Pedersen, 1995; Hickey & Doyle, 1977), perhaps within about 5 MYA (Moore, Bell, Soltis, & Soltis, 2007; Soltis, Bell, Kim, & Soltis, 2008). The two major angiosperm clades, eudicots (comprising about 75% of extant angiosperm species) and monocots (comprising about 22% of extant angiosperm species), diverged in a time window between 240 and 140 MYA (Bell et al., 2010; Crane et al., 1995; Soltis et al., 2008; Wolfe, Gouy, Yang, Sharp, & Li, 1989), a time period overlapping with the events of the Gondwanaland break-up, emergence of bees and diversification of insects, and the pancore eudicot γ paleohexaploidy, which may be key environmental and genetic factors driving early angiosperm and eudicot diversifications.

The first major diversification of dicotyledon plants likely occurred in early and mid-Cretaceous, involving many aspects of the organisms' lives such as floral structure, pollen structure, leaves, and pollination type (Crane et al., 1995; Friis et al., 2006; Hickey & Doyle, 1977). Evidence also revealed extensive diversification of core eudicots and monocots, respectively, beginning in the late Cretaceous (Crane et al., 1995; Friis et al., 2006). The phylogenetic relationships among basal eudicot lineages have not been fully resolved partly due to much diversity in their morphological and reproductive characters. More fundamentally, this is due to the antiquity and closeness of their divergence events and substantial variation in lineage evolutionary rates. Bearing these uncertainties in mind, it is currently generally agreed that Ranunculales, Sabiales, Proteales, Trochodendrales, and Buxales are sisters to the core eudicot common ancestor. Within the core eudicot clade phylogenetic analyses of 83 protein-coding and rRNA genes from seed plant, plastid genomes suggested that initial splits among *Dilleniaceae*, superrosids, and superasterids may have occurred as soon as 1 MYA after the divergence of Gunnerales from Pentapetalae (the rest of the core eudicots). Both superrosids and superasterids then continued to diversify rapidly to produce their respective major extant lineages in a few million years. Indeed, episodic rapid diversification is a theme throughout eudicot evolution (Moore, Soltis, Bell, Burleigh, & Soltis, 2010).

Many genes in floral development pathways seem to be duplicated in parallel time frames near early angiosperm and eudicot diversification events (Soltis et al., 2008), indicating that they may have been involved in polyploidization (whole-genome duplication) events rather than duplicated individually at the same time. It is now known that all angiosperms are paleopolyploids, having experienced one or more whole-genome duplications (WGDs) at some point(s) during their evolution (Blanc & Wolfe, 2004; Bowers, Chapman, Rong, & Paterson, 2003; Jiao et al., 2011; Tang, Bowers, et al., 2008). The widespread paleopolyploidy events in angiosperms and their coincident occurrences with several major species explosion periods support the hypotheses that paleopolyploidies may have been a crucial driving force in angiosperm evolution and diversification (Doyle et al., 2008; Fawcett, Maere, & Van de Peer, 2009; Lynch & Conery, 2000; Otto & Whitton, 2000; Paterson, Bowers, & Chapman, 2004; Soltis et al., 2009). In the sequenced portion of the eudicots, there are several examples of paleopolyploidies positioned on a clade's stem branch shortly preceding its radiation. In particular, a paleopolyploidy called 'gamma' (Bowers et al., 2003) is associated with early diversification of eudicots (Jaillon et al., 2007; Tang, Bowers, et al., 2008; Tang, Wang, et al., 2008).

Rates of genome evolution often vary greatly among plant lineages (Gaut, Yang, Takuno, & Eguiarte, 2011). For example, the *Vitis* lineage nuclear-gene nucleotide substitution rate is estimated to be about 20% less than that of *Populus* (Tang, Wang, et al., 2008), while *Nelumbo* is 30% slower than *Vitis* (Ming et al., 2013). Nucleotide substitution rates in plant organellar genomes also vary greatly, sometimes up to 100-fold or even more (Mower, Touzet, Gummow, Delph, & Palmer, 2007; Wolfe, Li, & Sharp, 1987). Although less explored, the frequency of genome rearrangements also varies among taxa, by at least 10-fold (Paterson et al., 1996; Zuccolo et al., 2011). One reason underlying these variations is the different generation time and life history of the organisms (Gaut et al., 2011; ;Smith & Donoghue, 2008 Tuskan et al., 2006; Young et al., 2011).

Abrupt origins, dynamic and often fast diversifications, widespread paleopolyploidies, and divergent lineage evolutionary rates are four key factors shaping the paths to modern eudicot plants. Having these in mind helps us better appreciate the conservation in modern eudicot genomes and understand the challenges in their comparative studies. Systematic evolutionary genome comparisons provide the necessary framework and building blocks to identify and dissect the functional conservations and innovations in this important group of organisms.

3. SEQUENCING OF EUDICOT GENOMES

A genome contains the genetic materials that are the blueprint for all aspects of an organism's development, functioning, interactions with other organisms as well as the environment, and its inheritance. The central functional units in a genome are genes (Beadle & Tatum, 1941; Morgan, 1910), which work together with each other and with regulatory and structural elements (Gilbert & Maxam, 1973; Jacob & Monod, 1961; Maniatis & Ptashne, 1973; McClintock, 1950). Knowing the information contained within genomes is a powerful and irreplaceable approach for biologists studying plants or other organisms. The initiation of sequencing of the first plant genome, that of the crucifer species *A. thaliana*, started in 1990, shortly after the Human Genome Project started (see Chapter 4). While *A. thaliana* has a small genome of about 135 Mb, many plant species possess huge genomes (e.g. see Table 6.1). Moreover, many plant genomes contain a repetitive fraction as much as or more than half of their total sizes and copies of highly similar (but not identical) sequences resulting from high intragenomic heterozygosity, polyploidy, and paleopolyploidy, posing challenges for

Table 6.1 Summary of genomic information from representative eudicot species with published genome assemblies—cont'd

Common name	Scientific name	Order	Chromosome # (1 ×)	WGD(s) (most recent first)	Estimated genome size (Mb)	Protein-coding gene #	TE (%)	Genome assembly level	Genome publication
Eurosid II									
Papaya	*Carica papaya*	Brassicales	9	γ	372	24,746	51.9	Scaffold	Ming et al. (2008)
Thale cress	*Arabidopsis thaliana*	Brassicales	5	α, β, γ	135	27,416	14	Chromosome	Arabidopsis Genome Initiative (2000) and Swarbreck et al. (2008)
	Arabidopsis lyrata	Brassicales	8	α, β, γ	207	32,670	29.7	Chromosome	Hu et al. (2011)
	Capsella rubella	Brassicales	8	α, β, γ	219	26,521	~50	Chromosome	Slotte et al. (2013)
Chinese cabbage	*Brassica rapa*	Brassicales	10	B, α, β, γ	284	41,174	39.5	Chromosome	Wang et al. (2011)
Salt cress	*Thellungiella parvula*	Brassicales	7	α, β, γ	160	28,901	7.5	Chromosome	Dassanayake et al. (2011)
Salt cress	*Thellungiella salsuginea*	Brassicales	7	α, β, γ	260	28,457	52	Chromosome	Wu et al. (2012)

Cotton	*Gossypium raimondii*	Malvales	13	C, γ	630–880	37,505	61	Chromosome	Paterson et al. (2012)
						40,976	57	Chromosome	Wang, Hobson, et al. (2012)
Cacoa	*Theobroma cacao*	Malvales	10	γ	430	28,798	24	Chromosome	Argout et al. (2011)
Neem	*Azadirachta indica*	Sapindales	14	NA	383	20,169	13	Scaffold	Krishnan et al. (2012)
Orange	*Citrus sinensis*	Sapindales	9	γ	367	29,445	20.5	Chromosome	Xu et al. (2013)
Basal rosids									
Grape	*Vitis vinifera*	Vitales	19	γ	475	26,346	41.4	Chromosome	Jaillon et al. (2007)
Asterids—Euasterid I									
Tomato	*Solanum lycopersicum*	Solanales	12	T, γ	900	34,727	63.3	Chromosome	Tomato Genome Consortium (2012)
Potato	*Solanum tuberosum*	Solanales	12	T, γ	844	39,031	62.2	Chromosome	Potato Genome Sequencing Consortium (2011)
Bladderwort	*Utricularia gibba*	Lamiales	14	Ua, Ub, Uc, γ	77	28,494	3	Scaffold	Ibarra-Laclette et al. (2013)
Basal eudicots									
Sacred lotus	*Nelumbo nucifera*	Proteales	8	λ	929	26,685	57	Megascaffold	Ming et al. (2013)

sequencing technologies relying on unique mappings between nucleotide sequences. Fortunately, although generally harder to sequence than animal genomes, advances in plant genome sequencing technologies in recent years (Hamilton & Buell, 2012; Schatz, Witkowski, & McCombie, 2012) have yielded rich data, tailored to studying many model and nonmodel plants of agricultural, economic, ecological, or theoretical interest. These data are key to people who want to understand and utilize specific plant species or botanical diversity. Goff et al. (Chapter 3) summarize plant genomes sequenced and published as of this writing, which include 35 eudicot species. There are also many more eudicot and plant genomes released (not yet published) or being sequenced (for a partial list, see http://genome.jgi.doe.gov/programs/plants/plant-projects.jsf). Table 6.1 gives an overview of genomic information from the 35 published eudicot genomes from species that have so far been selected to represent many more eudicot species.

4. THE GAMMA PALEOHEXAPLOIDY IN ANCESTRAL EUDICOT LINEAGES

Compared to more than 1000-fold difference in genome size, the chromosome number of plant species is relatively stable, varying only up to about 50-fold (Bennett & Smith, 1991; Bennett & Leitch, 2012). Sequenced angiosperm genomes typically have 25,000–45,000 protein-coding genes (Table 6.1). It had been long suspected that most angiosperms were paleopolyploids (Masterson, 1994; Otto & Whitton, 2000; Stebbins, 1966), but as noted earlier, it is now clear that all angiosperms have either single or compounded WGDs in their evolutionary history. Indeed, WGDs are especially widespread in plants (Blanc & Wolfe, 2004; Grant, Cregan, & Shoemaker, 2000; Paterson et al., 2000; Vision, Brown, & Tanksley, 2000) but are also found in animals (Jaillon et al., 2004; Ohno, 1970), fungi (Kellis, Birren, & Lander, 2004; Wolfe & Shields, 1997), and ciliates (Aury et al., 2006).

In particular, a paleohexaploidy (whole-genome triplication) has been detected in all core eudicot genomes sequenced so far (Table 6.1). This event, named gamma (γ) (Bowers et al., 2003), was hinted at in early studies (Ku, Vision, Liu, & Tanksley, 2000; Simillion, Vandepoele, Van Montagu, Zabeau, & Van de Peer, 2002; Vision et al., 2000), first unequivocally detected in *Arabidopsis* (Bowers et al., 2003) and fully profiled in *Vitis* (grape) (Jaillon et al., 2007; Tang, Bowers, et al., 2008; Tang, Wang, et al., 2008) and *Carica* (papaya) (Ming et al., 2008). Unlike the *Arabidopsis* genome, which was repeatedly duplicated, the grape genome has experienced no further polyploidization since gamma (Table 6.1). About 94.5% of grape genomic regions have up to two different paralogous regions, resulting from the original gamma triplets (Jaillon et al., 2007). By detecting and conducting multiple alignment of syntenic regions using a top-down algorithm (MCscan; see Chapter 8), Tang et al. were able to see triplication patterns in the common ancestor of papaya, *Populus*, and *Arabidopsis* (Ming et al., 2008; Tang, Bowers, et al., 2008; Tang, Wang, et al., 2008). When the alignments were further compared to the out-group grape genome, their triplication patterns correspond closely, revealing coalescence of up to four *Arabidopsis* regions, one papaya region, two *Populus* regions, and one grape region in each triplicated branch (Ming et al., 2008; Tang, Bowers, et al., 2008; Tang, Wang, et al., 2008). Both the grape and papaya genomes have experienced no further polyploidization since gamma. *Populus* has one additional duplication event (p) in its salicoid lineage, and *Arabidopsis* has two additional duplications (α and β) in its crucifer lineage.

The gamma paleohexaploidy occurred in eudicot ancestral lineages, likely being shortly before or during the earliest diversification of core eudicots. Although the event was first estimated to be shared by eudicots and monocots based on *Arabidopsis*–rice comparison (Bowers et al., 2003), comparison of grape genome to rice genome (Jaillon et al., 2007; Tang, Wang, et al., 2008) and 17 banana BACs (Tang, Wang, et al., 2008) indicated that gamma is more likely confined to eudicots.

Sacred lotus (*Nelumbo nucifera*) belongs to the basal eudicot order Proteales. Scrutinization of its genome revealed a paleotetraploidy event (λ), with each of the duplicated regions corresponding to the same set of gamma triplet regions in grape. Sacred lotus genes are typically diverged to similar degrees from their (up to) three orthologous grape genes, with the most similar ortholog distributed evenly among triplets of gamma regions. Molecular dating based on synonymous substitution rates between homeologous genes (Ks) positioned lambda at about 76–54 MYA. These results indicate that the ancestral *Nelumbo* lineage diverged from core eudicot ancestors before the gamma paleohexaploidy around 125 MYA and subsequently experienced a lineage-specific paleotetraploidy (Ming et al., 2013).

Tomato is an asterid species in which a second whole-genome triplication about 91–52 MYA was superimposed on gamma (Tomato Genome Consortium, 2012). Individual regions of the tomato genome correspond most closely to only one of the triplicated regions in grape, and no grape region is orthologous to more than one set of retriplicated regions in tomato.

abrupt diversifications of eudicots, making it statistically difficult to separate signals from individual events. This fact also makes difficult a more definitive test of allohexaploidy versus autohexaploidy, due to the lack of extant paleotetraploid parental species (Tang, Wang, et al., 2008). Better understanding of varied lineage evolutionary rates and taxon sampling on the deep branches are needed to further dissect the gamma event (Ming et al., 2013; Tang, Wang, et al., 2008). Being a genome-wide event that happened in the common ancestor of over 160,000 extant species and which has synteny conservation still readily detectable today, the evolutionary effects of gamma await further exploration.

5. STRUCTURAL COMPARISON OF EUDICOT GENOMES AND WIDESPREAD ANCIENT GENOME DUPLICATIONS

Despite early realization of structural similarity between many genomes (Bonierbale, Plaisted, & Tanksley, 1988; Vavilov, 1922), it is in fact often difficult, sometimes even impossible, to reliably detect conserved genomic regions. In well-assembled genomes, the main reason for this difficulty is repeated and nested alterations of genome structure through time. Genome structural mutations including insertions, deletions, inversions, translocations, recombinations, chromosome fissions and fusions, and, most dramatically, polyploidizations mask and even erase signatures of conservation of genome structure.

Although the phenomenon of ancient WGD is shared by plants and animals, studies of genome structural evolution are profoundly affected by key differences between the two kingdoms. Mammalian genomes, the main animal sequencing targets, have been free of polyploidization for and are thus much more conserved than plants over about 500 MYA (Nakatani, Takeda, Kohara, & Morishita, 2007;)Smith et al., 2013. Indeed, in plants, synteny conservation patterns across eudicots and monocots, which are separated by only 240–140 MYA, are extremely deteriorated (Salse et al., 2009) due in large part to paleopolyploidy and associated gene loss and rearrangement. Even when aligning a plant genome that experienced a lineage-specific paleopolyploidy with the genome of its closely related sister species lacking this event, such as *A. thaliana* versus *B. rapa* (Wang et al., 2011), severe gene loss and rearrangement result in complicated synteny maps. Most synteny detection software developed in mammalian studies, although excellent, is often designed to identify *single* best matching or orthologous regions (Bray & Pachter, 2004; Brudno et al., 2003;

Dubchak, Poliakov, Kislyuk, & Brudno, 2009; Kent, Baertsch, Hinrichs, Miller, & Haussler, 2003; Miller et al., 2007) and is therefore not suitable for comparing plant genomes with extensive intragenomic duplication.

In order to effectively compare divergent plant genomes across multiple WGDs, sensitive methods that can accommodate paralogous regions are prerequisite (Paterson, Freeling, Tang, & Wang, 2010). The 'bottom-up' approach, which iteratively interleaves gene positions on paralogous genomic segments, has been used to offset gene loss in the resultant merged profiles (Bowers et al., 2003). Inspired by early pairwise synteny detection algorithms ADHoRe (Vandepoele, Saeys, Simillion, Raes, & Van de Peer, 2002) and DiagHunter (Cannon, Kozik, Chan, Michelmore, & Young, 2003), MCscan implemented a novel 'top-down' approach of aligning multiple gene orders at once (e.g. A–B–C instead of A–B, B–C, A–C), possessing the distinct advantage of exploiting the transitive property of synteny (also see Chapter 8). This results in one-pass sensitive and accurate detection and alignment of synteny blocks across multiple genomes (Tang, Bowers, et al., 2008; Tang, Wang, et al., 2008). A complementary PAR (putative ancestral region) algorithm exhaustively identifies and hierarchically clusters homeologous regions in two genomes into concentrated sets. Individual clusters of PARs can then be statistically evaluated, aligned, and further analysed (Tang, Bowers, Wang, & Paterson, 2010). Segments duplicated in one polyploidy event have two distinct characters: They have same 'birth time' and they are nonoverlapping with each other. Each of these characters has been used to sort duplicated regions as belonging to the same (or different) polyploidy events. Based on both structural and sequence divergence criteria, a binary integer programming algorithm, QUOTA-ALIGN, screens synteny blocks into separate WGD events (Tang et al., 2011). This is especially valuable in analysing genomes bearing more than one round of paleopolyploidy. Some of the aforementioned software has been incorporated into user-friendly interfaces of online comparative genomics platforms such as PGDD (Lee, Tang, Wang, & Paterson, 2013) and CoGe (Lyons & Freeling, 2008).

Through more than a decade of comparative studies, paleopolyploidy events spanning many eudicot lineages have been characterized. Besides the deepest 'gamma' paleohexaploidy, at least 11 paleotetraploidies, 2 paleohexaploidies, and 1 paleo(do)decaploidy have been profiled in sequenced eudicot genomes (Fig. 6.2). We give an introductory description of each event here, for readers to use as a starting point to further explore the events.

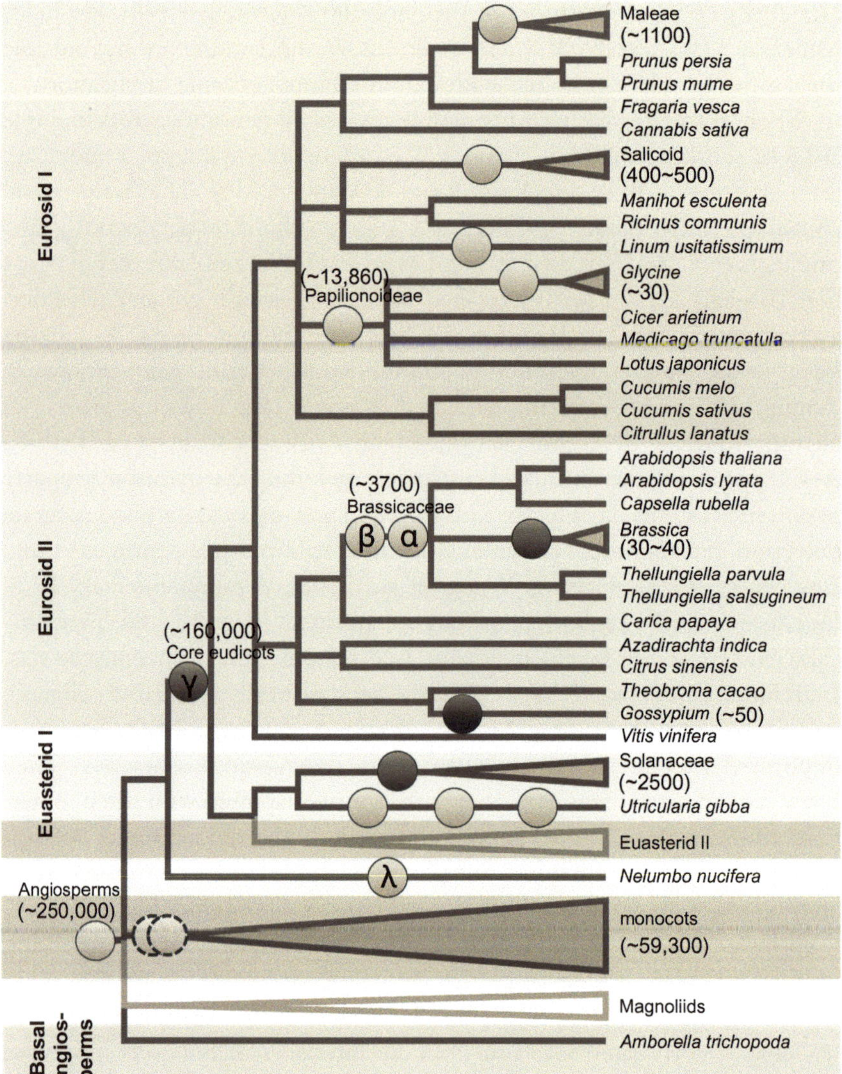

Figure 6.2 Paleopolyploidy events mapped to the phylogeny of major eudicot lineages. Since this chapter focuses on dicots, monocots and other angiosperm clades are abstracted. Inside the parentheses are numbers of known extant species in the respective clades. Light pink circles represent paleotetraploidies. Dark pink circles represent paleohexaploidies. Red circle represents the paleodecaploidy or dodecaploidy in *Gossypium*. The two overlapping light pink circles with dashed outline are an abstract representation of WGD events identified in monocot lineages. Clades that have no genome sequenced yet are omitted or in light grey (euasterid II and magnoliids). The cladogram is based on the NCBI Taxonomy database. (See colour plate.)

A WGD occurred in the common ancestor of the apple (*Malus*) and pear (*Pyrus*) lineages. Strong colinearity patterns exist between pairs of 17 apple chromosomes. Following duplication of the *Maleae* ancestor, a few genome rearrangements including one chromosome fusion and three translocations brought the ancestral *Maleae* genome back to 17 chromosomes (Velasco et al., 2010). All apple and pear chromosomes are highly colinear (Wu et al., 2013). This event is not shared with other species in the *Rosaceae* family (Wu et al., 2013).

A WGD occurred in the common ancestor of the *Papilionoideae* clade of legumes, which includes *Medicago*, chickpea (*Cicer arietinum*), *Lotus japonicus*, soybean (*Glycine max*), and pigeon pea (*Cajanus cajan*), shortly before the papilionoid radiation (Young et al., 2011). Subsequent to this WGD, the *Medicago* genome experienced higher levels of genome rearrangement and proximal duplications than soybean and *L. japonicus*, although macrosynteny is still relatively well conserved among the genetic regions of the three genomes (Young et al., 2011), despite an estimated separation of the millettioid (containing pigeon pea and soybean) and galegoid clades (including *Medicago*, *L. japonicus*, and chickpea) ~54 MYA (Lavin, Herendeen, & Wojciechowski, 2005). This indicated that major genome rearrangements following the Papilionoideae WGD took place quickly, before the separation of millettioids and galegoids. The soybean genome experienced an additional lineage-specific WGD, which is likely an allotetraploidy (Gill et al., 2009; Schmutz et al., 2010). Consistent with the relatively slow synonymous substitution rate of soybean compared to many other eudicots, levels of gene loss and structural mutations in soybean are also noticeably lower (Schmutz et al., 2010).

Besides the pancore eudicot γ event, the two most recent paleopolyploidies affecting *Arabidopsis*, α and β, following the usage in Bowers et al. (2003), appear to have occurred within the crucifer (*Brassicaceae*) lineage after its separation with the *Caricaceae* family about 72 MYA (Jaillon et al., 2007; Ming et al., 2008). No lineage sequenced so far originated between the two WGD events. Following the shared events, ten major rearrangements, including two reciprocal translocations and three chromosomal fusions, differentiated the *A. thaliana* karyotype of five chromosomes from the more ancestral eight-chromosome karyotype found in *A. lyrata* (Yogeeswaran et al., 2005), although 90% of the two genomes remained syntenic (Hu et al., 2011). There is an additional whole-genome triplication in the *Brassica* lineage after its separation from the remaining crucifers (Wang et al., 2011). In this event, the three paleo-subgenomes likely were merged

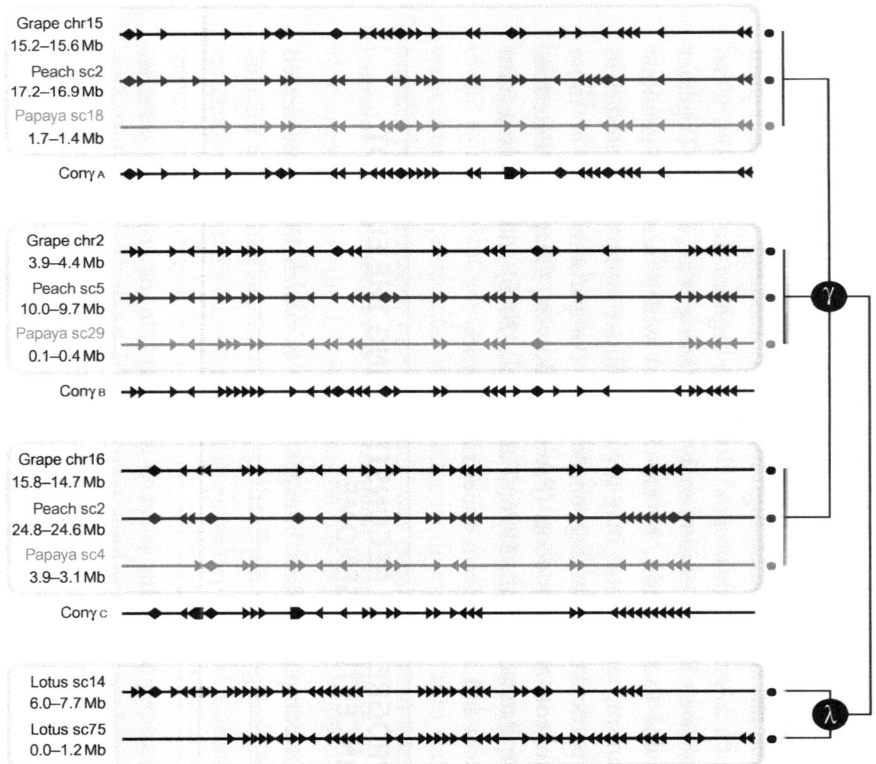

Figure 6.3 Multiple alignments of a set of syntenic regions in papaya, peach, grape and sacred lotus. Triangles represent individual genes and their transcriptional orientations. Genes with no syntenic matches are not plotted. The event γ is the paleohexaploidy that occurred in ancestral eudicots and is shared by the grape, peach, and papaya lineages. The event λ is the paleotetraploidy in the *Nelumbo* (sacred lotus) lineage after it diverged from the rest of eudicot lineages. The two events are also shown in Fig. 6.2. The γ regions are grouped into three γ subgenomes based on parsimony principles. Aligned genes within each γ subgenome are merged into a consensus order (Con γ_A, γ_B, and γ_C, respectively). Ancestral genes with uncertain orientations are represented by squares. The pair of sister λ regions in lotus is displayed at the bottom. (See colour plate.)

often affect paleopolyploidy and paralogy detection as some syntenic matching regions in a poorly assembled genome may be missed. This artefact can be mitigated by incorporating additional information from the ancestral genome and sister genomes. Therefore, although paleopolyploidies are often apparent in carefully performed genome alignments or age groupings of homeologous genes, the reconstruction of the inferred ancestral genome is the best way to recover syntenic mappings among homeologous regions. The reconstructed ancestral genome may not be the same as the true ancestral genome, but it likely has high structural similarity and is an irreplaceable bridge in whole-genome alignments including complicated combinations of global and local alignments (Tang, Bowers, et al., 2008). This is especially true when a 'clean' (i.e. having no WGD) out-group genome is not available. Moreover, the simulation has shown that reconstructed ancestral sequence may be a better predictor of extant sequence than using closest extant neighbouring sequences (Paten et al., 2008).

Experiments to compare genome structure preceded molecular biology (Dobzhansky & Sturtevant, 1938; Dobzhansky & Tan, 1936; Dubinin, Sokolov, & Tiniakov, 1936). Before the era of whole-genome sequencing, hybridization-based techniques such as chromosome painting (Ried, Schrock, Ning, & Wienberg, 1998) were used to compare genomes from different species at low resolution. With more than 40 plant genomes now sequenced and even more animal genomes, it has become increasingly interesting and possible to reconstruct the ancestral genomes that preceded speciations and/or genome duplications. Extant genomes differ from their ancestral genome by gene gain/loss, nucleotide substitution, insertion, deletion, translocation, inversion, chromosome fusion/fission, and duplications. Early attempts to solve the ancestral genome reconstruction problem used viral or organellar genomes (Blanchette, Kunisawa, & Sankoff, 1999; Cosner et al., 2000; Hannenhalli, Chappey, Koonin, & Pevzner, 1995; Moret, Wang, Warnow, & Wyman, 2001). Since then, manual and computational methods have been developed for this problem. Although manual reconstruction has its own advantages as being somewhat more tractable and curated (Gordon, Byrne, & Wolfe, 2009), it fails to handle or make use of the wealth of growing genome data.

Computational ancestral reconstruction is essential for genomics now and in the future, with a typical pipeline outlined in Fig. 6.4. Gene order is traditionally modelled as the permutation of integer sequences, with homeologous genes represented with the same integer. Rearrangement distance is then defined as the minimum number of rearrangements (such as inversions and translocations) required to convert one permutation into the other (Sankoff, 1992). The ancestral reconstruction problem for the case of three genomes can then be formulated as the 'genome median' problem, minimizing the sum of rearrangement distances between the ancestor and each of the descendent sequences (Sankoff & Blanchette, 1997). The genome median problem was later extended to more than three genomes, known as the 'small phylogeny problem' (Sankoff & Blanchette, 1998; Zheng & Sankoff, 2012). Alternatively, the equivalent problem can be approached by adopting heuristic rules to select 'good reversals' (i.e. that reduce total reversal distance) iteratively until all genomes are 'devolved' to an identical genome, which is claimed as the most likely ancestral median (Bourque & Pevzner, 2002). A simpler and in some sense more general 'breakpoint distance' (Blanchette, Bourque, & Sankoff, 1997), measuring the number of different adjacent pairs between permutations, is nonetheless more ambiguous in identifying the ancestor, probably because there are no unified mappings between real (biological) gene order rearrangements and number of breakpoints.

When a reconstruction involves multiple species, it is useful to know their phylogenetic relationships. The most parsimonious rearrangement

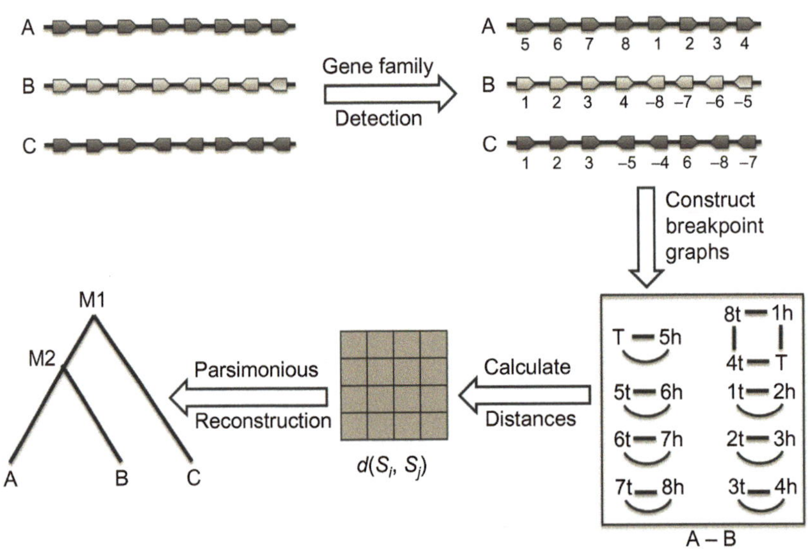

Figure 6.4 Pipeline for computational reconstruction of the ancestral genome. A toy example of three extant genomic regions (A, B, and C) and their ancestral regions (M1 and M2) is used for illustration. Oriented blocks represent genes, which are usually converted to signed integers in computation. (See colour plate.)

scenario can then be inferred on the species tree. Adjacencies of varied reliability can be modelled as edges of a weighted directed graph. The reconstruction algorithm then finds a set of paths that maximize the total weight (Ma et al., 2006). A related approach used 'travelling salesman' algorithms instead to chain the adjacencies (Bertrand, Gagnon, Blanchette, & El-Mabrouk, 2010). Alternatively, probabilistic models, such as the TKF91 model (Thorne, Kishino, & Felsenstein, 1991) or transducers (Paten et al., 2008), can be used to infer the evolutionary history of structural alterations.

Because of typically large genomes, high repetitive sequence content, and complicated paralogy relationships due to recurring genome and small-scale duplications, reconstructions of ancestral genomes in plants are much more difficult and therefore have lagged behind yeast and animal studies. Although several reconstruction algorithms have been developed in vertebrate studies, none is directly applicable to plant genomes. The reconstruction of plant ancestral genomes has been formulated by several computational models. The 'genome aliquoting' problem is the problem of finding a genome with one copy of every gene given a genome with exactly p copies of every gene (when $p=2$, it is called the genome halving problem; El-Mabrouk, Nadeau, & Sankoff, 1998), such that the number of rearrangements necessary to convert the reconstructed ancestral genome into the observed genome is minimized (Warren & Sankoff, 2009). While finding exact solutions to the problems is NP-hard, heuristic algorithms (with restrictive assumptions) have been implemented to reduce the time complexity to polynomial or even linear. Recently, the PATHGROUPS algorithm has been developed to relax the restriction of equal gene complement so that information from incomplete homeolog groups can also be incorporated (Zheng & Sankoff, 2012). Because WGDs are often followed by extensive chromosome rearrangements (Otto, 2007; Semon & Wolfe, 2007; Song, Lu, Tang, & Osborn, 1995), species with fewer historical WGDs, such as grape, papaya, peach, and sacred lotus, are often preferentially chosen for the reconstruction of the eudicot ancestral genome.

It should be noted that in most cases, there is uncertainty in the reconstruction analyses. This is on one hand due to more than one solution being equally probable in some computational algorithms, while on the other hand, it is due to the lack of a true ancestral genome to evaluate reconstructions (simulation studies are currently unable to model all aspects of genome structural evolution). Besides, many heuristic techniques that have been used lack immediate biological reasons and may not be justified

Edger, Pires, & Conant, 2011). The newly created 'redundant' copies are often relieved of selective pressure and sometimes experience functional modifications via subfunctionalization or neofunctionalization (Kellis et al., 2004; Lynch & Force, 2000; Ohno, 1970). The gene balance (or gene dosage) theory constrains changes in some duplicated genes coding for interacting products to fulfil stoichiometric balance (Birchler, Bhadra, Bhadra, & Auger, 2001; Papp, Pal, & Hurst, 2003; Thomas et al., 2006), possibly driving them to different post-WGD evolutionary paths. Moreover, the cohort of whole-genome duplicates greatly increases the 'buffer capacity' of a genome, making it more genetically robust (Chapman, Bowers, Feltus, & Paterson, 2006; Gu et al., 2003; Paterson et al., 2006). Polyploids often have increased regulatory and morphological complexity (Freeling & Thomas, 2006) and a higher chance of obtaining new gene combinations and hybrid vigour (De Bodt, Maere, & Van de Peer, 2005; Rieseberg et al., 2003). All these consequences, some of which are relatively dramatic and may quickly follow WGDs, often foreshadow the development of derived or novel phenotypes and the diversification of plant species (Otto & Whitton, 2000; Paterson et al., 2000; Soltis et al., 2009). The far-ranging, multilayered, and possibly compounded effects of polyploidies in plant genome evolution need further elucidation.

8. PERSPECTIVE

Compared to two rounds of WGDs (2R) thought to have occurred in primitive vertebrate lineages (Dehal & Boore, 2005;)Smith et al., 2013, plant lineages exhibit continuous propensity for polyploidies. Since the first plant genome, that of the thale cress *A. thaliana*, was sequenced in 2000 (see Chapter 4), both a few previously suspected and many completely unknown paleopolyploidy events have been unravelled, pervading angiosperm phylogeny (Fig. 6.2). Widespread paleopolyploidies in plants may have originated from more flexible and diverse reproductive schemes in plants and in turn drive the enormous genetic and phenotypic diversity observed in this kingdom. Being a major evolutionary force, recurring paleopolyploidization underlies the origin and diversification of extant angiosperm species, contributing to speciation, biological diversity, crop domestication and agricultural applications, and perhaps survival in extreme environments such as those associated with the Cretaceous–Tertiary mass extinction (Freeling & Thomas, 2006; Paterson et al., 2010; Soltis et al., 2009; Van de Peer, Maere, & Meyer, 2009). On the other hand, in plant genome comparative studies, the timing and effects of paleopolyploidies are two most important factors. In addition, studying the fates and interactions of homeologous genes (duplicated genes created in polyploidy events) is a major approach to delineate the structural and functional architecture of plant genomes and evolution of specific gene families of interest. Therefore, for botanists, evolutionists, or genomicists, it is crucial to understand the polyploidization history of plant lineages.

One particular interesting feature in eudicots is that they are presently the only clade known to be affected by paleohexaploidy (ancient whole-genome triplication) events. These events have been identified in the core eudicot stem lineage (gamma), the asterid *Solanaceae* family (T), the rosid *Brassica* genus (B), and perhaps in *Gossypium* (rosids). Although some wild monocot plants such as the grass 'timothy' and crops such as the bread wheat are neohexaploids, paleohexaploidy has not been found in any monocot genome studied so far. This raises curious questions about possible reasons and consequences associated with these events through eudicot evolution or, alternatively, possible suppression of such events in the evolution of other lineages.

At the centre of comparative genomics is the need to effectively detect sequence and functional correspondences among related genomes. Systematic comparison and reconstruction of genome structural evolution have started to show great power in this research area. For example, in the past, molecular evolution studies have mostly relied on analyses of individual gene families. However, inference of relationships among members of a gene family based on sequence alignment can be complicated by gene loss, different mutation rates among taxa and genes, gene conversion, horizontal gene transfer, and other processes. Alternatively, evolutionary inferences based on genome-wide synteny data have proven to be less confusing and more accurate in both animal (Dehal & Boore, 2005) and plant (Tang, Bowers, et al., 2008) studies.

Intergenomic alignment is the fundamental method to reveal similarities and differences among genomes. In particular, the alignment with orthologous regions in the genomes of model organisms facilitates accurate knowledge transfer to and accelerates studies of nonmodel organisms. Multiway syntenic mapping aided by inferred 'intermediate' genomes provides an effective framework for genome alignment, with the unique advantage of tolerating long evolutionary distance and extensive genome rearrangement. Nucleotide-level alignments can in turn be conducted to enable more evolutionary analyses, such as nucleotide-level conservation, categorization of indels in coding and regulatory regions, discovery of candidates for new and lineage-specific genes and other functional elements, recovery of ancestral functional elements that are lost in extant sequences, and detection of genomic selection patterns.

The last decade has seen exponential increase in the number of published (plant) genomes. Rapidly developing sequencing technology and plummeting cost seem to assure a continuing explosion of genome data. We are entering the golden age of genome informatics, with unprecedented challenges and opportunities. Novel tools and solid knowledge of polyploidization history have greatly facilitated systematic comparisons among plant genomes that were once thought to be beyond the reach of such comparisons. There is a clear need and opportunity for future comparative genomic studies to take more depth. For example, the frequency of structural alterations is not uniform throughout a genome. Hotspots of microrearrangements are found near centromeres, telomeres, duplications, and interspersed repeats (Eichler & Sankoff, 2003; Murphy et al., 2005). The rate of recombination, a major cause of genome rearrangement, varies along genomic regions of different nucleotide composition, gene density, repeat content, chromatin packing, and other potential factors (Cirulli, Kliman, & Noor, 2007; Giraut et al., 2011; McVean et al., 2004; Pal & Hurst, 2003). Arrangement and diversification of functional elements in the genome are also different in organisms with different reproductive types, lifestyles, habitats, and other factors. More detailed studies of genome structural evolution will promise better understanding of the fundamental questions of the rules and consequences of genome organization in plants and other organisms.

ACKNOWLEDGEMENTS

This work was supported by grants from the U.S. National Science Foundation to A. H. P. (DBI 0849896, MCB 0821096, and MCB 1021718) and to R. M., Q. Yu, P. H. Moore, J. Jiang, and A. H. P. (Award Nos. DBI0553417 and DBI-0922545). This study was supported in part by resources and technical expertise from the University of Georgia, the Georgia Advanced Computing Resource Center, and a partnership between the Office of the Vice President for Research and the Office of the Chief Information Officer.

REFERENCES

Arabidopsis Genome Initiative, (2000). Analysis of the genome sequence of the flowering plant Arabidopsis thaliana. *Nature*, *408*(6814), 796–815. http://dx.doi.org/10.1038/35048692.

Arabidopsis Interactome Mapping Consortium, (2011). Evidence for network evolution in an Arabidopsis interactome map. *Science, 333*(6042), 601–607. http://dx.doi.org/10.1126/science.1203877.

Argout, X., Salse, J., Aury, J. M., Guiltinan, M. J., Droc, G., Gouzy, J., et al. (2011). The genome of Theobroma cacao. *Nature Genetics, 43*(2), 101–108. http://dx.doi.org/10.1038/ng.736.

Aury, J. M., Jaillon, O., Duret, L., Noel, B., Jubin, C., Porcel, B. M., et al. (2006). Global trends of whole-genome duplications revealed by the ciliate Paramecium tetraurelia. *Nature, 444*(7116), 171–178. http://dx.doi.org/10.1038/nature05230.

Beadle, G. W., & Tatum, E. L. (1941). Genetic control of biochemical reactions in Neurospora. *Proceedings of the National Academy of Sciences of the United States of America, 27*, 499–506.

Bekaert, M., Edger, P. P., Pires, J. C., & Conant, G. C. (2011). Two-phase resolution of polyploidy in the Arabidopsis metabolic network gives rise to relative and absolute dosage constraints. *Plant Cell, 23*(5), 1719–1728. http://dx.doi.org/10.1105/tpc.110.081281.

Bell, C. D., Soltis, D. E., & Soltis, P. S. (2010). The age and diversification of the angiosperms re-revisited. *American Journal of Botany, 97*(8), 1296–1303. http://dx.doi.org/10.3732/ajb.0900346.

Bennett, M. D., & Leitch, I. J. (2012). *Plant DNA C-values database release 6.0.* http://www.kew.org/cvalues/.

Bennett, M. D., & Smith, J. B. (1991). Nuclear DNA amounts in angiosperms. *Philosophical Transactions of the Royal Society of London Series B, Biological Sciences, 334*(1271), 309–345.

Bennetzen, J. L. (2005). Transposable elements, gene creation and genome rearrangement in flowering plants. *Current Opinion in Genetics & Development, 15*(6), 621–627. http://dx.doi.org/10.1016/j.gde.2005.09.010.

Bertrand, D., Gagnon, Y., Blanchette, M., & El-Mabrouk, N. (2010). Reconstruction of ancestral genome subject to whole genome duplication, speciation, rearrangement and loss. In V. Moulton & M. Singh (Eds.), *Algorithms in bioinformatics*, Vol. 6293, (pp. 78–89). Berlin: Springer.

Birchler, J. A., Bhadra, U., Bhadra, M. P., & Auger, D. L. (2001). Dosage-dependent gene regulation in multicellular eukaryotes: Implications for dosage compensation, aneuploid syndromes, and quantitative traits. *Developmental Biology, 234*(2), 275–288. http://dx.doi.org/10.1006/dbio.2001.0262.

Blanc, G., & Wolfe, K. H. (2004). Widespread paleopolyploidy in model plant species inferred from age distributions of duplicate genes. *Plant Cell, 16*(7), 1667–1678. http://dx.doi.org/10.1105/tpc.021345.

Blanchette, M., Bourque, G., & Sankoff, D. (1997). Breakpoint phylogenies. *Genome Informatics Workshop on Genome Informatics, 8*, 25–34.

Blanchette, M., Kunisawa, T., & Sankoff, D. (1999). Gene order breakpoint evidence in animal mitochondrial phylogeny. *Journal of Molecular Evolution, 49*(2), 193–203.

Bonierbale, M. W., Plaisted, R. L., & Tanksley, S. D. (1988). RFLP maps based on a common set of clones reveal modes of chromosomal evolution in potato and tomato. *Genetics, 120*(4), 1095–1103.

Bourque, G., & Pevzner, P. (2002). Genome-scale evolution: Reconstructing gene orders in the ancestral species. *Genome Research, 12*(1), 26–36.

Bowers, J. E., Arias, M. A., Asher, R., Avise, J. A., Ball, R. T., Brewer, G. A., et al. (2005). Comparative physical mapping links conservation of microsynteny to chromosome structure and recombination in grasses. *Proceedings of the National Academy of Sciences of the United States of America, 102*(37), 13206–13211. http://dx.doi.org/10.1073/pnas.0502365102.

Hannenhalli, S., Chappey, C., Koonin, E. V., & Pevzner, P. A. (1995). Genome sequence comparison and scenarios for gene rearrangements: A test case. *Genomics*, *30*(2), 299–311. http://dx.doi.org/10.1006/geno.1995.9873.

Hawkins, J. S., Kim, H., Nason, J. D., Wing, R. A., & Wendel, J. F. (2006). Differential lineage-specific amplification of transposable elements is responsible for genome size variation in Gossypium. *Genome Research*, *16*(10), 1252–1261. http://dx.doi.org/10.1101/gr.5282906.

Hedges, S. B., & Kumar, S. (2009). *The timetree of life*. Oxford: OUP.

Hickey, L. J., & Doyle, J. A. (1977). Early cretaceous fossil evidence for angiosperm evolution. *The Botanical Review*, *43*(1), 3–104. http://dx.doi.org/10.1007/BF02860849.

Hu, T. T., Pattyn, P., Bakker, E. G., Cao, J., Cheng, J. F., Clark, R. M., et al. (2011). The Arabidopsis lyrata genome sequence and the basis of rapid genome size change. *Nature Genetics*, *43*(5), 476–481. http://dx.doi.org/10.1038/ng.807.

Huang, S., Li, R., Zhang, Z., Li, L., Gu, X., Fan, W., et al. (2009). The genome of the cucumber, Cucumis sativus L. *Nature Genetics*, *41*(12), 1275–1281. http://dx.doi.org/10.1038/ng.475.

Ibarra-Laclette, E., Lyons, E., Hernandez-Guzman, G., Perez-Torres, C. A., Carretero-Paulet, L., Chang, T. H., et al. (2013). Architecture and evolution of a minute plant genome. *Nature*http://dx.doi.org/10.1038/nature12132.

Jacob, F., & Monod, J. (1961). Genetic regulatory mechanisms in the synthesis of proteins. *Journal of Molecular Biology*, *3*, 318–356.

Jaillon, O., Aury, J. M., Brunet, F., Petit, J. L., Stange-Thomann, N., Mauceli, E., et al. (2004). Genome duplication in the teleost fish Tetraodon nigroviridis reveals the early vertebrate proto-karyotype. *Nature*, *431*(7011), 946–957.

Jaillon, O., Aury, J. M., Noel, B., Policriti, A., Clepet, C., Casagrande, A., et al. (2007). The grapevine genome sequence suggests ancestral hexaploidization in major angiosperm phyla. *Nature*, *449*(7161), 463–467. http://dx.doi.org/10.1038/nature06148.

Jain, M., Misra, G., Patel, R. K., Priya, P., Jhanwar, S., Khan, A. W., et al. (2013). A draft genome sequence of the pulse crop chickpea (Cicer arietinum L.). *Plant Journal*, *74*(5), 715–729. http://dx.doi.org/10.1111/tpj.12173.

Jiao, Y., Leebens-Mack, J., Ayyampalayam, S., Bowers, J. E., McKain, M. R., McNeal, J., et al. (2012). A genome triplication associated with early diversification of the core eudicots. *Genome Biology*, *13*(1), R3. http://dx.doi.org/10.1186/gb-2012-13-1-r3.

Jiao, Y., Wickett, N. J., Ayyampalayam, S., Chanderbali, A. S., Landherr, L., Ralph, P. E., et al. (2011). Ancestral polyploidy in seed plants and angiosperms. *Nature*, *473*(7345), 97–100. http://dx.doi.org/10.1038/nature09916.

Kellis, M., Birren, B. W., & Lander, E. S. (2004). Proof and evolutionary analysis of ancient genome duplication in the yeast Saccharomyces cerevisiae. *Nature*, *428*(6983), 617–624.

Kent, W. J., Baertsch, R., Hinrichs, A., Miller, W., & Haussler, D. (2003). Evolution's cauldron: Duplication, deletion, and rearrangement in the mouse and human genomes. *Proceedings of the National Academy of Sciences of the United States of America*, *100*(20), 11484–11489. http://dx.doi.org/10.1073/pnas.1932072100.

Khawaja, H. I., Ellis, J. R., & Sybenga, J. (1995). Cytogenetics of Lathyrus palustris, a natural autohexaploid. *Genome*, *38*(4), 827–831.

Krishnan, N. M., Pattnaik, S., Jain, P., Gaur, P., Choudhary, R., Vaidyanathan, S., et al. (2012). A draft of the genome and four transcriptomes of a medicinal and pesticidal angiosperm Azadirachta indica. *BMC Genomics*, *13*, 464. http://dx.doi.org/10.1186/1471-2164-13-464.

Ku, H. M., Vision, T., Liu, J., & Tanksley, S. D. (2000). Comparing sequenced segments of the tomato and Arabidopsis genomes: Large-scale duplication followed by selective gene loss creates a network of synteny. *Proceedings of the National Academy of Sciences of the United States of America*, *97*(16), 9121–9126. http://dx.doi.org/10.1073/pnas.160271297.

Lavin, M., Herendeen, P. S., & Wojciechowski, M. F. (2005). Evolutionary rates analysis of Leguminosae implicates a rapid diversification of lineages during the tertiary. *Systematic Biology*, *54*(4), 575–594. http://dx.doi.org/10.1080/10635150590947131.

Lee, T. H., Tang, H., Wang, X., & Paterson, A. H. (2013). PGDD: A database of gene and genome duplication in plants. *Nucleic Acids Research*, *41*(Database Issue), D1152–D1158. http://dx.doi.org/10.1093/nar/gks1104.

Lynch, M., & Conery, J. S. (2000). The evolutionary fate and consequences of duplicate genes. *Science*, *290*(5494), 1151–1155. http://dx.doi.org/10.1126/science.290.5494.1151.

Lynch, M., & Force, A. (2000). The probability of duplicate gene preservation by subfunctionalization. *Genetics*, *154*(1), 459–473.

Lyons, E., & Freeling, M. (2008). How to usefully compare homologous plant genes and chromosomes as DNA sequences. *Plant Journal*, *53*(4), 661–673. http://dx.doi.org/10.1111/j.1365-313X.2007.03326.x.

Lyons, E., Pedersen, B., Kane, J., & Freeling, M. (2008). The value of nonmodel genomes and an example using SynMap within CoGe to dissect the hexaploidy that predates the rosids. *Tropical Plant Biology*, *1*(3), 181–190. http://dx.doi.org/10.1007/s12042-008-9017-y.

Ma, J., Zhang, L., Suh, B. B., Raney, B. J., Burhans, R. C., Kent, W. J., et al. (2006). Reconstructing contiguous regions of an ancestral genome. *Genome Research*, *16*(12), 1557–1565. http://dx.doi.org/10.1101/gr.5383506.

Maniatis, T., & Ptashne, M. (1973). Multiple repressor binding at the operators in bacteriophage lambda. *Proceedings of the National Academy of Sciences of the United States of America*, *70*(5), 1531–1535.

Masterson, J. (1994). Stomatal size in fossil plants—Evidence for polyploidy in majority of angiosperms. *Science*, *264*(5157), 421–424.

Matsuoka, Y. (2011). Evolution of polyploid triticum wheats under cultivation: The role of domestication, natural hybridization and allopolyploid speciation in their diversification. *Plant & Cell Physiology*, *52*(5), 750–764. http://dx.doi.org/10.1093/pcp/pcr018.

McClintock, B. (1950). The origin and behavior of mutable loci in maize. *Proceedings of the National Academy of Sciences of the United States of America*, *36*(6), 344–355.

McVean, G. A., Myers, S. R., Hunt, S., Deloukas, P., Bentley, D. R., & Donnelly, P. (2004). The fine-scale structure of recombination rate variation in the human genome. *Science*, *304*(5670), 581–584. http://dx.doi.org/10.1126/science.1092500.

Miller, W., Rosenbloom, K., Hardison, R. C., Hou, M., Taylor, J., Raney, B., et al. (2007). 28-way vertebrate alignment and conservation track in the UCSC Genome Browser. *Genome Research*, *17*(12), 1797–1808. http://dx.doi.org/10.1101/gr.6761107.

Ming, R., Hou, S., Feng, Y., Yu, Q., Dionne-Laporte, A., Saw, J. H., et al. (2008). The draft genome of the transgenic tropical fruit tree papaya (Carica papaya Linnaeus). *Nature*, *452*(7190), 991–996. http://dx.doi.org/10.1038/nature06856.

Ming, R., Vanburen, R., Liu, Y., Yang, M., Han, Y., Li, L. T., et al. (2013). Genome of the long-living sacred lotus (Nelumbo nucifera Gaertn.). *Genome Biology*, *14*(5), R41. http://dx.doi.org/10.1186/gb-2013-14-5-r41.

Moore, M. J., Bell, C. D., Soltis, P. S., & Soltis, D. E. (2007). Using plastid genome-scale data to resolve enigmatic relationships among basal angiosperms. *Proceedings of the National Academy of Sciences of the United States of America*, *104*(49), 19363–19368. http://dx.doi.org/10.1073/pnas.0708072104.

Moore, M. J., Soltis, P. S., Bell, C. D., Burleigh, J. G., & Soltis, D. E. (2010). Phylogenetic analysis of 83 plastid genes further resolves the early diversification of eudicots. *Proceedings of the National Academy of Sciences of the United States of America*, *107*(10), 4623–4628. http://dx.doi.org/10.1073/pnas.0907801107.

Moret, B. M., Wang, L. S., Warnow, T., & Wyman, S. K. (2001). New approaches for reconstructing phylogenies from gene order data. *Bioinformatics*, *17*(Suppl. 1), S165–S173.

Morgan, T. H. (1910). Sex limited inheritance in Drosophila. *Science, 32*(812), 120–122. http://dx.doi.org/10.1126/science.32.812.120.

Morgante, M. (2006). Plant genome organisation and diversity: The year of the junk! *Current Opinion in Biotechnology, 17*(2), 168–173. http://dx.doi.org/10.1016/j.copbio.2006.03.001.

Morgante, M., De Paoli, E., & Radovic, S. (2007). Transposable elements and the plant pan-genomes. *Current Opinion in Plant Biology, 10*(2), 149–155. http://dx.doi.org/10.1016/j.pbi.2007.02.001.

Mower, J. P., Touzet, P., Gummow, J. S., Delph, L. F., & Palmer, J. D. (2007). Extensive variation in synonymous substitution rates in mitochondrial genes of seed plants. *BMC Evolutionary Biology*, 7, 135. http://dx.doi.org/10.1186/1471-2148-7-135.

Murphy, W. J., Larkin, D. M., Everts-van der Wind, A., Bourque, G., Tesler, G., Auvil, L., et al. (2005). Dynamics of mammalian chromosome evolution inferred from multispecies comparative maps. *Science, 309*(5734), 613–617. http://dx.doi.org/10.1126/science.1111387.

Nakatani, Y., Takeda, H., Kohara, Y., & Morishita, S. (2007). Reconstruction of the vertebrate ancestral genome reveals dynamic genome reorganization in early vertebrates. *Genome Research, 17*(9), 1254–1265. http://dx.doi.org/10.1101/gr.6316407, gr.6316407 [pii].

NordenskiÖLd, H. (1953). A genetical study in the mode of segregation in hexaploid phleum pratense. *Hereditas, 39*(3–4), 469–488. http://dx.doi.org/10.1111/j.1601-5223.1953.tb03431.x.

Ohno, S. (1970). *Evolution by gene duplication.* Berlin: Springer.

Otto, S. P. (2007). The evolutionary consequences of polyploidy. *Cell, 131*(3), 452–462. http://dx.doi.org/10.1016/j.cell.2007.10.022.

Otto, S. P., & Whitton, J. (2000). Polyploid incidence and evolution. *Annual Review of Genetics, 34,* 401–437. http://dx.doi.org/10.1146/annurev.genet.34.1.401.

Pal, C., & Hurst, L. D. (2003). Evidence for co-evolution of gene order and recombination rate. *Nature Genetics, 33*(3), 392–395. http://dx.doi.org/10.1038/ng1111.

Papp, B., Pal, C., & Hurst, L. D. (2003). Dosage sensitivity and the evolution of gene families in yeast. *Nature, 424*(6945), 194–197. http://dx.doi.org/10.1038/nature01771.

Paten, B., Herrero, J., Fitzgerald, S., Beal, K., Flicek, P., Holmes, I., et al. (2008). Genome-wide nucleotide-level mammalian ancestor reconstruction. *Genome Research, 18*(11), 1829–1843. http://dx.doi.org/10.1101/gr.076521.108.

Paterson, A. H., Bowers, J. E., Bruggmann, R., Dubchak, I., Grimwood, J., Gundlach, H., et al. (2009). The Sorghum bicolor genome and the diversification of grasses. *Nature, 457*(7229), 551–556. http://dx.doi.org/10.1038/nature07723.

Paterson, A. H., Bowers, J. E., Burow, M. D., Draye, X., Elsik, C. G., Jiang, C. X., et al. (2000). Comparative genomics of plant chromosomes. *Plant Cell, 12*(9), 1523–1540.

Paterson, A. H., Bowers, J. E., & Chapman, B. A. (2004). Ancient polyploidization predating divergence of the cereals, and its consequences for comparative genomics. *Proceedings of the National Academy of Sciences of the United States of America, 101*(26), 9903–9908. http://dx.doi.org/10.1073/pnas.0307901101.

Paterson, A. H., Bowers, J. E., Chapman, B. A., Peterson, D. G., Rong, J. K., & Wicker, T. M. (2004). Comparative genome analysis of monocots and dicots, toward characterization of angiosperm diversity. *Current Opinion in Biotechnology, 15*(2), 120–125.

Paterson, A. H., Chapman, B. A., Kissinger, J. C., Bowers, J. E., Feltus, F. A., & Estill, J. C. (2006). Many gene and domain families have convergent fates following independent whole-genome duplication events in Arabidopsis, Oryza, Saccharomyces and Tetraodon. *Trends in Genetics, 22*(11), 597–602.

Paterson, A. H., Freeling, M., Tang, H., & Wang, X. (2010). Insights from the comparison of plant genome sequences. *Annual Review of Plant Biology, 61*, 349–372. http://dx.doi.org/10.1146/annurev-arplant-042809-112235.

Paterson, A. H., Lan, T. H., Reischmann, K. P., Chang, C., Lin, Y. R., Liu, S. C., et al. (1996). Toward a unified genetic map of higher plants, transcending the monocot-dicot divergence. *Nature Genetics, 14*(4), 380–382.

Paterson, A. H., Wendel, J. F., Gundlach, H., Guo, H., Jenkins, J., Jin, D., et al. (2012). Repeated polyploidization of Gossypium genomes and the evolution of spinnable cotton fibres. *Nature, 492*(7429), 423–427. http://dx.doi.org/10.1038/nature11798.

Potato Genome Sequencing Consortium (2011). Genome sequence and analysis of the tuber crop potato. *Nature, 475*(7355), 189–195. http://dx.doi.org/10.1038/nature10158.

Prochnik, S., Marri, P. R., Desany, B., Rabinowicz, P. D., Kodira, C., Mohiuddin, M., et al. (2012). The cassava genome: Current progress, future directions. *Tropical Plant Biology, 5*(1), 88–94. http://dx.doi.org/10.1007/s12042-011-9088-z.

Rahman, A. Y., Usharraj, A. O., Misra, B. B., Thottathil, G. P., Jayasekaran, K., Feng, Y., et al. (2013). Draft genome sequence of the rubber tree Hevea brasiliensis. *BMC Genomics, 14*, 75. http://dx.doi.org/10.1186/1471-2164-14-75.

Ried, T., Schrock, E., Ning, Y., & Wienberg, J. (1998). Chromosome painting: A useful art. *Human Molecular Genetics*, 7(10), 1619–1626.

Rieseberg, L. H., Raymond, O., Rosenthal, D. M., Lai, Z., Livingstone, K., Nakazato, T., et al. (2003). Major ecological transitions in wild sunflowers facilitated by hybridization. *Science, 301*(5637), 1211–1216. http://dx.doi.org/10.1126/science.1086949.

Salse, J., Abrouk, M., Bolot, S., Guilhot, N., Courcelle, E., Faraut, T., et al. (2009). Reconstruction of monocotelydoneous proto-chromosomes reveals faster evolution in plants than in animals. *Proceedings of the National Academy of Sciences of the United States of America, 106*(35), 14908–14913. http://dx.doi.org/10.1073/pnas.0902350106.

Sankoff, D. (1992). Edit distance for genome comparison based on non-local operations. In A. Apostolico, M. Crochemore, Z. Galil, & U. Manber (Eds.), *Combinatorial pattern matching*, Vol. 644, (pp. 121–135). Berlin: Springer.

Sankoff, D., & Blanchette, M. (1997). The median problem for breakpoints in comparative genomics. In T. Jiang & D. T. Lee (Eds.), *Computing and combinatorics*, Vol. 1276, (pp. 251–263). Berlin: Springer.

Sankoff, D., & Blanchette, M. (1998). Multiple genome rearrangement and breakpoint phylogeny. *Journal of Computational Biology, 5*(3), 555–570.

Sato, S., Nakamura, Y., Kaneko, T., Asamizu, E., Kato, T., Nakao, M., et al. (2008). Genome structure of the legume, Lotus japonicus. *DNA Research, 15*(4), 227–239. http://dx.doi.org/10.1093/dnares/dsn008.

Schatz, M. C., Witkowski, J., & McCombie, W. R. (2012). Current challenges in de novo plant genome sequencing and assembly. *Genome Biology, 13*(4), 243. http://dx.doi.org/10.1186/gb4015.

Schmutz, J., Cannon, S. B., Schlueter, J., Ma, J., Mitros, T., Nelson, W., et al. (2010). Genome sequence of the palaeopolyploid soybean. *Nature, 463*(7278), 178–183. http://dx.doi.org/10.1038/nature08670.

Schnable, J. C., Springer, N. M., & Freeling, M. (2011). Differentiation of the maize subgenomes by genome dominance and both ancient and ongoing gene loss. *Proceedings of the National Academy of Sciences of the United States of America, 108*(10), 4069–4074. http://dx.doi.org/10.1073/pnas.1101368108.

Semon, M., & Wolfe, K. H. (2007). Rearrangement rate following the whole-genome duplication in teleosts. *Molecular Biology and Evolution, 24*(3), 860–867. http://dx.doi.org/10.1093/molbev/msm003.

Wang, Z., Hobson, N., Galindo, L., Zhu, S., Shi, D., McDill, J., et al. (2012). The genome of flax (Linum usitatissimum) assembled de novo from short shotgun sequence reads. *Plant Journal*, *72*(3), 461–473. http://dx.doi.org/10.1111/j.1365-313X.2012.05093.x.

Wang, K., Wang, Z., Li, F., Ye, W., Wang, J., Song, G., et al. (2012). The draft genome of a diploid cotton Gossypium raimondii. *Nature Genetics*, *44*(10), 1098–1103. http://dx.doi.org/10.1038/ng.2371.

Wang, X., Wang, H., Wang, J., Sun, R., Wu, J., Liu, S., et al. (2011). The genome of the mesopolyploid crop species Brassica rapa. *Nature Genetics*, *43*(10), 1035–1039. http://dx.doi.org/10.1038/ng.919.

Warren, R., & Sankoff, D. (2009). Genome aliquoting with double cut and join. *BMC Bioinformatics*, *10*, S2. http://dx.doi.org/10.1186/1471-2105-10-s1-s2.

Wikstrom, N., Savolainen, V., & Chase, M. W. (2001). Evolution of the angiosperms: Calibrating the family tree. *Proceedings of the Biological Sciences*, *268*(1482), 2211–2220. http://dx.doi.org/10.1098/rspb.2001.1782.

Wolfe, K. H., Gouy, M., Yang, Y. W., Sharp, P. M., & Li, W. H. (1989). Date of the monocot-dicot divergence estimated from chloroplast DNA sequence data. *Proceedings of the National Academy of Sciences of the United States of America*, *86*(16), 6201–6205.

Wolfe, K. H., Li, W. H., & Sharp, P. M. (1987). Rates of nucleotide substitution vary greatly among plant mitochondrial, chloroplast, and nuclear DNAs. *Proceedings of the National Academy of Sciences of the United States of America*, *84*(24), 9054–9058.

Wolfe, K. H., & Shields, D. C. (1997). Molecular evidence for an ancient duplication of the entire yeast genome. *Nature*, *387*(6634), 708–713.

Wu, J., Wang, Z., Shi, Z., Zhang, S., Ming, R., Zhu, S., et al. (2013). The genome of the pear (Pyrus bretschneideri Rehd.). *Genome Research*, *23*(2), 396–408. http://dx.doi.org/10.1101/gr.144311.112.

Wu, H. J., Zhang, Z., Wang, J. Y., Oh, D. H., Dassanayake, M., Liu, B., et al. (2012). Insights into salt tolerance from the genome of Thellungiella salsuginea. *Proceedings of the National Academy of Sciences of the United States of America*, *109*(30), 12219–12224. http://dx.doi.org/10.1073/pnas.1209954109.

Xu, Q., Chen, L. L., Ruan, X., Chen, D., Zhu, A., Chen, C., et al. (2013). The draft genome of sweet orange (Citrus sinensis). *Nature Genetics*, *45*(1), 59–66. http://dx.doi.org/10.1038/ng.2472.

Yogeeswaran, K., Frary, A., York, T. L., Amenta, A., Lesser, A. H., Nasrallah, J. B., et al. (2005). Comparative genome analyses of Arabidopsis spp.: Inferring chromosomal rearrangement events in the evolutionary history of A. thaliana. *Genome Research*, *15*(4), 505–515. http://dx.doi.org/10.1101/gr.3436305.

Young, N. D., Debelle, F., Oldroyd, G. E., Geurts, R., Cannon, S. B., Udvardi, M. K., et al. (2011). The Medicago genome provides insight into the evolution of rhizobial symbioses. *Nature*, *480*(7378), 520–524. http://dx.doi.org/10.1038/nature10625.

Zhang, Q., Chen, W., Sun, L., Zhao, F., Huang, B., Yang, W., et al. (2012). The genome of Prunus mume. *Nature Communications*, *3*, 1318. http://dx.doi.org/10.1038/ncomms2290.

Zheng, C., & Sankoff, D. (2012). Gene order in rosid phylogeny, inferred from pairwise syntenies among extant genomes. *BMC Bioinformatics*, *13*(Suppl. 10), S9. http://dx.doi.org/10.1186/1471-2105-13-S10-S9.

Zuccolo, A., Bowers, J. E., Estill, J. C., Xiong, Z., Luo, M., Sebastian, A., et al. (2011). A physical map for the Amborella trichopoda genome sheds light on the evolution of angiosperm genome structure. *Genome Biology*, *12*(5), R48. http://dx.doi.org/10.1186/gb-2011-12-5-r48.

CHAPTER SEVEN

Insights into the Common Ancestor of Cereals

Xiyin Wang[*,†,‡,1], **Hui Guo**[*,§], **Jingpeng Wang**[†,‡]

[*]Plant Genome Mapping Laboratory, University of Georgia, Athens, Georgia, USA
[†]Center for Genomics and Computational Biology, Hebei United University, Tangshan, PR China
[‡]College of Life Sciences, Hebei United University, Tangshan, PR China
[§]Department of Plant Biology, University of Georgia, Athens, Georgia, USA
[1]Corresponding author: e-mail address: wang.xiyin@gmail.com

Contents

Abstract

Their economic and agricultural importance has motivated whole-genome sequencing efforts of a diverse sampling of cereals, facilitating new research to understand the evolution of their common ancestor. Our analyses of the genome sequences of rice, sorghum, maize, and a non-cultivated pooid, *Brachypodium*, have revealed the occurrence of polyploidizations, genome structural changes, illegitimate recombination between homoeologous chromosomal regions, biological pathway evolution, evolution of gene repertoire, and other important dimensions of evolution in cereal common ancestor(s) after and even before a whole-genome duplication occurring tens of million years ago in a common ancestor of all these plants. One pair of grass chromosomes duplicated in this event has been greatly affected by illegitimate recombination, facilitating ongoing gene conversion and resulting in a stratification of divergence patterns along their length from the centromeres to the telomeres. These findings will possibly

Advances in Botanical Research, Volume 69
ISSN 0065-2296
http://dx.doi.org/10.1016/B978-0-12-417163-3.00007-X

help to decipher mysteries regarding the ecological and agricultural success of cereals and other members of the grass family. Increasing fundamental knowledge of the cereal common ancestor may contribute to understanding botanical diversity and applying that knowledge to sustainable improvement of crop productivity.

1. THE ECONOMIC AND AGRICULTURAL IMPORTANCE OF CEREALS

Cereal plants are from the grass family Poaceae of monocotyledonous flowering plants. The grass family has about 600 genera and ~10,000 species, including members that are among the most economically important plants, such as rice, wheat, maize, sorghum, and sugar cane (Watson & Dallwitz, 1992). Cereals account for about 70% of crops. Grown for their edible seeds, cereals are the primary source of human nutrition, providing more than half of all our calories and appreciable protein (Kellogg, 2001). Rice is a staple food in southern and eastern Asia; maize in central and south America; wheat and barley in Europe, northern Asia, and the Americas; and sorghum in some African countries. These four cereals constitute the most important global agricultural commodities by quantity (Global Perspective Studies Unit, 2006). In addition, sugar cane is the major source of sugar production. Grasses are also grown for forage and fodder. Cow's milk, the sole animal product in the top 10 agricultural commodities by quantity, largely comes from grass-fed animals (Bevan, Garvin, & Vogel, 2010). Grasses are also used for house construction (throughout East Asia (bamboo) and sub-Saharan Africa (sorghum)), papermaking (*Miscanthus*), water treatment, wetland habitat preservation, and land reclamation. Moreover, grasses with C4 photosynthesis, including *Miscanthus*, switchgrass, sugar cane, and sorghum, are attractive for biofuel production. A growing human population (predicted to be 9 billion by 2050) and expected increase in living standards will require the ongoing sustainable exploitation of cereal resources.

2. GENOME SEQUENCING OPENS A NEW ERA OF GRASS RESEARCH

In view of their central importance to humanity, four cereal plants and one relative have been sequenced, including rice (Ehrhartoideae) (International Rice Genome Sequencing Project, 2005; Yu et al., 2005), sorghum (Paterson et al., 2009) and maize (Panicoideae) (Schnable et al., 2009), and *Brachypodium*

(Pooideae) (The International Brachypodium Initiative, 2010) (Fig. 7.1). During the writing of this chapter, draft genome sequences of barley, one wheat chromosome (Jia et al., 2013; Ling et al., 2013), two diploid wheat relatives, and foxtail millet (Devos, 2010) were published (Mayer et al., 2011; Wicker et al., 2011).

Rice was sequenced by four independent groups, including the Beijing Genomics Institute (BGI) (Yu et al., 2002, 2005), International Rice Genome Sequencing Project (IRGSP) (2005), Syngenta (Goff et al., 2002), and Monsanto (see Chapter 3). BGI analysed the genotypes 93–11 and PA64s, the parental strains for a popular land race of super hybrid rice, LYP9. Both subgenomes were sequenced up to ~6× coverage. IRGSP presented a map-based, finished quality sequence that covers 95% of the estimated 389 Mb genome, virtually including all the euchromatic regions and even two complete centromeres (International Rice Genome Sequencing Project, 2005). Both BGI and IRGSP reported ~38,000 rice genes. Syngenta also used a WGS method and published a 10× draft that incorporates the Syngenta data.

Sorghum was the second grass genome sequence that was decoded (Paterson et al., 2009). Despite a repeat content of 61%, a high-quality sequence of its ~730 Mb genome was assembled from homozygous sorghum genotype BTx623 by using a modified WGS approach and incorporating the following: (1) 8.5 genome equivalents of paired-end reads from genomic libraries spanning a 100-fold range of insert sizes, resolving many repetitive regions, and (2) a high-quality read length averaging 723 bp, facilitating assembly. The sorghum sequence was carefully validated by genetic, physical, and syntenic information, and a comparison with 27 finished bacterial artificial chromosomes (BACs) showed the assembly to be 98.46% complete and accurate to ~1 error per 10 kb.

The maize genome of 2.3 Gb was, like sorghum, assembled with support of genetic and physical maps (Schnable et al., 2009). Over 32,000 genes were predicted, of which 99.8% were placed on reference chromosomes. Nearly 85% of the genome is composed of hundreds of families of transposable elements, dispersed non-uniformly across the genome.

As a representative of the Pooideae subfamily, which contains the large and complex hexaploid genome of bread wheat, the wild grass *Brachypodium distachyon* was sequenced (The International Brachypodium Initiative, 2010). The five compact *Brachypodium* pseudochromosomes comprise 272 Mbp, and the assembly was validated by cytogenetic analysis and alignment with two physical maps and sequenced BACs.

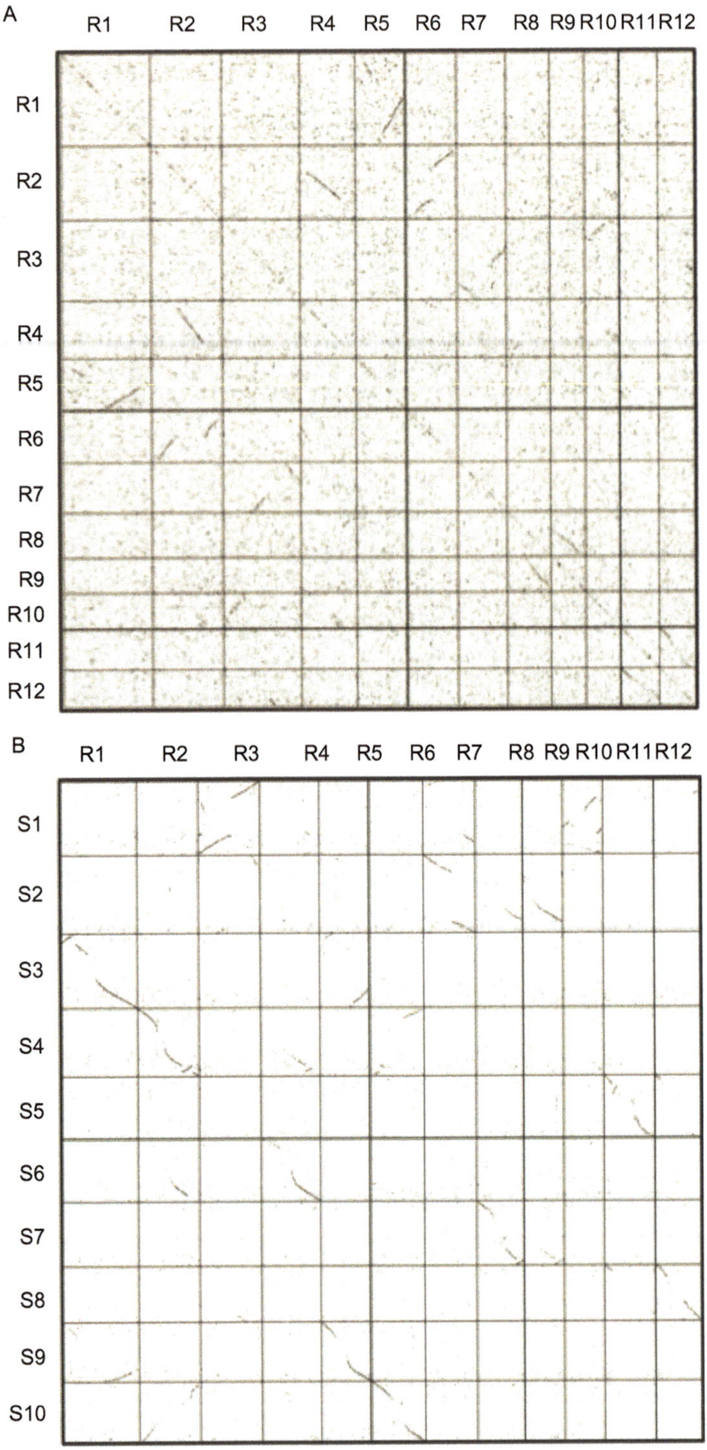
A
R1 R2 R3 R4 R5 R6 R7 R8 R9 R10 R11 R12
R1 R2 R3 R4 R5 R6 R7 R8 R9 R10 R11 R12
B
R1 R2 R3 R4 R5 R6 R7 R8 R9 R10 R11 R12
S1 S2 S3 S4 S5 S6 S7 S8 S9 S10

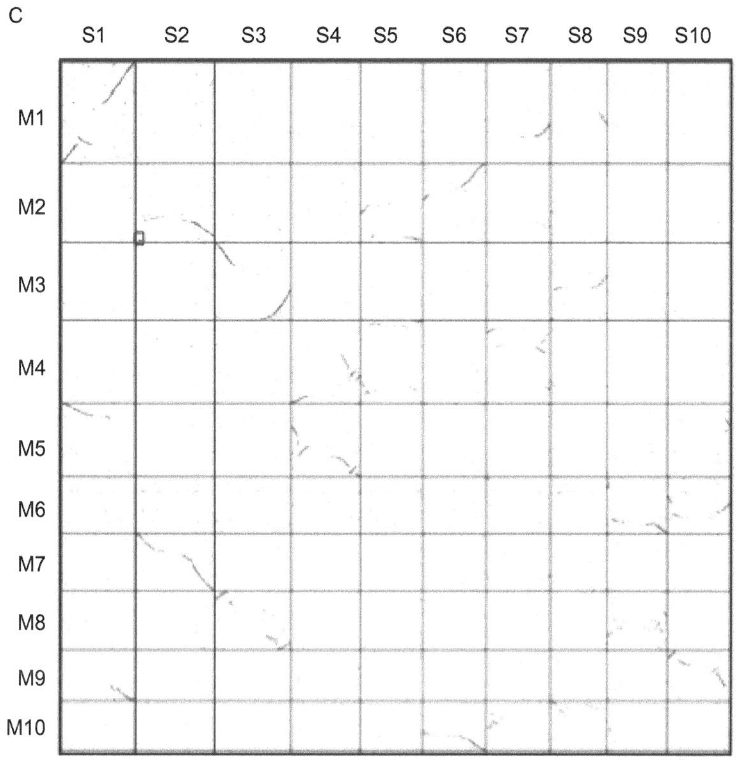

Figure 7.1 Homoeologous gene dot plots. Chromosomes from a grass are arranged horizontally or vertically. R: rice, S: sorghum, M: maize. (A) A dot plot of rice genome reveals a pattern of whole-genome duplication. (B) A dot plot between rice and sorghum shows a pattern of whole-genome duplication shared by both of them, resulting major and minor homoeologous blocks for a genomic region. (C) A dot plot between sorghum and maize shows a pattern of whole-genome duplication common to them and one specific to maize, resulting major and minor blocks. With the major ones, there is a 1:2 ratio between sorghum and maize. (For color version of this figure, the reader is referred to the online version of this chapter.)

3. GENE COLINEARITY CONTRIBUTES TO DECIPHER GENOME STRUCTURE

Comparative genomic analysis of different cereal genomes will help to understand their common ancestor. Such analysis has heavily relied on efficiently detecting intra- and intergenomic homoeologous DNA segments (Van de Peer, 2004), which often preserve a considerable proportion of colinear genes, reflecting gene composition and genome structure of their common ancestor (Fig. 7.1). Gene colinearity is often referred to as gene synteny, when genes are on corresponding homoeologous chromosomes in different taxa, but strict order is not conserved or not known. In this text, we only mention gene colinearity when strictly referring to corresponding gene order.

The inference of gene colinearity is often the starting point for many further biological and evolutionary explorations. Several approaches have been proposed to infer genome-wide gene homology with underlying algorithms to find gene colinearity or gene synteny (Salse, Abrouk, Murat, Quraishi, & Feuillet, 2009; Vandepoele, Simillion, & Van de Peer, 2002). The utilization of colinearity in genome alignments reflects the fact that many genes in extant genomes have remained at their ancestral locations on chromosomes despite widespread genomic repatterning and gene losses, especially after WGDs (see Tang et al., Chapter 8). To cope with the added complexity in plant genomes due to WGDs and subsequent repatterning, we developed MCSCAN software to infer gene colinearity among multiple genomes (Tang, Wang, et al., 2008). MCSCAN was based on elements of two previous pairwise gene colinearity searching tools, DAGchainer (Haas, Delcher, Wortman, & Salzberg, 2004) and ColinearScan (Wang et al., 2006). From the former, it borrowed a scoring scheme and partial searching approach, while from the latter, it borrowed the statistical methods to evaluate the significance of chromosomal homology represented by gene colinearity, which has been well established in theory and tested by computationally simulating datasets. The usefulness of multiple alignments has been proved in several large-scale chromosomal homology searches involving different combinations of genomes (Ming et al., 2008; Tang, Bowers, et al., 2008). An updated version of MCSCAN, MCSCANX, was recently published and is characterized by additional useful elements to explore gene colinearity and gene evolution (Wang et al., 2012). Some initial alignments of genomes have been performed, and related results including gene colinearity information are available in the Plant Genome Duplication Database (http://chibba.agtec.uga.edu/duplication/) (Lee, Tang, Wang, & Paterson, 2013).

3.1. Gene colinearity facilitates paleogenomic exploration

Considerable gene colinearity is shared between rice, sorghum, and *Brachypodium*, with less but still readily discernible and useful colinearity with maize (Fig. 7.1). This fact means that after even tens of millions of years divergence, many genes have preserved their ancestral location, which is of central importance to profound comparative genomics analysis. The sequencing efforts for these grasses will contribute to understanding the domestication and agricultural improvements of staple crops. Here, by using ColinearScan (Wang et al., 2006), homoeologous regions were revealed within each grass and between any two of them.

3.2. Intragenomic gene colinearity

Rice, sorghum, and *Brachypodium* have similar numbers of homoeologous blocks within each genome. There are 175 homoeologous blocks in rice, containing 3946 genes in total. Some blocks can be quite large, with 85 of them having more than 10 colinear genes (paralogs), and 12 having more than 50 colinear genes. The largest block is between rice chromosomes 1 and 5 (Os01 and Os5), including 432 paralogs in colinear positions. There are 170 homoeologous blocks in sorghum, containing 3505 genes in total. A total of75 and 16 blocks have more than 10 and 50 colinear genes, respectively. The largest block is between chromosomes Sb03 and Sb09 and contains 402 colinear paralogs. There are 181 homoeologous blocks in *Brachypodium*, containing 3100 genes, with 82 and 11 blocks having more than 10 and 50 colinear genes, respectively. The largest block is located in chromosome Bd02 and contains 402 colinear paralogs. Comparatively, the homoeologous blocks in maize are smaller in size. There are 332 homoeologous blocks in maize, containing 3505 genes in total. A total of 142 and 1 blocks have more than 10 and 50 colinear genes, respectively. The largest block is located in Zm02 and Zm10 and contains 69 genes.

3.3. Intergenomic gene colinearity

Intergenomic gene colinearity is much better than intragenomic colinearity. Between rice and *Brachypodium*, there are 273 homoeologous blocks, including 15,461 genes in colinearity. 160 and 48 Blocks have 10 and 50 colinear genes, respectively. The largest block sprawls nearly the whole chromosomal arm of Os01 and the corresponding region on Bd02, including 1243 colinear genes. Between rice and sorghum, there are 212 homoeologous blocks, including 15,955 genes in colinearity. 120 and 53 Blocks have 10 and 50 colinear genes, respectively. The largest block sprawls nearly the whole chromosomal arm of Os01 and Sb03, including 1281 colinear genes. Between *Brachypodium* and sorghum, there are 344 homoeologous blocks, including 15,441 genes in colinearity. 210 and 56 Blocks have 10 and 50 colinear genes, respectively. The largest block sprawls nearly the whole chromosomal arm of Br01 and Sb03, including 1398 colinear genes. Between sorghum and maize, there are 577 homoeologous blocks, containing 20,013 genes. 390 and 56 Blocks have 10 and 50 colinear genes, respectively. The longest block is located between Sb03 and Zm03, containing 684 colinear genes.

4. AN ANCESTRAL POLYPLOIDIZATION PRESIDES THE DIVERGENCE OF MAJOR CEREALS

Characterization of gene colinearity (or gene synteny) among four grasses reveals a 'whole-genome duplication' pattern due to a shared polyploidy event (Fig. 7.1). Regions in which duplication can still be discerned cover ~70% of the genome in rice, sorghum, and *Brachypodium* (Paterson, Bowers, & Chapman, 2004). Noting that heterochromatic regions often have insufficient gene densities and more extensive rearrangement that precludes the detection of ancient genome duplications, the observed pattern indicates that these genomes have been affected by large-scale genomic duplication(s). Most chromosomal segments correspond closely to one and only one other region, implying a whole-genome duplication (cWGD). By characterizing synonymous nucleotide substitution rates (Ks) between colinear paralogs and orthologs, we find major peaks of Ks in each grass, which show that the cWGD is more ancient than the divergence of these grasses from a common ancestor (Fig 7.2), as reported previously (Paterson et al., 2004; Wang, Shi, Hao, Ge, & Luo, 2005; Yu et al., 2005). That is, the common ancestor of cereals is inferred to have been a tetraploid. The duplication event was estimated to have occurred ~70 million years ago (mya) by assuming a synonymous nucleotide evolutionary rate of 6.5×10^{-9} substitutions per site per year (Gaut, 1998).

Maize was affected by another lineage-specific whole-genome duplication (mWGD) (Gaut & Doebley, 1997; Schnable et al., 2009). One sorghum genomic region always corresponds to two regions in maize, while one maize region corresponds to only a single sorghum region. Based on the sequence divergence between colinear maize paralogs, this event occurred about ~20 mya, likely quite near the split of sorghum and maize (Fig. 7.1C), consistent with previous reports (Swigonova et al., 2004a, 2004b).

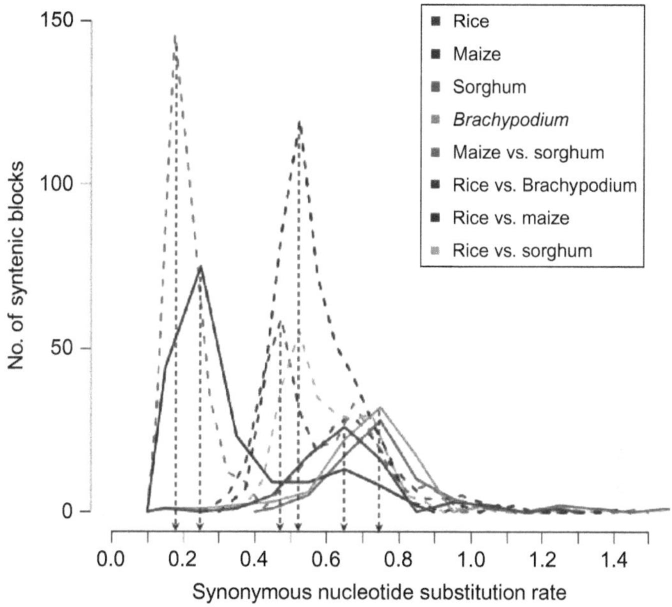

Figure 7.2 Distribution of synonymous nucleotide substitution rates (Ks) between homoeologous genes in grasses. Solid lines show Ks between paralogs within each genome, and broken lines show Ks between homologs (a mixture of othologs and paralogs) between two genomes. (For color version of this figure, the reader is referred to the online version of this chapter.)

4.1. Genomic fractionation after the common polyploidization

Large-scale gene losses followed the polyploidization in the cereal common ancestral genome (Paterson et al., 2009; Wang et al., 2005), with most of these losses completed before the divergence of the major cereal lineages. In rice, for example, 30–65% of duplicated genes have lost at least one duplicated copy. Different cereal genomes have highly similar patterns of gene loss, characterized by more colinear genes between orthologous blocks in different species than between paralogous blocks within a single grass. For example, between rice and sorghum, more than 96% genes have colinear orthologs in the other genome, indicating only 1.7–3% gene loss after their divergence ~50 mya (Paterson et al., 2009). The ~20 million year interval between genome duplication and lineage divergence is thought to be substantially longer than the half-life of duplicated genes (Lynch & Conery, 2003), consistent with the notion that post-duplication gene loss was largely completed before these grass' radiation.

of 70 mya, would have been much diverged than the orthologs formed by species (lineage) divergence ~50 mya. Unusual cases in which the paralogs are much more similar to one another than the orthologs indicate gene conversion or some other mechanism of concerted evolution, perhaps as a result of homoeologous and illegitimate recombination. In both rice and sorghum (and other grasses), cases of gene conversion have been inferred at appreciable frequencies (Fig. 7.3A–D).

Illegitimate recombination may have affected not only individual genes but also sizable chromosomal regions. Different blocks of colinearity within a genome or between genomes that trace to the same duplication event, such as the pancereal duplication of 70 mya, can have very different degrees of divergence (e.g. Ks) between duplicated genes. Such differences may suggest that restriction of illegitimate recombination was accomplished at different times following the duplication (Wang et al., 2009), with earlier restriction resulting in larger Ks.

Notably, one pair of grass chromosomes has been much affected by illegitimate recombination (Wang, Tang, & Paterson, 2011). A duplicated block at the very end of the short arms of homoeologous rice chromosomes 11 and 12, and their respective sorghum orthologous chromosomes 5 and 8, has remarkably small Ks at one end (Wang et al., 2011), and so do the corresponding regions in *Brachypodium* and maize, with a singular pattern of stratification, becoming progressively more similar from centromeres to telomeres along their short arm(s). This observation is reminiscent of the human Y chromosome but singular in all plant genomes sequenced to date and appears to be explained by the temporal, gradual, and segmental restriction of recombination between two duplicated chromosomes produced by the cWGD (Fig. 7.3E). At the very end of the short arms on rice chromosomes 11 and 12, duplicated genes are nearly identical, showing ongoing homoeologous recombination, which has independent supporting evidence from *Oryza* phylogenetic analysis (Jacquemin, Laudie, & Cooke, 2009; Wang, Tang, Bowers, Feltus, & Paterson, 2007).

7. ALIGNMENT OF MULTIPLE GENOMES

An alignment of multiple genomes would help transfer genetic information among them and understand genomic evolution in both large-scale and detailed levels (Abeel, Van Parys, Saeys, Galagan, & Van de Peer, 2012; Abrouk et al., 2010; Salse, Abrouk, Bolot, et al., 2009; Salse et al., 2008). By checking DNA and gene similarity between two genomes, the orthologous

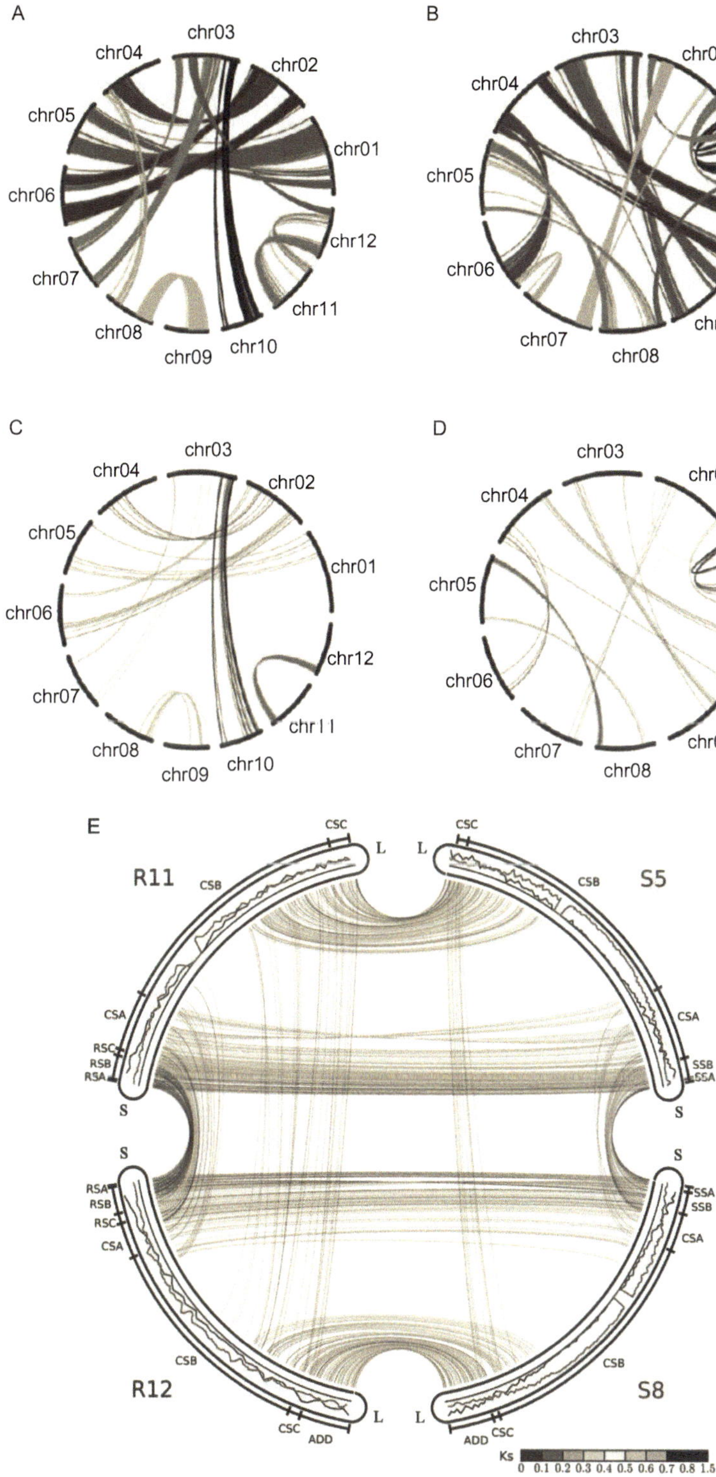

Figure 7.3 (See legend on next page)

and 'outparalogous' blocks can be separated. The outparalogous blocks are those for which homology was established by cWGD, whereas orthologous blocks are those for which homology was established due to speciation. By referring to the rice genome and by integrating gene colinearity information within and between genomes, multiple alignments of genomes were constructed in a stepwise manner, ultimately constructing a multiple alignment of all four genomes (Fig. 7.4). If the rice chromosomes were taken to represent the ancestral genome of grass-common ancestor (as is most frequently inferred), from the alignment pattern in Fig. 7.4, chromosomal breakage points, fissions, and fusions can be easily read out. A detailed alignment of a local genomic region is displayed (Fig. 7.5) to show large-scale gene loss in homoeologous regions from different grasses, affected by cWGD and mWGD.

8. INFERENCE OF THE GENE COMPOSITION OF ANCESTRAL GENOMES

Checking the alignment of different grass genomes sheds light on ancestral genome composition (Tang, Wang, et al., 2008). Here, orthologous gene groups were checked, totalling 10,631 for rice–sorghum, 10,150 for rice–*Brachypodium*, 9782 for sorghum–maize, and 8802 for rice–maize. The lower numbers of orthologous gene groups in comparisons with

Figure 7.3—cont'd Gene conversion between duplicated genes produced by whole-genome duplication common to major grasses. Chromosomes are arranged in circles. Major duplicated blocks are displayed, and curvy lines are used to link duplicated genes. Lines in panels (A–D) are coloured to separate different blocks in rice (A, C) and sorghum (B, D), respectively. Panels (A) and (B) show information of all duplicated and colinear genes, whereas (C) and (D) show converted genes. In panel (E), we show rice homoeologous chromosomes 11 and 12 and their respective sorghum orthologous chromosomes, and two homoeologous chromosomes have been affected by very recent and even ongoing recombination, leading to gene conversion. Curvy lines linking duplicated and orthologous genes are coloured as to Ks values between the duplicated genes (see colour scheme in the panel). As to Ks, two homoeologous chromosomes from rice or sorghum are divided into strata (RSA–RSC, SSA and SSB, and CSA–CSC), showing rice-specific, sorghum-specific, and common and ancestral strata as a result of temporal and segmental restriction of recombination. Red and blue curvy lines in chromosome boxes show distribution of repetitive sequences and genes along chromosomes. L: long arms; S: short arms; ADD: additional chromosome segment. (For color version of this figure, the reader is referred to the online version of this chapter.)

maize are consistent with erosion of synteny due to gene loss following the mWGD and its much closer relationship to sorghum than to rice.

This alignment permitted us to estimate gene numbers before and after the cWGD, requiring that the existence of an ancestral gene must be supported by extant gene colinearity in paralogous blocks within any grass genome or in orthologous blocks between any two genomes. Excluding maize due to the gene losses associated with the mWGD, we found that the duplicated ancestral genome contained at least 20,708 genes. If two paralogous genes were merged into one node, we estimated that there were at least 10,885 genes before the cWGD. We emphasize that these estimates

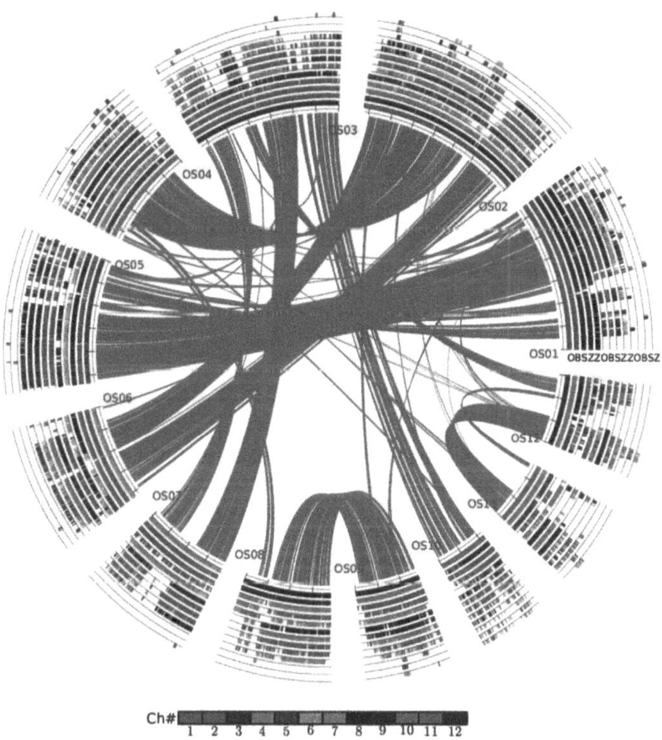

Figure 7.4 Alignment of grass chromosomes. By using gene colinearity, chromosomes from rice (O or Os), sorghum (S), *Brachypodium* (B), and maize (Z) are aligned using rice as reference. A common whole-genome duplication in all these grasses causes rice, sorghum, and *Brachypodium* to have two circles of chromosomes, and an additional lineage-specific whole-genome duplication causes maize to have four. Genes are coloured according to their chromosome number in rice. For example, genes on chromosome 1 from all grasses are in the same colour. (For color version of this figure, the reader is referred to the online version of this chapter.)

Chen, J. M., Cooper, D. N., Chuzhanova, N., Ferec, C., & Patrinos, G. P. (2007). Gene conversion: Mechanisms, evolution and human disease. *Nature Reviews Genetics*, *8*(10), 762–775. http://dx.doi.org/10.1038/nrg2193, nrg2193 [pii].
Devos, K. M. (2010). Grass genome organization and evolution. *Current Opinion in Plant Biology*, *13*(2), 139–145. http://dx.doi.org/10.1016/j.pbi.2009.12.005, S1369-5266(09)00181-2 [pii].
Freeling, M. (2001). Grasses as a single genetic system: Reassessment 2001. *Plant Physiology*, *125*(3), 1191–1197.
Gaut, B. S. (1998). Molecular clocks and nucleotide substitution rates in higher plants. *Evolutionary Biology*, *30*, 93–120.
Gaut, B. S., & Doebley, J. F. (1997). DNA sequence evidence for the segmental allotetraploid origin of maize. *Proceedings of the National Academy of Sciences of the United States of America*, *94*(13), 6809–6814.
Global Perspective Studies Unit, Food and Agriculture Organization of the United Nations. (2006). *FAQ: World agriculture: towards 2030/2050. Interim Report*. Rome, Italy.
Goff, S. A., Ricke, D., Lan, T. H., Presting, G., Wang, R., Dunn, M., et al. (2002). A draft sequence of the rice genome (*Oryza sativa* L. ssp. *japonica*). *Science*, *296*(5565), 92–100.
Haas, B. J., Delcher, A. L., Wortman, J. R., & Salzberg, S. L. (2004). DAGchainer: A tool for mining segmental genome duplications and synteny. *Bioinformatics*, *20*(18), 3643–3646.
Hilu, K. W. (2004). Phylogenetics and chromosomal evolution in the Poaceae (grasses). *Australian Journal of Botany*, *52*, 10.
Hurles, M. E. (2001). Gene conversion homogenizes the CMT1A paralogous repeats. *BMC Genomics*, *2*(1), 11.
International Rice Genome Sequencing Project (2005). The map-based sequence of the rice genome. *Nature*, *436*(7052), 793–800.
Jacquemin, J., Laudie, M., & Cooke, R. (2009). A recent duplication revisited: Phylogenetic analysis reveals an ancestral duplication highly-conserved throughout the Oryza genus and beyond. *BMC Plant Biology*, *9*, 146. http://dx.doi.org/10.1186/1471-2229-9-146, 1471-2229-9-146 [pii].
Jia, J., Zhao, S., Kong, X., Li, Y., Zhao, G., He, W., et al. (2013). Aegilops tauschii draft genome sequence reveals a gene repertoire for wheat adaptation. *Nature*, *496*(7443), 91–95. http://dx.doi.org/10.1038/nature12028.
Kellogg, E. A. (2001). Evolutionary history of the grasses. *Plant Physiology*, *125*(3), 1198–1205.
Lee, T. H., Tang, H., Wang, X., & Paterson, A. H. (2013). PGDD: A database of gene and genome duplication in plants. *Nucleic Acids Research*, *41*(Database issue), D1152–D1158. http://dx.doi.org/10.1093/nar/gks1104.
Ling, H. Q., Zhao, S., Liu, D., Wang, J., Sun, H., Zhang, C., et al. (2013). Draft genome of the wheat A-genome progenitor Triticum urartu. *Nature*, *496*(7443), 87–90. http://dx.doi.org/10.1038/nature11997.
Lynch, M., & Conery, J. S. (2003). The evolutionary demography of duplicate genes. *Journal of Structural and Functional Genomics*, *3*(1–4), 35–44.
Mayer, K. F., Martis, M., Hedley, P. E., Simkova, H., Liu, H., Morris, J. A., et al. (2011). Unlocking the barley genome by chromosomal and comparative genomics. *Plant Cell*, *23*(4), 1249–1263. http://dx.doi.org/10.1105/tpc.110.082537, tpc.110.082537 [pii].
Ming, R., Hou, S., Feng, Y., Yu, Q., Dionne-Laporte, A., Saw, J. H., et al. (2008). The draft genome of the transgenic tropical fruit tree papaya (Carica papaya Linnaeus). *Nature*, *452*(7190), 991–996.
Murat, F., Xu, J. H., Tannier, E., Abrouk, M., Guilhot, N., Pont, C., et al. (2010). Ancestral grass karyotype reconstruction unravels new mechanisms of genome shuffling as a source of plant evolution. *Genome Research*, *20*(11), 1545–1557. http://dx.doi.org/10.1101/gr.109744.110, gr.109744.110 [pii].

Paterson, A. H., Bowers, J. E., Bruggmann, R., Dubchak, I., Grimwood, J., Gundlach, H., et al. (2009). The Sorghum bicolor genome and the diversification of grasses. *Nature, 457*(7229), 551–556.

Paterson, A. H., Bowers, J. E., & Chapman, B. A. (2004). Ancient polyploidization predating divergence of the cereals, and its consequences for comparative genomics. *Proceedings of the National Academy of Sciences of the United States of America, 101*(26), 9903–9908.

Salse, J., Abrouk, M., Bolot, S., Guilhot, N., Courcelle, E., Faraut, T., et al. (2009a). Reconstruction of monocotelydoneous proto-chromosomes reveals faster evolution in plants than in animals. *Proceedings of the National Academy of Sciences of the United States of America, 106*(35), 14908–14913. http://dx.doi.org/10.1073/pnas.0902350106, 0902350106 [pii].

Salse, J., Abrouk, M., Murat, F., Quraishi, U. M., & Feuillet, C. (2009b). Improved criteria and comparative genomics tool provide new insights into grass paleogenomics. *Briefings in Bioinformatics, 10*(6), 619–630. http://dx.doi.org/10.1093/bib/bbp037, bbp037 [pii].

Salse, J., Bolot, S., Throude, M., Jouffe, V., Piegu, B., Quraishi, U. M., et al. (2008). Identification and characterization of shared duplications between rice and wheat provide new insight into grass genome evolution. *Plant Cell, 20*, 11–24.

Schnable, J. C., Springer, N. M., & Freeling, M. (2011). Differentiation of the maize subgenomes by genome dominance and both ancient and ongoing gene loss. *Proceedings of the National Academy of Sciences of the United States of America, 108*(10), 4069–4074. http://dx.doi.org/10.1073/pnas.1101368108, 1101368108 [pii].

Schnable, P. S., Ware, D., Fulton, R. S., Stein, J. C., Wei, F., Pasternak, S., et al. (2009). The B73 maize genome: Complexity, diversity, and dynamics. *Science, 326*(5956), 1112–1115. http://dx.doi.org/10.1126/science.1178534, 326/5956/1112 [pii].

Soderstrom, T. R. H., Hilu, K. W., Campbell, C. S., & Barkworth, M. A. (1987). *Grass systematics and evolution*. Washington, DC: Smithsonian Institution Press.

Swigonova, Z., Lai, J. S., Ma, J. X., Ramakrishna, W., Llaca, M., Bennetzen, J. L., et al. (2004a). On the tetraploid origin of the maize genome. *Comparative and Functional Genomics, 5*(3), 281–284.

Swigonova, Z., Lai, J. S., Ma, J. X., Ramakrishna, W., Llaca, V., Bennetzen, J. L., et al. (2004b). Close split of sorghum and maize genome progenitors. *Genome Research, 14*(10A), 1916–1923.

Tang, H., Bowers, J. E., Wang, X., Ming, R., Alam, M., & Paterson, A. H. (2008a). Synteny and colinearity in plant genomes. *Science, 320*, 486–488.

Tang, H. B., Wang, X. Y., Bowers, J. E., Ming, R., Alam, M., & Paterson, A. H. (2008b). Unraveling ancient hexaploidy through multiply-aligned angiosperm gene maps. *Genome Research, 18*(12), 1944–1954.

The Arabidopsis Genome Initiative, (2000). Analysis of the genome sequence of the flowering plant *Arabidopsis thaliana*. *Nature, 408*(6814), 796 815.

The International Brachypodium Initiative, (2010). Genome sequencing and analysis of the model grass Brachypodium distachyon. *Nature, 463*(7282), 763–768. http://dx.doi.org/10.1038/nature08747, nature08747 [pii].

Thomas, B. C., Pedersen, B., & Freeling, M. (2006). Following tetraploidy in an Arabidopsis ancestor, genes were removed preferentially from one homeolog leaving clusters enriched in dose-sensitive genes. *Genome Research, 16*(7), 934–946. http://dx.doi.org/10.1101/gr.4708406, gr.4708406 [pii].

Van de Peer, Y. (2004). Computational approaches to unveiling ancient genome duplications. *Nature Review Genetics, 5*(10), 752–763.

Vandepoele, K., Simillion, C., & Van de Peer, Y. (2002). Detecting the undetectable: Uncovering duplicated segments in *Arabidopsis* by comparison with rice. *Trends in Genetics, 18*(12), 606–608.

Wang, X., Shi, X., Hao, B., Ge, S., & Luo, J. (2005). Duplication and DNA segmental loss in the rice genome: Implications for diploidization. *New Phytologist, 165*(3), 937–946.

Wang, X., Shi, X., Li, Z., Zhu, Q., Kong, L., Tang, W., et al. (2006). Statistical inference of chromosomal homology based on gene colinearity and applications to Arabidopsis and rice. *BMC Bioinformatics*, 7(1), 447.

Wang, X., Tang, H., Bowers, J. E., Feltus, F. A., & Paterson, A. H. (2007). Extensive concerted evolution of rice paralogs and the road to regaining independence. *Genetics*, *177*(3), 1753–1763.

Wang, X., Tang, H., Bowers, J. E., & Paterson, A. H. (2009). Comparative inference of illegitimate recombination between rice and sorghum duplicated genes produced by polyploidization. *Genome Research*, *19*(6), 1026–1032.

Wang, Y., Tang, H., Debarry, J. D., Tan, X., Li, J., Wang, X., et al. (2012). MCScanX: A toolkit for detection and evolutionary analysis of gene synteny and collinearity. *Nucleic Acids Research*, *40*(7), e49. http://dx.doi.org/10.1093/nar/gkr1293.

Wang, X., Tang, H., & Paterson, A. H. (2011). Seventy million years of concerted evolution of a homoeologous chromosome pair, in parallel, in major Poaceae lineages. *Plant Cell*, *23*(1), 27–37. http://dx.doi.org/10.1105/tpc.110.080622.

Watson, L., & Dallwitz, M. J. (1992). *The grass genera of the world*. Wallingford: CAB International.

Wicker, T., Mayer, K. F., Gundlach, H., Martis, M., Steuernagel, B., Scholz, U., et al. (2011). Frequent gene movement and pseudogene evolution is common to the large and complex genomes of wheat, barley, and their relatives. *Plant Cell*, *23*(5), 1706–1718. http://dx.doi.org/10.1105/tpc.111.086629, tpc.111.086629 [pii].

Wiese, C., Pierce, A. J., Gauny, S. S., Jasin, M., & Kronenberg, A. (2002). Gene conversion is strongly induced in human cells by double-strand breaks and is modulated by the expression of BCL-x(L). *Cancer Research*, *62*(5), 1279–1283.

Woodhouse, M. R., Schnable, J. C., Pedersen, B. S., Lyons, E., Lisch, D., Subramaniam, S., et al. (2010). Following tetraploidy in maize, a short deletion mechanism removed genes preferentially from one of the two homologs. *PLoS Biology*, *8*(6), e1000409. http://dx.doi.org/10.1371/journal.pbio.1000409.

Yu, J., Hu, S., Wang, J., Wong, G. K., Li, S., Liu, B., et al. (2002). A draft sequence of the rice genome (*Oryza sativa* L. ssp. *indica*). *Science*, *296*(5565), 79–92.

Yu, J., Wang, J., Lin, W., Li, S. G., Li, H., Zhou, J., et al. (2005). The genomes of Oryza sativa: A history of duplications. *PLoS Biology*, *3*(2), 266–281.

CHAPTER EIGHT

Early History of the Angiosperms

Haibao Tang[*,†,1], **Eric Lyons**[‡], **James C. Schnable**[§]
[*]FAFU and UIUC-SIB Joint Center for Genomics and Biotechnology, Fujian Agriculture and Forestry University, Fuzhou, Fujian Province, China
[†]J. Craig Venter Institute, Rockville, Maryland, USA
[‡]School of Plant Sciences, iPlant Collaborative, University of Arizona, Tucson, Arizona, USA
[§]Donald Danforth Plant Science Center, St. Louis, Missouri, USA
[1]Corresponding author: e-mail address: htang@jcvi.org

Contents

Abstract

The flowering plants, also known as the angiosperms, are the most diverse group of plants. The basal flowering plant lineages diverged at very early stage in flowering plant evolution, followed by rapid diversification of the magnoliids, the eudicots and the monocots. Genomic comparisons within and across plant lineages help identify the critical events that occurred during the evolution of major groups of flowering plants. In this chapter, we first review the basic concepts and analytic methods for studying ancient polyploidy—a prominent feature during plant evolution. We then highlight recent progress on the dating of deep polyploidies in the eudicot and monocot lineage, respectively. With a clear knowledge of genomic history, we can effectively compare the eudicot genomes to monocot genomes, which promise to bridge functional equivalence between genes of the two well-studied groups. Finally, we deduce the composition and structure of the 'ancestral genome' on the basis of the arrangements of genes in the extant species. The *in silico* reconstruction of the ancestral genome provides an integrated framework under which conservation of modern plant genomes can be systematically studied.

Advances in Botanical Research, Volume 69
ISSN 0065-2296
http://dx.doi.org/10.1016/B978-0-12-417163-3.00008-1

genomes—from various ecotypes that have adapted to their local environments to morphotypes that have modified their developmental programmes and species that are distinctive in their characteristics. Finding commonalities as well as differences among various genomes allows us to identify key patterns of evolution that contribute to creating such rich biological diversity. Comparing various plant genomes from an evolutionary perspective holds the key to answering many of these questions.

2. PLANT GENOME COMPARISONS AND PALEOPOLYPLOIDY EVENTS

Plant genomes offer opportunities at the frontiers of the study of the evolution of genome structure and dynamics and the underlying processes and principles. Plant genomes have high turnover rates of genetic material and a dynamic history of chromosome evolution. To catalogue these genomic changes, researchers have developed methods to identify relatively small-scale genomic mutations, such as point mutations, indels and gene transpositions, and large-scale changes involving chromosomes or entire genomes such as inversions, translocations, chromosomal fusions or fissions and segmental or genome duplications. It is thought that large-scale changes often lead to speciation by creating chromosomal incompatibilities and thus reproductive barriers between the mutant (i.e. the 'hopeful monster') and the rest of its source population, permitting new species to more rapidly diverge from one another (Chouard, 2010).

Arguably, the largest genomic changes that can be imposed on a genome are polyploidy events—where in one generation, all the genetic information is duplicated. Polyploidy events such as whole-genome duplications (WGDs) are evident in lineages across the spectrum of eukaryotes, including fungi (Kellis, Birren, & Lander, 2004), animals (Aury et al., 2006; Jaillon et al., 2004) and plants (Bowers, Chapman, Rong, & Paterson, 2003; Cui et al., 2006). WGDs offer rich opportunities for the evolution of new or modified gene functions and create new gene arrangements. In addition to creating reproductive barriers with parental diploid species, reciprocal gene loss following WGDs can contribute to reproductive isolation between polyploid populations through divergent resolution of duplicate copies (Bikard et al., 2009) and foreshadow the diversification of species (Lynch & Force, 2000; Scannell, Byrne, Gordon, Wong, & Wolfe, 2006;

Soltis et al., 2009). Some have suggested a possible link between polyploidy and the likelihood of surviving mass extinction events, noting that many ancient polyploidy events observed in the genomes of extant plant species appeared to date to times close to the Cretaceous–Tertiary (K–T) extinction (Fawcett, Maere, & Van de Peer, 2009; Van de Peer, Maere, & Meyer, 2009). Additionally, events following polyploidy may create heterogeneity among different parts within the same genome, in the extreme creating one 'dominant' subgenome as well as facilitating the rewiring of regulatory networks (Bekaert, Edger, Pires, & Conant, 2011; Schnable, Springer, & Freeling, 2011; Tang et al., 2012).

One important aspect of comparisons between plant genomes is how to deal with the complications derived from the post-WGD degeneration or loss of duplicate genes (Tang, Bowers, et al., 2008, 2011; Thomas, Pedersen, & Freeling, 2006). Traces of past WGD events can often be detected from pairwise segments through conserved synteny—pairs of paralogous genes retained in the same order in different regions of a modern genome. Gene loss at duplicated loci 'fractionates' the duplicate gene pairs and reduces the density of colinear gene pairs, making the signatures of ancient 'paleopolyploidy' more difficult to detect. Depending on the level of gene loss, the residual signals of paleopolyploidy are sometimes so eroded that homologous segments can no longer be identified based only on similarity to one another. The problem is worsened when the species in question has undergone several genome duplication events, with recent duplications obscuring synteny from more ancient events (Van de Peer, 2004).

In order to handle paleopolyploidy events in plant genome comparisons, two methods are repeatedly used—'bottom-up' (intragenomic) and 'top-down' (intergenomic) methods (Paterson, Freeling, Tang, & Wang, 2010; Tang, Bowers, et al., 2008). The 'bottom-up' approach is based on an iterative step that interleaves the most recently duplicated segments within a single genome and then generates intermediate 'profiles' that are further recursively merged (Aury et al., 2006; Bowers et al., 2003). The alternative 'top-down' approach requires only one cycle of deduction by simultaneously searching for and aligning all structurally similar segments across multiple genomes and subgenomes. The dual methods are complementary and are both useful for identifying evidence of ancient duplications. For deep genome comparisons, these approaches may be used in conjunction with one other.

3. ANALYTIC TOOLS AND WORKFLOW FOR GENOME COMPARISONS

Before delving into polyploidy inference and dating in early flowering plant lineages, here, we offer a quick introduction and review of some basic but frequently used analytic tools. Many of these tools are specifically designed or tuned for comparisons between plant genomes. Much of the additional algorithmic efforts are in place to cope with differential ploidy levels between genomes that are common in comparison of plant lineages.

A genomic dot plot is a basic but popular visualization tool for genome comparisons. Almost all major genomic changes between two genomes or within one genome can be viewed on a genomic dot plot. A two-dimensional genomic dot plot is designed to show a 'pairwise' all-against-all comparison—with chromosomes from one genome on the *x*-axis and chromosomes from another genome on the *y*-axis. A dot indicates sequence similarity between pairs of regions from the two genomes. If the two genome sequences are identical or very similar, the result is a straight line along the diagonal. Any patterns that deviate from the diagonal are indicative of structural changes, for example, inversions and duplications.

Within this type of genomic dot plots, we are mostly interested in cluster of dots called 'synteny blocks'. The pipeline to identify synteny blocks for comparing genomes contains a number of steps to extract, enrich and visualize the signal contained in a dot plot. Figure 8.2 shows a current curation pipeline used by the Plant Genome Duplication Database (PGDD) that is a database to store and query synteny blocks across many sequenced plant genomes (Lee, Tang, Wang, & Paterson, 2013). The typical steps of a synteny block pipeline are as follows: (1) generate the anchors (dots), (2) filter the anchors to make the blocks more easily extractable, (3) generate the blocks, (4) filter the blocks, (5) curate the blocks and (6) visualize the blocks.

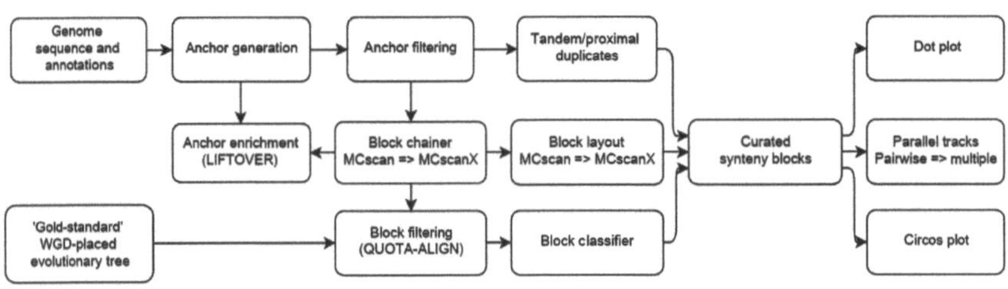

Figure 8.2 Genome comparison workflow connected through individual components.

In order to find good synteny blocks, one needs to have good anchors to start with. To generate the anchors, a number of sequence alignment software can be used, with the most popular tools including TBLASTX, BLASTN, LASTZ and LAST (Camacho et al., 2009; Kielbasa, Wan, Sato, Horton, & Frith, 2011; Schwartz et al., 2003). Differences between alignment algorithms are mostly in sensitivity and computational efficiency. To filter the matching sequences (anchors), some researchers use high stringency (percent identity and percent coverage) (Salse et al., 2009), while others use adaptive filtering such as variants of reciprocal best hit or C-score filtering (Tang, Bowers, Wang, & Paterson, 2010).

After the anchors are generated and filtered, synteny blocks can be extracted between a pair of genomes. There have been a number of different synteny block methods of two major classes—density-based (Catchen, Conery, & Postlethwait, 2009; Hampson, McLysaght, Gaut, & Baldi, 2003) and colinearity-based methods (Haas, Delcher, Wortman, & Salzberg, 2004; Proost et al., 2012; Tang, Wang, et al., 2008). The key idea is to specify a distance cut-off between the anchors and assign all the anchors close to one another to the same block. These blocks are then screened to contain a minimum number of anchors. Isolated anchors outside the blocks, or blocks with fewer than the minimum number of anchors, are usually noise derived from transposed genes or high-copy gene families. After the synteny blocks are extracted, there is a second filtering of blocks to remove artefacts. Two classes of methods can be used to filter the synteny blocks—divergence-based and depth-based filtering. Divergence-based filtering assumes a mode of genetic distances (e.g. nucleotide substitutions per synonymous site (Ks)) for each block based on the average distance between gene pairs within the block. Blocks of significantly high or low values are then filtered or differentially labelled (Lyons, Pedersen, Kane, & Freeling, 2008). Another type of block-level filtering is based on syntenic depth. QUOTA-ALIGN can be used to clean up artefacts as well as classify blocks resulting from speciation and/or different WGD events (Tang et al., 2011). QUOTA-ALIGN is useful even when conventional Ks-based filtering is less effective due to overlapping peaks and for older events when substitutions at the third codon position are saturated. QUOTA-ALIGN has been adopted by a few online comparative genomics platforms, including Legume Information System (Gonzales, Gajendran, Farmer, Archuleta, & Beavis, 2007), PGDD (Lee et al., 2013) and CoGe (Lyons & Freeling, 2008).

regions were then constructed and correlated back to these regions (Sampedro, Lee, Carey, dePamphilis, & Cosgrove, 2005).

Historically, the names for inferred polyploidy events were largely ad hoc, a system that has become undesirable due to increasing numbers of events discovered and codiscovery of events in different lineages. Standardization of the names of paleopolyploidy events is much needed for easy reference and communication among researchers. Several naming systems have been used by different groups working on polyploidy, including three different types—precedence-based, clade-based and sequential naming systems. Following a number of successive group discussions first initiated during Plant and Animal Genome meeting (PAG) in 2010, two naming approaches were proposed: (1) relative naming, starting from the leaf organism and moving up in the phylogenetic tree and calling successive events in Greek letter order, for example, Brassica-α and Brassica-β, and (2) unique naming, use a unique integer for each new discovered event. The unique naming was the preferred method for standardizing the names and the events that have been discovered so far are tracked in a public website (http://polyploidy.iplantc.org/). Each event will then track its phylogenetic placement (including uncertain placements) on the tree of life, as well as the evidence in support of the inference of the event. Supporting evidence types can include cytological observation, Ks distribution of paralogs, phylogenomic studies or structural (synteny) patterns, as previously discussed.

Over the past few years, various polyploidy events were mapped to the phylogeny of flowering plants, and some earlier dating results based on limited data have been rectified (Fig. 8.3). The dates of many relatively recent paleopolyploidy events, in particular events that are limited to a family or genus, are quite clear. For example, the two most recent paleopolyploidies affecting *Arabidopsis*, α and β, following the usage in Bowers et al. (2003), now appear to have occurred within the crucifer lineage (Jaillon et al., 2007; Ming et al., 2008). *Populus trichocarpa* (poplar) underwent a duplication specific to its own salicoid lineage (Tuskan et al., 2006) and shares only one of the three paleopolyploidies (γ) affecting *Arabidopsis*. *Vitis vinifera* (grape) (Jaillon et al., 2007) and *Carica papaya* (papaya) (Ming et al., 2008), the latter within the same taxonomic order (Brassicales) as *Arabidopsis*, each have no subsequent polyploidies since γ (Fig. 8.3). Many lineages have experienced not just one, but several events. By tracing how many events one species experienced from root to leaf, we can then calculate the expected ancestral ploidy level for each modern species. For example, the genome of the model plant Arabidopsis contains a total ploidy of $12\times$ $(2\times2\times3)$ relative to the common ancestor of angiosperms, and the maize genome contains a total ploidy of 16× (2×2×2×2) (Fig. 8.3).

The accurate dating of polyploidy is impacted by the duration of the interval to subsequent speciation events. When polyploidy and speciation events are closely spaced (so that very few shared gene losses occurred), then a shared polyploidy may appear as independent events in two lineages, especially if each lineage experienced substantial independent gene fractionation. Accurate dating becomes more challenging if the two events were very recent (little gene loss) or very ancient (massive gene loss). Both of these cases further reduce the information available from which to infer their relative timing.

Deep paleopolyploidy events in major plant groups are particularly important since any deep genomic comparisons will inevitably require knowledge of such events. In eudicots, the most ancient duplication (called γ by Bowers et al., 2003) is now known to be a hexaploidy (Jaillon et al., 2007; Ming et al., 2008). In the monocots, the two most ancient duplications are called σ and τ (Tang et al., 2010). A full understanding of when and how these more ancient events occurred has many important implications in studying flowering plant evolution and is instrumental to uncovering genomic and functional correspondence between deeply divergent clades.

5. CIRCUMSCRIBING THE PANEUDICOT PALEOHEXAPLOIDY

Gene order alignments revealed a high degree of colinearity between *Arabidopsis*, *Carica* (papaya) and *Populus* (poplar), and a triplicated structure of a common *Arabidopsis–Carica–Populus* ancestor was suggested (Ming et al., 2008). Comparison to the genome of *Vitis*, a eudicot out-group for *Arabidopsis*, *Carica* and *Populus*, validated the reconstructed order and the inferred triplicated structure (Tang, Bowers, et al., 2008). This genome triplication, or paleohexaploidy event, corresponded to the γ event first inferred in Bowers et al. (2003).

Following the *Vitis* genome analysis, the γ triplication event was initially dated to have occurred after the monocot–eudicot separation but before the radiation of the rosids (Jaillon et al., 2007), with a rather wide range of possible time points for it to occur (Soltis et al., 2009). Recent efforts resulting in an increasing level of precision on the placement of the γ event show a classic example of how phylogenetic 'circumscription' worked in dating a paleopolyploidy event, by successfully closing the gap between the lower-bound (lineages that do have this event) and upper-bound estimates (lineages that do *not* have the event). The first line of evidence came from comparisons of local tomato regions to the corresponding regions in grape. Tang et al. analysed 78 *Solanum* BACs with more than 10 genes (i.e. from gene-rich regions) (Tang, Wang, et al., 2008). Out of the 78 BACs, 72 have more than half of their genes showing primary synteny to a single *Vitis* chromosome. Each individual tomato BAC corresponds closely to only one of the triplicate regions rather than showing equal matches to each of the three γ triplicated chromosomes in *Vitis*. This evidence strongly supports the hypothesis that the γ triplication occurred in a common ancestor of asterids and rosids, also suggesting that each *Solanum* segment would be expected to have up to four primary syntenic segments in *Arabidopsis* (due to α and β) (Tang, Wang, et al., 2008). Comparison of the completed tomato genome sequence to the grape genome supported the hypothesis that the γ triplication occurred in a common core eudicot ancestor (The Tomato Genome Consortium, 2012). Analyses from a few other recently sequenced asterid genomes further supported that the γ triplication did indeed occur in a common ancestor of eurosids and asterids (Ibarra-Laclette et al., 2013; Xu et al., 2011). This had delimited the time frame of γ earlier to just prior to divergence of the core eudicots but not addressing the various lineages of basal eudicots (Fig. 8.4).

To further delimit the time of the γ triplication, Jiao et al. employed a phylogenomic approach using transcriptome sequences in basal eudicots. Most gene duplications were before divergence of the rosids and asterids but after the divergence of the Ranunculales (Jiao et al., 2012), thereby giving a tighter upper-bound. Another phylogenomic study found only rare evidence of γ duplicates within the transcriptome of Gunnera and Pachysandra (Vekemans et al., 2012), although being inconclusive due to limited data.

The sacred lotus genome, representing the basal eudicot group Proteales, showed very little evidence of the γ triplication in lotus, although revealing a lineage-specific duplication (λ). Therefore, comparison between grape and lotus shows mostly three (grape) to two (lotus) matches (Ming et al., 2013). While the inference from the spatial analysis is in favour of the hypothesis that lotus does *not* have γ event, there are some gene pairs supporting a 'shared' model. This incongruence prompted the speculative hypothesis of a 'two-step' model for the γ triplication (Ming et al., 2013), with one event forming a tetraploid and subsequent addition of another genome forming the hexaploid. The third genome was added some time later, during

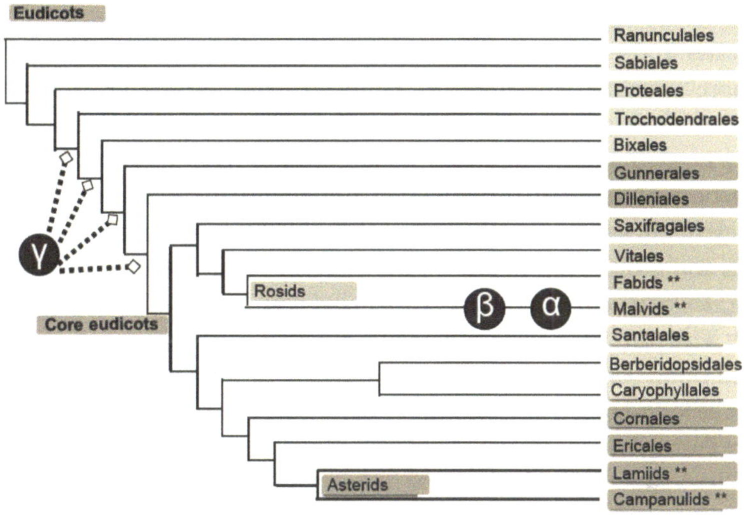

Figure 8.4 Phylogeny of selected families of eudicots. Circles represent paleopolyploidy events discovered in the Arabidopsis lineage (Bowers et al., 2003; Jaillon et al., 2007; Tang, Wang, et al., 2008). Arrows indicate possible placement for certain events. Phylogeny is based on APG III (The Angiosperm Phylogeny Group, 2009). **indicates unranked groups instead of regular taxonomic orders. (See colour plate.)

which time the gene fractionation had already occurred within the first two genomes (Lyons et al., 2008). A possible separation of the two rounds of mergers could partially explain the incongruence but this hypothesis remains to be clearly demonstrated.

In partial summary, while the γ triplication has unanimous support as being prior to the rosid–asterid split (i.e. ancestral to the core eudicots), its relationship to the basal eudicot lineages is less clear due to lack of genomic data and some uncertainties associated with phylogenomic inference. The most likely range of dates for the γ event based on present data is after the divergence of the Proteales and core eudicots. Figure 8.4 reflects the most specific dates so far obtained for the core eudicot γ event. Complete genome sequences of the basal eudicot taxa, in particular those that have close-to-ancestral karyotypes (no subsequent WGDs or chromosomal rearrangements), might further circumscribe the dating of the γ event.

Curiously, the median Ks among *Vitis* γ paleologs (1.22) is much lower than that of *Carica* (1.76) and *Populus* (1.54) (Tang, Wang, et al., 2008). While γ was in a common ancestor of the three species, the difference in the Ks values must indicate a difference in the nucleotide substitution rates between lineages. The median Ks values between *Arabidopsis* β and γ

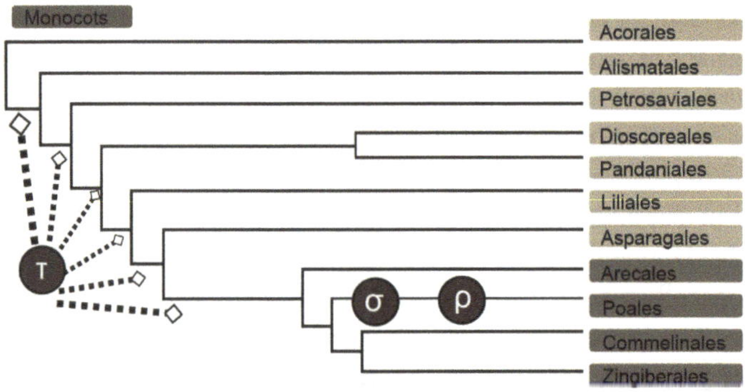

Figure 8.5 Phylogeny of selected families of monocots. Circles represent duplication events discovered in the cereal lineages (Paterson et al., 2004; Tang et al., 2010). Arrows indicate possible placement for certain events. Phylogeny is based on APG III (The Angiosperm Phylogeny Group, 2009). (See colour plate.)

have orthologous relationships with the same *Musa* regions. Such a pattern suggested that σ event is *not* shared with *Musa*; otherwise, the ρ blocks will be orthologous to different *Musa* regions (D'Hont et al., 2012). In addition, three separate duplication events were inferred in the Zingiberales lineage (D'Hont et al., 2012), indicating that *Musa* has a higher paleoploidy level than rice. The higher complexity in terms of ploidy level as well as higher base substitution rates suggested that *Musa* might not be an ideal genome for monocot comparisons.

A sequence of a second nongrass monocot genome, date palm, is also available. Date palm belongs to Arecales order, which is out-group to Zingiberales–Poales divergence (Al-Dous et al., 2011). However, the draft assembly of date palm is quite fragmented, which further degrades spatial synteny analysis and requires additional analysis to more effectively compare their genome structures. Perhaps, a careful reanalysis of some large contigs or scaffolds in the date palm genome is necessary for clarifying duplication history in this lineage.

There was clear spatial evidence of a third round of paleoduplication event (τ event) preceding both ρ and σ duplications could not be clearly resolved based on the Ks distribution, suggesting a much earlier date when Ks has become saturated (Ks of values greater than 2 are rarely informative) (Tang et al., 2010). As the *Musa* genome analysis suggested, if ρ and σ duplications occurred within the Poales lineage, the τ event might have occurred in lineages prior to Zingiberales–Poales divergence, with a rather wide range of possible dates (Fig. 8.5). Additional evidence from transcriptome sequences is already ongoing, and several basal monocot genome sequencing projects are under way. For example, asparagus and orchids (Asparagales), or duckweed and eelgrass (Alismatales), will help circumscribe the τ event. In addition, both ρ and σ have been dated to have occurred within the Poales, and there is so far no lineage that can separate these two events. Sequences from basal Poales lineages such as the genome of pineapple may add resolution to the timings of these two events.

7. EFFECTIVE COMPARISONS BETWEEN EUDICOT AND MONOCOT GENOMES THROUGH PARs

The 'holy grail' of plant comparative genomics is to effectively compare eudicot genomes to monocot genomes, with the promise to bridge functional equivalence between genes of the two well-studied groups of flowering plants. Alignments on the whole-genome scale between eudicots and monocots were suspected long before (Paterson et al., 1996), but early efforts (Liu, Sachidanandam, & Stein, 2001; Salse et al., 2009) were not successful.

The main difficulty arising in such deep comparison is that the syntenic regions between eudicot and monocot genomes are often small and interrupted by large numbers of noncolinear genes (Liu et al., 2001). Similarities between monocot and eudicot genomes resulting from common ancestry tend to have been obscured by many rounds of paleopolyploidy and numerous genome rearrangements along each lineage (Jaillon et al., 2007; Liu et al., 2001). Thus, direct comparisons between them are complicated by these deep events. For example, when comparing Arabidopsis to maize, patterns of 12(Arabidopsis):16(maize) are expected (Fig. 8.3). It is more tractable to consider multiple-to-multiple alignments for eudicot–monocot comparison while recognizing that each individual alignment remains sparse. To compile multiple-to-multiple alignments, one must take into full consideration all prior evidence of ancient WGDs that occurred earlier in the eudicot and monocot lineages.

To identify eudicot–monocot synteny alignments, a hierarchical clustering method can be used (Tang et al., 2010). The hierarchical clustering approach partially circumvents many of the difficulties to identify scattered synteny by consolidating related regions into sets. Briefly, the chromosomes were first cut into small segments and comparisons were made between every pair of rice and grape segments. For example, assume we have rice segments *O1* and *O2* and grape segment *V1* and comparisons *O1–V1* and *O2–V1* show a significant number of homologs. Based on this information, *O1* and *O2* can be clustered together, because they both match the same grape region(s). This clustering can be performed in both directions, ultimately recruiting multiple correlated rice and grape regions in one set, or 'putative ancestral region' (PAR) (Tang et al., 2010).

Following this unique approach, only the 'dense' (syntenic) portions of the whole-genome dot plot are clustered, assembled and interpreted; the 'sparse' (nonsyntenic) portions are ignored in further analyses (Fig. 8.6). Within each set of regions, duplicated segments retained in grape following the eudicot γ hexaploidy event (Jaillon et al., 2007), and homologous segments retained in rice following at least three rounds of duplication (ρ, σ and τ), contain a total of 38 PARs. The PARs collectively explain 19.1% of all observed homolog pairs and 31.0% of reciprocal best hits between grape and rice genes. The PARs interleave multiple grape and rice genomic regions, collectively explaining around 70% of each genome.

The patterns exemplified by PARs often show threefold redundancies on the grape axis and four- to eightfold redundancies on the rice axis. In 22 of the 38 PARs, grapevine–rice colinearity is clear, which allows us to evaluate the level of redundancies in both genomes. These redundancies reflect the number of genome duplication events observable in both lineages. Among the 22 PARs, 12 are threefold redundant in grapevine, consistent with hexaploidy. The level of redundancy in rice is more variable, ranging from as little as twofold (1 PAR) to sevenfold (3 PARs) and eightfold (5 PARs). The high redundancies suggested that the rice lineage likely experienced three rounds of WGDs, explaining the maximum of eightfold redundancy observed (Tang et al., 2010). Through knowledge of deep paleoduplications in eudicots and monocots as well as the clustering and sorting approach, we can now group and collate such 'scattered' synteny patterns, presenting compelling evidence that regions of good synteny conservation indeed exist between a eudicot and a monocot genome.

8. PANANGIOSPERM COMPARISONS THROUGH ANCESTRAL GENOME RECONSTRUCTION

While the PAR reconstruction mentioned earlier is an elegant way of visualizing a typical panangiosperm comparison, a better and generic solution is to directly infer the 'ancestral genome' for both eudicot and monocot

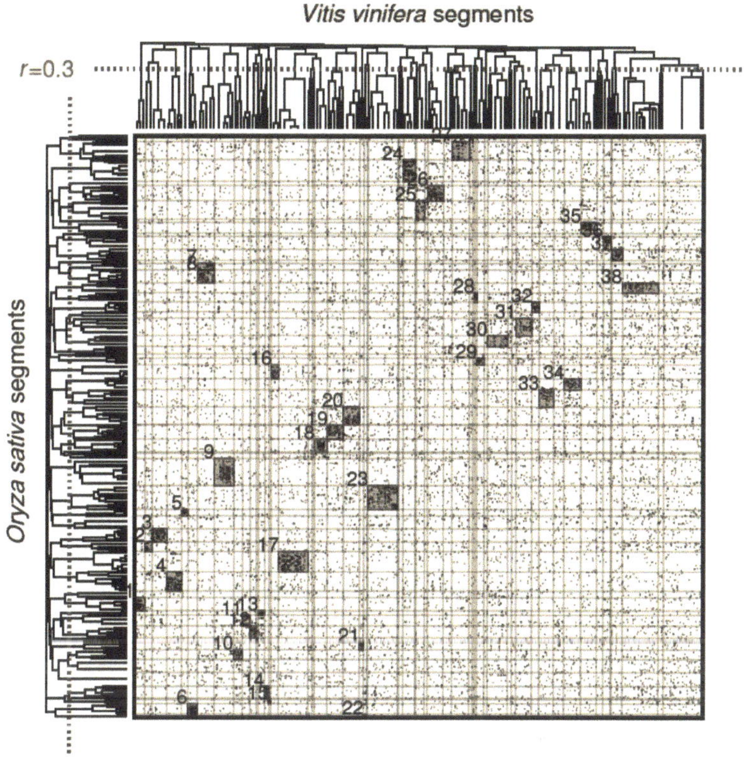

Figure 8.6 Hierarchical clustering method for constructing putative ancestral regions (PARs). Dots represent pairs of homologous genes between grape and rice. The chromosomal segments in the two genomes are reordered by the hierarchical clusters and concatenated. The trees on the top and left of the plot represent average correlation coefficient (*r*) among clustered grape and rice segments. Horizontal and vertical lines separate clusters of grape and rice segments, as defined by having an average correlation coefficient of distribution of hits to the other genome greater than 0.3. The squares highlighted in yellow are the 38 PARs that show high density of gene pairs between grape and rice clusters, with the PAR identifier shown in the upper-left corner of each highlighted block. (See colour plate.)

genomes. The goal of ancestral genome reconstruction is to determine the composition and structure of a putative genome in the common ancestor based on the arrangements of genes within the genomes of extant species. The heart of this problem is to 'unscramble' large-scale rearrangements and genome duplications. Ancestral genomes, if reconstructed properly, are better proxies for genome comparisons since they significantly reduce the evolutionary distance between clades (relative to comparisons between the genomes of exemplar species within each clade) (Fig. 8.7).

dot plot and mostly obscure and small (Fig. 8.8D). We argue that comparison between the ancestral genomes, rather than comparison between extant genomes directly, has substantially enriched the signal of synteny. In the meantime, full tracking of the segments derived from lineage-specific WGDs is provided during the reconstruction of the ancestral genomes.

The reconstructed genomes have created the potential to bridge chromosomal regions from eudicots and monocots. This bridge allows direct comparison between many species within the flowering plant clade, mediated through the ancestral genomes (Fig. 8.7). Along with CoGe's SynFind and GEvo tools, we can now compile a panangiosperm homologous set by having all the extant genomes, comparing the ancestral genomes in a streamlined fashion (Lyons & Freeling, 2008).

9. DEEPLY CONSERVED NONCODING SEQUENCES IN FLOWERING PLANTS

While more studies have focused on mapping the genic contents across taxa, researchers have also been tracking the evolution of conserved noncoding sequences (CNSs) in flowering plant genomes. In contrast to the mammalian CNSs, the elements in plant genomes tend to be shorter and less numerous (Freeling & Subramaniam, 2009). Plant CNSs around genes appear to have a faster turnover rate although many short CNSs (or 'motifs') are still known to have binding properties. The genes that are associated with more CNSs are typically transcription factors or other regulatory genes, while genes associated with few or no CNSs often encode enzymes or structural proteins (Inada et al., 2003). More generally, there is a strong correlation of CNS clustering with proteins that have high connectivity in regulatory networks (Baxter et al., 2012). A systematic search for long identical ultraconserved elements from six plant species across eudicots and monocots suggested the number is ~70-fold lower than those from six vertebrate genomes with a similar degree of divergence (Reneker et al., 2012). In contrast to those of vertebrates, plant ultraconserved elements are all in nonsyntenic locations after the exclusion of sequences that are repetitive in nature. One type of ultraconserved element in the nonsyntenic regions was often derived from mitochondrial sequences suggesting a possible origin by horizontal transfer (Freeling & Subramaniam, 2009; Reneker et al., 2012). Another major type of long ultraconserved sequences appears to be associated with rRNAs and was hypothesized to have maintained a very low polymorphism level due to concerted evolution between rRNA loci (Reneker et al., 2012).

Other studies have examined the evolution of microRNAs that are involved in posttranscriptional gene silencing. Like protein-coding genes, microRNAs can follow a similar birth and death process, and new members can arise through tandem, segmental, or WGDs (Maher, Stein, & Ware, 2006). Comparison among several sequenced grass genomes suggested that the conservation of microRNAs is closely related to miRNA–target interactions. Retention of some microRNA family members following the pancereal tetraploidy is in support of the gene dosage balance hypothesis, because overretained microRNAs are correlated with the number of target genes they regulate (Abrouk et al., 2012).

10. FUTURE PROSPECTS

The plant community is still in need of an application where researchers can query cross-species sequence conservation easily through a comparative genome browser. CoGe has been a useful tool for comparing gene contents, but it may be useful to highlight genomic regions that resist changes across the comparisons (i.e. under purifying selection) and regions that are under accelerated evolution. Perhaps, a 'conservation' module of CoGe can be developed to address this need. The popular UCSC genome browser lists the human genome as reference that well suits animal genome comparisons. Comparisons among vertebrates are easier to render, because the human genome is karyotypically similar to most other vertebrates with the most recent genome duplications shared by all vertebrates (Nakatani, Takeda, Kohara, & Morishita, 2007). The challenge in plants is that even commonly used plant models are poor reference genomes for comparisons—they often have incredibly high ploidy numbers when taking into account all WGDs since the divergence of the angiosperm lineages. Using ancestral genomes as a reference holds much promise as they enable the alignment of the structures of modern genomes with ease and use synteny transitivity to infer evolutionary history between extant genomes. Each high ploidy modern genome can show multiple tracks derived from genome duplications, not just a single track as typically seen in vertebrate genome comparisons (Miller et al., 2007). Such a browser would be very useful in plant genomes by allowing researchers to capture and visualize the complexities of modern plant genomes.

REFERENCES

Abrouk, M., Zhang, R., Murat, F., Li, A., Pont, C., Mao, L., et al. (2012). Grass microRNA gene paleohistory unveils new insights into gene dosage balance in subgenome partitioning after whole-genome duplication. *Plant Cell, 24*(5), 1776–1792.

Al-Dous, E. K., George, B., Al-Mahmoud, M. E., Al-Jaber, M. Y., Wang, H., Salameh, Y. M., et al. (2011). De novo genome sequencing and comparative genomics of date palm (Phoenix dactylifera). *Nature Biotechnology, 29*(6), 521–527.

Aury, J. M., Jaillon, O., Duret, L., Noel, B., Jubin, C., Porcel, B. M., et al. (2006). Global trends of whole-genome duplications revealed by the ciliate Paramecium tetraurelia. *Nature, 444*(7116), 171–178.

Baxter, L., Jironkin, A., Hickman, R., Moore, J., Barrington, C., Krusche, P., et al. (2012). Conserved noncoding sequences highlight shared components of regulatory networks in dicotyledonous plants. *Plant Cell, 24*(10), 3949–3965.

Bekaert, M., Edger, P. P., Pires, J. C., & Conant, G. C. (2011). Two-phase resolution of polyploidy in the Arabidopsis metabolic network gives rise to relative and absolute dosage constraints. *Plant Cell, 23*(5), 1719–1728.

Bikard, D., Patel, D., Le Mette, C., Giorgi, V., Camilleri, C., Bennett, M. J., et al. (2009). Divergent evolution of duplicate genes leads to genetic incompatibilities within A. thaliana. *Science, 323*(5914), 623–626.

Blanc, G., & Wolfe, K. H. (2004). Widespread paleopolyploidy in model plant species inferred from age distributions of duplicate genes. *Plant Cell, 16*(7), 1667–1678.

Bowers, J. E., Chapman, B. A., Rong, J., & Paterson, A. H. (2003). Unravelling angiosperm genome evolution by phylogenetic analysis of chromosomal duplication events. *Nature, 422*(6930), 433–438.

Camacho, C., Coulouris, G., Avagyan, V., Ma, N., Papadopoulos, J., Bealer, K., et al. (2009). BLAST+: Architecture and applications. *BMC Bioinformatics, 10*, 421.

Catchen, J. M., Conery, J. S., & Postlethwait, J. H. (2009). Automated identification of conserved synteny after whole-genome duplication. *Genome Research, 19*(8), 1497–1505.

Chase, M. W., Soltis, D. E., Olmstead, R. G., Morgan, D., Les, D. H., Mishler, B. D., et al. (1993). Phylogenetics of seed plants: An analysis of nucleotide sequences from the plastid gene rbcL. *Annals of the Missouri Botanical Garden, 80*(3), 528–580.

Chouard, T. (2010). Evolution: Revenge of the hopeful monster. *Nature, 463*(7283), 864–867.

Chunfang, Z., Albert, V. A., Lyons, E., & Sankoff, D. (2012). Ancient angiosperm hexaploidy meets ancestral eudicot gene order. In *Paper presented at the 2012 IEEE 2nd international conference on computational advances in bio and medical sciences (ICCABS)*.

Cui, L., Wall, P. K., Leebens-Mack, J. H., Lindsay, B. G., Soltis, D. E., Doyle, J. J., et al. (2006). Widespread genome duplications throughout the history of flowering plants. *Genome Research, 16*(6), 738–749.

Davies, T. J., Barraclough, T. G., Chase, M. W., Soltis, P. S., Soltis, D. E., & Savolainen, V. (2004). Darwin's abominable mystery: Insights from a supertree of the angiosperms. *Proceedings of the National Academy of Sciences of the United States of America, 101*(7), 1904–1909.

D'Hont, A., Denoeud, F., Aury, J. M., Baurens, F. C., Carreel, F., Garsmeur, O., et al. (2012). The banana (Musa acuminata) genome and the evolution of monocotyledonous plants. *Nature, 488*(7410), 213–217.

Fares, M. A., Byrne, K. P., & Wolfe, K. H. (2006). Rate asymmetry after genome duplication causes substantial long-branch attraction artifacts in the phylogeny of Saccharomyces species. *Molecular and Biological Evolution, 23*(2), 245–253.

Fawcett, J. A., Maere, S., & Van de Peer, Y. (2009). Plants with double genomes might have had a better chance to survive the Cretaceous-Tertiary extinction event. *Proceedings of the National Academy of Sciences of the United States of America, 106*(14), 5737–5742.

Felsenstein, J. (2004). *Inferring phylogenies*. Sunderland, MA: Sinauer.

Freeling, M., & Subramaniam, S. (2009). Conserved noncoding sequences (CNSs) in higher plants. *Current Opinion in Plant Biology*, *12*(2), 126–132.

Gaut, B. S., Morton, B. R., McCaig, B. C., & Clegg, M. T. (1996). Substitution rate comparisons between grasses and palms: Synonymous rate differences at the nuclear gene Adh parallel rate differences at the plastid gene rbcL. *Proceedings of the National Academy of Sciences of the United States of America*, *93*(19), 10274–10279.

Gonzales, M. D., Gajendran, K., Farmer, A. D., Archuleta, E., & Beavis, W. D. (2007). Leveraging model legume information to find candidate genes for soybean sudden death syndrome using the legume information system. *Methods in Molecular Biology*, *406*, 245–259.

Haas, B. J., Delcher, A. L., Wortman, J. R., & Salzberg, S. L. (2004). DAGchainer: A tool for mining segmental genome duplications and synteny. *Bioinformatics*, *20*(18), 3643–3646.

Hampson, S., McLysaght, A., Gaut, B., & Baldi, P. (2003). LineUp: Statistical detection of chromosomal homology with application to plant comparative genomics. *Genome Research*, *13*(5), 999–1010.

Ibarra-Laclette, E., Lyons, E., Hernandez-Guzman, G., Perez-Torres, C. A., Carretero-Paulet, L., Chang, T. H., et al. (2013). Architecture and evolution of a minute plant genome. *Nature*, *498*(7452), 94–98.

Inada, D. C., Bashir, A., Lee, C., Thomas, B. C., Ko, C., Goff, S. A., et al. (2003). Conserved noncoding sequences in the grasses. *Genome Research*, *13*(9), 2030–2041.

Jaillon, O., Aury, J. M., Brunet, F., Petit, J. L., Stange-Thomann, N., Mauceli, E., et al. (2004). Genome duplication in the teleost fish Tetraodon nigroviridis reveals the early vertebrate proto-karyotype. *Nature*, *431*(7011), 946–957.

Jaillon, O., Aury, J. M., Noel, B., Policriti, A., Clepet, C., Casagrande, A., et al. (2007). The grapevine genome sequence suggests ancestral hexaploidization in major angiosperm phyla. *Nature*, *449*(7161), 463–467.

Jiao, Y., Leebens-Mack, J., Ayyampalayam, S., Bowers, J. E., McKain, M. R., McNeal, J., et al. (2012). A genome triplication associated with early diversification of the core eudicots. *Genome Biology*, *13*(1), R3.

Jiao, Y., Wickett, N. J., Ayyampalayam, S., Chanderbali, A. S., Landherr, L., Ralph, P. E., et al. (2011). Ancestral polyploidy in seed plants and angiosperms. *Nature*, *473*(7345), 97–100.

Kellis, M., Birren, B. W., & Lander, E. S. (2004). Proof and evolutionary analysis of ancient genome duplication in the yeast Saccharomyces cerevisiae. *Nature*, *428*(6983), 617–624.

Kielbasa, S. M., Wan, R., Sato, K., Horton, P., & Frith, M. C. (2011). Adaptive seeds tame genomic sequence comparison. *Genome Research*, *21*(3), 487–493.

Koch, M. A., Haubold, B., & Mitchell-Olds, T. (2000). Comparative evolutionary analysis of chalcone synthase and alcohol dehydrogenase loci in Arabidopsis, Arabis, and related genera (Brassicaceae). *Molecular and Biological Evolution*, *17*(10), 1483–1498.

Lee, T. H., Tang, H., Wang, X., & Paterson, A. H. (2013). PGDD: A database of gene and genome duplication in plants. *Nucleic Acids Research*, *41*(Database issue), D1152–D1158.

Leebens-Mack, J., Raubeson, L. A., Cui, L., Kuehl, J. V., Fourcade, M. H., Chumley, T. W., et al. (2005). Identifying the basal angiosperm node in chloroplast genome phylogenies: Sampling one's way out of the Felsenstein zone. *Molecular Biology and Evolution*, *22*(10), 1948–1963.

Lescot, M., Piffanelli, P., Ciampi, A. Y., Ruiz, M., Blanc, G., Leebens-Mack, J., et al. (2008). Insights into the Musa genome: Syntenic relationships to rice and between Musa species. *BMC Genomics*, *9*, 58.

Liu, H., Sachidanandam, R., & Stein, L. (2001). Comparative genomics between rice and Arabidopsis shows scant collinearity in gene order. *Genome Research*, *11*(12), 2020–2026.

Lynch, M., & Force, A. G. (2000). The origin of interspecific genomic incompatibility via gene duplication [review]. *American Naturalist*, *156*(6), 590–605.

Wang, X., Shi, X., Hao, B., Ge, S., & Luo, J. (2005). Duplication and DNA segmental loss in the rice genome: Implications for diploidization. *The New Phytologist*, *165*(3), 937–946.
Wang, X., Tang, H., Bowers, J. E., Feltus, F. A., & Paterson, A. H. (2007). Extensive concerted evolution of rice paralogs and the road to regaining independence. *Genetics*, *177*(3), 1753–1763.
Wang, X., Tang, H., Bowers, J. E., & Paterson, A. H. (2009). Comparative inference of illegitimate recombination between rice and sorghum duplicated genes produced by polyploidization. *Genome Research*, *19*(6), 1026–1032.
Xu, X., Pan, S., Cheng, S., Zhang, B., Mu, D., Ni, P., et al. (2011). Genome sequence and analysis of the tuber crop potato. *Nature*, *475*(7355), 189–195.
Zheng, C., Swenson, K., Lyons, E., & Sankoff, D. (2011). OMG! Orthologs in multiple genomes—Competing graph-theoretical formulations. In T. Przytycka & M.-F. Sagot (Eds.), *Algorithms in bioinformatics*, Vol. 6833, (pp. 364–375). Berlin: Springer.

CHAPTER NINE

Prehistory of the Angiosperms: Characterization of the Ancient Genomes

Yuannian Jiao[1], Hui Guo
Plant Genome Mapping Laboratory, University of Georgia, Athens, Georgia, USA
[1]Corresponding author: e-mail address: jiaoyn@uga.edu

Contents

Abstract

Angiosperms are by far the largest group of land plants with more than 30,000 living species. Genomics approaches have been most widely applied in the angiosperms, where significant genomics resources, including whole-genome sequences, have been developed. The genomics of land plant lineages that branched off earlier than the angiosperms provide valuable references for comparative evolutionary analyses. In this chapter, we therefore first reviewed the evolutionary history of plant lineages from green algae to angiosperms and current knowledge discovered from the chloroplast genomes in major ancestral groups. We then reviewed the genomic features and resources from three representative outgroup species suitable for comparison to angiosperms (*Chlamydomonas reinhardtii*, *Physcomitrella patens* and *Selaginella moellendorffii*). In ferns, we discussed extensive RNA editing in plastid genomes and the very high chromosome numbers in homosporous ferns. Despite the lack of fully sequenced genomes to date in gymnosperms, rich expressed sequence tag and BAC resources are available, with several genome sequencing projects in progress. We discuss one of the most controversial issues in gymnosperm systematics, which is the phylogenetic position of Gnetales. Finally, we discuss recent studies elucidating two ancestral polyploidy events before the origin of angiosperms—one in the common ancestor of extant seed plants

Advances in Botanical Research, Volume 69
ISSN 0065-2296
http://dx.doi.org/10.1016/B978-0-12-417163-3.00009-3

and the other in the common ancestor of angiosperms. These two very ancient polyploidy events could help to provide better insight into the rise and eventual dominance of seed plants and angiosperms.

1. BRIEF EVOLUTIONARY HISTORY FROM ALGAE TO FLOWERING PLANTS

The embryophytes (land plants), including bryophytes and vascular plants, descended from early water-dwelling algae (Fig. 9.1) about 480 million years ago (mya) (Becker & Marin, 2009; Karol, McCourt, Cimino, & Delwiche, 2001; Kenrick & Crane, 1997; Lewis & McCourt, 2004). Bryophytes, comprising hornworts, mosses and liverworts, are the most primitive terrestrial plants surviving today. Fossil

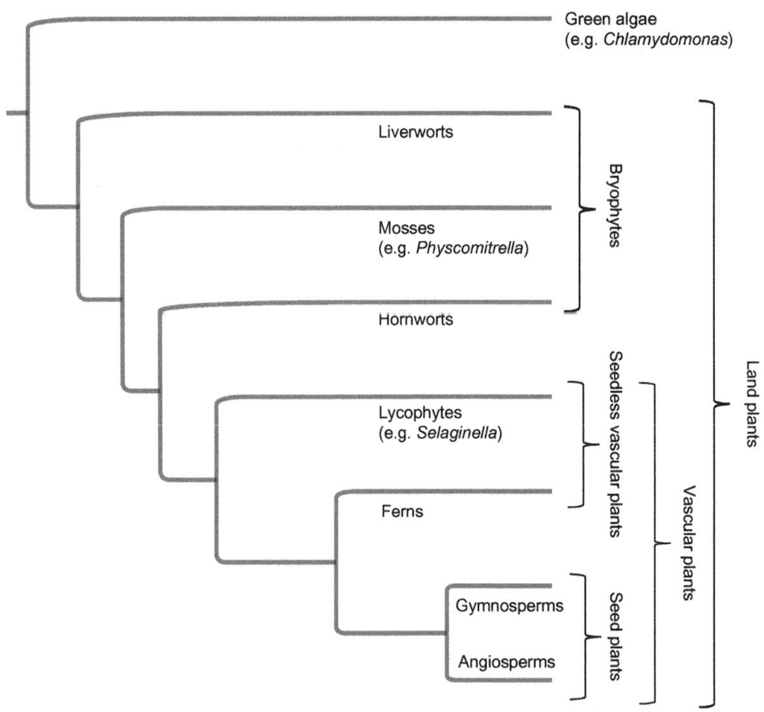

Figure 9.1 Simplified phylogenetic tree showing relationships among the major groups of extant land plants.

evidence suggests that early land plants were structurally similar to extant bryophytes (Kenrick & Crane, 1997). They probably had a dominant haploid phase and required very moist conditions for survival, especially for sexual reproduction.

Ferns and the related lycophytes are among the earliest vascular plants, with the earliest fossils dating from nearly 443 mya (Steemans et al., 2009). Vascular plants switched to sporophyte-dominated life cycles, colonized drier habitats and were no longer restricted to damper areas, which led to an explosion in their diversity (Bateman et al., 1998). In addition, the development of an internal vascular system for transportation of water, minerals and nutrients has allowed vascular plants to evolve erect forms and larger sizes than the non-vascular plants around them (Graham, Cook, & Busse, 2000).

The advent of seed plants is thought to have changed the way that the world looked around 309 mya (Miller, 1999). The evolution of seed plays an important role in the reproduction and spread of gymnosperms and angiosperms relative to the more primitive bryophytes, ferns and lycophytes. Seed substantially reduced reliance on moist conditions for growing and reproduction, providing a durable structure that can remain quiescent until suitable growing conditions stimulate it to germinate and develop into a new plant. Gymnosperms were the first seed-bearing plants to appear on Earth, including four main groups: conifers, cycads, Gnetales and Ginkgo (a single living species). Gymnosperms had become the dominant vegetation on Earth before the rapid diversification of angiosperms.

Angiosperms are flowering plants, evolving from gymnosperms during the Mesoz, which include about 85–90% of all living plant species. Angiosperms have seeds contained within a fruit, unlike gymnosperms that have naked seeds (no fruit). They have also evolved flowers, which actually are organs for reproduction. Fruits can attract animals to help with dispersal of seeds, and flowers can attract animal pollinators to facilitate outcrossing by carrying pollen to other individuals of the same species. The fruits, flowers and other characters of angiosperms are thought to have contributed to their emergence as the most diverse group of land plants.

2. CHLOROPLAST GENOME EVOLUTION IN LAND PLANTS

Chloroplast genomes (cpDNA) are circular and relatively conserved among land plants in terms of size, structure and gene content

has been expanded to host two genes (*ndhB* and *rps7*) and the 3′ exon of *rps12*. A striking finding in the *A. formosae* chloroplast genome is that more than half of protein-coding genes have nonsense codons, which are converted to sense codons by RNA editing (Kugita et al., 2003). A high RNA-editing rate of genes in the chloroplast genome is also found in ferns (Wolf, Rowe, & Hasebe, 2004).

Analysis of the chloroplast genome from a lycophyte, *Selaginella uncinata*, revealed some unique features: (1) A unique 20 kb inversion located in the large single copy region between the two inverted repeats. (2) Much gene loss, for example, only 12 tRNA genes are identified in contrast to an average of >30 tRNA genes in other chloroplast genomes. In addition, four genes, *rps15*, *rps16*, *rpl32* and *ycf10*, which are conserved in other land plants, are not present in *S. uncinata*. (3) Transposition of two segments from the large single copy region to the small single copy region (Tsuji et al., 2007). In addition, the *Selaginella* chloroplast genome has much higher GC content (54.8%) than the 36.2% average of other sequenced plastid genomes (Smith, 2009). A number of factors causing the inflated GC content are proposed, such as GC-biased gene conversion, natural selection on high GC in the first two codon positions and fewer tRNA genes (Smith, 2012).

In conifers, due to lack of two conserved inverted repeats, the size of the chloroplast genome is relatively small among land plants (Strauss et al., 1988). Chloroplast genomes are extensively rearranged in conifers (Strauss et al., 1988; Wakasugi et al., 1994), arguably the result of the deletion of inverted repeats. Wakasugi et al. also show that all 11 *ndh* genes (coding subunits of NADH dehydrogenase) are missing in black pine, which implies that these genes are either transferred to the nuclear genome or not essential in the physiology of black pine.

3. AN ANCIENT GREEN ALGAE MODEL ORGANISM: *C. REINHARDTII*

Unicellular green algae diverged from land plants over 1 billion years ago when a cyanobacterium was captured by a heterotrophic protist and incorporated as an endosymbiont, giving rise to the first eukaryotic alga (Hackett et al., 2007). Green algae can be found all over the world, in soil, fresh water, oceans and elsewhere. By the time of writing, many genomes of algae have been published, such as *C. reinhardtii* (Merchant et al., 2007),

Volvox carteri (Prochnik et al., 2010), *Micromonas pusilla* (Worden et al., 2009) and *Ostreococcus lucimarinus* (Worden et al., 2009) with *Coccomyxa subellipsoidea* to be published soon.

Chlamydomonas has been a model organism with a well-established genetic background for a long time. In 2007, Merchant et al. published a draft nuclear genome sequence of *C. reinhardtii* (Merchant et al., 2007) of about 120 megabases (Mb) covering around 95% of the genome, separated into 17 linkage groups (chromosomes). The GC content of the *Chlamydomonas* genome is unusually high at about 64%, compared to *Arabidopsis thaliana* with 36% (Arabidopsis Genome Initiative, 2000) and human with 41% (Venter et al., 2001). Merchant et al. annotated a *Chlamydomonas* reference set of 15,143 protein-coding genes, with most proteins slightly more similar to *Arabidopsis* than to human proteins. Flagellar and basal body proteins in *Chlamydomonas* tend to have greater similarity to animal homologues, consistent with losses of these genes in angiosperms. In general, the *Chlamydomonas* genome has approximately uniform densities of genes, simple sequence repeats and transposable elements. Most genes (92%) have at least one intron, which is a relatively high percentage. Merchant et al. proposed that *Chlamydomonas* introns are subject to creation or invasion by transposable elements (Merchant et al., 2007). The *Chlamydomonas* genome is a critical reference to study the evolution of key features of animals and plants and ultimately sheds lights on the nature of the common ancestor of animals and plants.

4. THE FIRST MODEL MOSS GENOME SEQUENCE: *P. PATENS*

Mosses are non-vascular plants with about 12,000 species classified in the Bryophyta. Unlike vascular plants, mosses lack xylem and absorb water and nutrients mainly through their leaves. As is true of the other two groups of Bryophyta (liverworts and hornworts), the haploid gametophyte generation is the dominant phase of the life cycle in mosses, while vascular plants have the diploid sporophyte generation as the dominant phase. *Physcomitrella* is the first, and so far the only, species with a completed genome sequence in the Bryophyta (Rensing et al., 2008). The *Physcomitrella* genome sequence allows us to begin to unravel evolutionary changes that occurred during the colonization of land.

Whole-genome shotgun sequencing was used to sequence *P. patens*. The draft genome sequence was assembled into 480 Mb pairs of scaffold sequence with an average depth of ~8.6 × per nucleotide (nt). The total of 35,938 gene models predicted and annotated in *Physcomitrella* (Rensing et al., 2008) is more than twice the number in *Chlamydomonas* (15,143) (Merchant et al., 2007; Rensing et al., 2008), suggesting a general increase in gene family complexity. Genome duplication seems to have contributed to the expanded gene number in *Physcomitrella*. A Ks distribution plot of *Physcomitrella* genes shows a clear peak at around 0.5–0.9, indicating a whole-genome duplication (WGD) event in evolutionary history. Previous expressed sequence tag (EST)-based analysis also suggested that a genome duplication occurred between 30 and 60 mya (Rensing et al., 2007). A relatively small percentage (~1%) of genes have tandem duplicates (Ks < 0.1) in *Physcomitrella*, compared to *Arabidopsis* (~16%), *Oryza* (~14%) and poplar (~11%) (Arabidopsis Genome Initiative, 2000; International Rice Genome Sequencing Project, 2005; Rensing et al., 2008; Tuskan et al., 2006). Photosynthesis proteins, particularly antenna proteins, are significantly enriched among the recent tandem duplicated genes in *Physcomitrella*.

Comparison with the genome sequences of green algae (*Chlamydomonas*) and several angiosperms (*Arabidopsis*, *Populus* and *Oryza*) shows that genes associated with aquatic environments (e.g. flagellar components for gametic motility) and dynein-mediated transport have been lost in *Physcomitrella* since it shared common ancestry with angiosperms (Rensing et al., 2008). In contrast, *Physcomitrella* gained or increased family size of genes associated with signal transduction (e.g. through gibberellic acid, jasmonic acid, ethylene and brassinosteroids), transport capabilities and tolerance for abiotic stresses. Adaptation to land required plants to adapt to a much greater variation in temperature, light and water availability than was experienced in aquatic environments. An example of such adaptation may be the heat shock protein 70 (HSP70) gene family—all algal genomes sequenced to date have only one cytosolic HSP70 gene (Wang, Vinocur, Shoseyov, & Altman, 2004), while there are nine in *P. patens* (Rensing et al., 2008). Moreover, the light-harvesting complex proteins have significantly expanded in *P. patens*, which might contribute to robustness of the photosynthetic antenna to deal with high light intensities. The photoprotective early light-induced proteins also expanded greatly in *P. patens*, putatively associated with avoidance of photo-oxidative damage. Some of these expansions may be results of WGDs (Rensing et al., 2008).

5. THE FIRST LYCOPHYTE GENOME SEQUENCE: *SELAGINELLA MOELLENDORFFII*

Lycophytes, together with euphyllophytes (ferns and seed plants), are the only two surviving lineages after the divergence of vascular plants. Lycophytes are an ancient lineage that quickly branched out after land plants evolved vascular tissues about 400 mya and has features typical of vascular plants such as a dominant and complex sporophyte generation. *S. moellendorffii*, in the lycophyte family Selaginellaceae (spikemosses), is the first non-seed vascular plant with its genome sequenced (Banks et al., 2011), providing a resource for identifying genes that have been important in early evolution of developmental and metabolic processes specific to vascular plants.

The *Selaginella* genome was sequenced by whole-genome shotgun sequencing. Two haplotypes were assembled with a genome size of 212.6 Mb, about twice the 88–127 Mb determined by flow cytometry (Wang et al., 2005). The DAGchainer program (Haas, Delcher, Wortman, & Salzberg, 2004) was used to identify duplicated regions to enable whole-genome analysis of genes in a deduced haplotype genome (Banks et al., 2011). A total of 22,285 protein-coding genes were annotated, fewer than *Arabidopsis* (26,207) and *Physcomitrella* (27,949). The average intron number is 5.7 per spliced gene with a median of 4 per spliced gene (minimum of 1 and maximum of 78), and the average intron size is smaller than other species (*Selaginella*, 103 bp; *Arabidopsis*, 164 bp; and *Physcomitrella*, 310 bp). The age distribution of duplicate genes did not reveal any evidence of ancient WGD in the *Selaginella* lineage.

Selaginella transposable element content (37.5% of the genome and 42.7% of the raw genomic shotgun sequence data) is higher than that of *Arabidopsis* (15%). In total, 59,748 LTR retrotransposons, 46 LINEs, 5394 *Helitrons*, 2386 MITEs, 11 *hAT* elements and 37 DNA transposons were identified in the *Selaginella* genome (Banks et al., 2011). Several unique aspects of epigenetic or post-transcriptional gene regulation found in *Selaginella* (Banks et al., 2011) are as follows: (1) the proportion of sRNAs 23–24 nt in length is extraordinarily small, while three-quarters of the *Selaginella* sRNA mapped to miRNA loci are 21 nt in length; (2) no *DCL4*, *RDR6* and *MIR390* loci were found; and (3) extraordinarily large numbers of RNA-edited sites are found in the *Selaginella* plastome.

Newly originated or expanded gene families coinciding with evolutionary innovations and losses in land plants may be identified by comparative

Angiosperms are the most diverse and widely studied seed plants. Large-scale phylogenetic analyses have identified complex patterns of diversification (Bell, Soltis, & Soltis, 2010; Magallon & Castillo, 2009; Smith, Beaulieu, Stamatakis, & Donoghue, 2011), and numerous genomes have been fully, or at least partially, sequenced. Gymnosperms are a group of seed-producing plants that include conifers, cycads, Ginkgo and Gnetales, with fewer than 1000 extant species (compared to about 300,000 extant angiosperms). By far, the largest group of living gymnosperms is the conifers (pines, cypresses and relatives). Compared to angiosperms, little is known about the patterns of diversification and genome evolution in gymnosperms, and there is no sequenced genome in this clade so far.

Many gymnosperms have exceptionally large genomes, partly accounting for their limited genomic resources. For example, conifer genome sizes range from 18 to 35 Gb (Murray, Leitch, & Bennett, 2012), which has hindered whole-genome sequencing. The huge genome size is of interest, because it has been suggested that polyploidy is rare among gymnosperms (Delevoryas, 1979). Recent efforts have elucidated that the large genome size might be associated with rapid expansion of retrotransposons and may be limited to conifers, Pinaceae (Grotkopp, Rejmanek, Sanderson, & Rost, 2004; Hall, Dvorak, Johnston, Price, & Williams, 2000; Kovach et al., 2010; Morse et al., 2009; Wakamiya, Newton, Johnston, & Price, 1993). A recent study suggested elevated rates of genome size and diversification within the last 100 million years, especially in *Pinus* (Burleigh, Barbazuk, Davis, Morse, & Soltis, 2012).

Although there is not yet a completed gymnosperm genome sequence, large-scale transcriptome data have been generated and deposited in public databases such as GenBank and PlantGDB. Large-insert BAC genomic libraries have been constructed for *P. pinaster*, *P. glauca* and *P. taeda* (Bautista et al., 2007; Hamberger et al., 2009; Magbanua et al., 2011). By sequencing BACs, it has been proposed that pseudogene formation may be a frequent feature within conifer genomes (Kovach et al., 2010; Magbanua et al., 2011). Very large amounts of repetitive DNAs are in conifer genomes as well (Magbanua et al., 2011; Morse et al., 2009), consistent with their large genome sizes. Given the taxonomic position, ecological and economic significance of conifers, several research groups have been funded to sequence whole conifer genomes, including pines, spruces and Douglas fir in the last few years. A European consortium led by Sweden is sequencing the genome of Norway spruce (Nystedt et al., 2013). A USDA-funded project was launched to sequence the genomes of loblolly pine (*P. taeda*), Douglas fir (*P. menziesii*) and sugar pine (*P. lambertiana*) (http://www.pinegenome.

org/pinerefseq/). A Genome Canada project is sequencing the white spruce (*P. glauca*) genome (http://www.smartforests.ca/en-ca/home.aspx). These genomes will provide important resources for better understanding of plant evolution and function, enhancing and protecting the word's conifer forests.

One of the most long-standing and controversial issues in gymnosperm systematics is the phylogenetic position of Gnetales (Burleigh & Mathews, 2004; Chaw, Parkinson, Cheng, Vincent, & Palmer, 2000; Donoghue & Doyle, 2000; Mathews, 2009; Zhong et al., 2011; Zhong, Yonezawa, Zhong, & Hasegawa, 2010), a morphologically and ecologically diverse group of gymnosperms. The gnetophytes have vessel elements such as those found in flowering plants, which transport water within the plant. The Gnetales were initially thought to be the nearest relatives of flowering plants (angiosperms) based on morphological similarities (Fig. 9.2A), an idea called the 'anthophyte' hypothesis (Crane, 1985; Doyle & Donoghue, 1986; Rothwell & Serbet, 1994). However, all recent molecular evolutionary evidence is against this hypothesis, although failing to reach a final conclusion about the phylogenetic placement of Gnetales (Burleigh & Mathews, 2004; Zhong et al., 2011). There are three different hypotheses somewhat supported by molecular analysis for the position of Gnetales: (1) as sister group to all conifers (the 'Gnetifer' hypothesis—Fig. 9.2B; Chaw et al., 2000); (2) within conifers, close to Pinaceae (the 'Gnepine' hypothesis—Fig. 9.2C; Bowe, Coat, & dePamphilis, 2000; Chaw et al., 2000; Hajibabaei, Xia, & Drouin, 2006; Wu, Wang, Liu, & Chaw, 2007; Zhong et al., 2010); (3) within conifers, but sister to Cupressophyta (non-Pinaceae conifers; the 'Gnecup' hypothesis—Fig. 9.2D; Doyle, 2006; Nickrent, Parkinson, Palmer, & Duff, 2000). Zhong et al. evaluated the robustness of several systematic errors in seed plant phylogenomic inferences, including taxon sampling, long-branch attraction (LBA) (Felsenstein, 1978; Hendy & Penny, 1989) and parallel substitutions. It has been proposed that improved taxon sampling was not sufficient to overcome LBA between Curessophytes and Gnetales (Wu, Wang, Hsu, Lin, & Chaw, 2011; Zhong et al., 2011). These controversial results from chloroplast genomes might be solved by the conifer nuclear genomes that are being sequenced as noted in the preceding text.

8. ANCESTRAL POLYPLOIDY EVENTS BEFORE THE ORIGIN OF ANGIOSPERMS

WGD followed by gene loss and diploidization has long been recognized as an important evolutionary force in animals, fungi and other

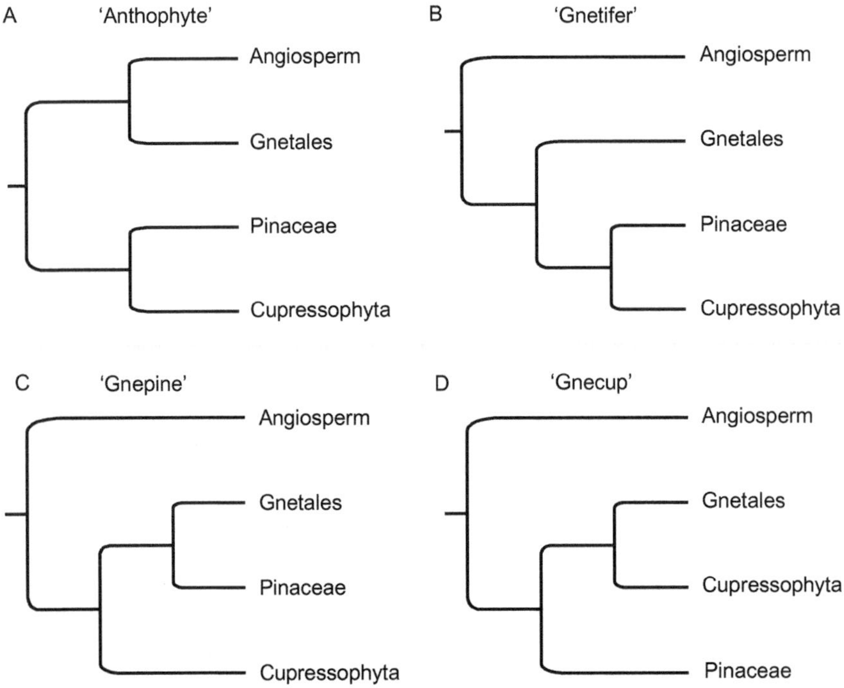

Figure 9.2 Four different hypotheses regarding the phylogenetic position of Gnetales. (A) The 'anthophyte' hypothesis: Gnetales is sister to angiosperms. (B) The 'Gnetifer' hypothesis: Gnetales is sister to conifers as a whole. (C) The 'Gnepine' hypothesis: Gnetales is sister to Pinaceae, which has relatively more support from molecular analyses. (D) The 'Gnecup' hypothesis: Gnetales is sister to Cupressophyta (non-Pinaceae conifers).

organisms (Edger & Pires, 2009; Lynch, 2007; Ohno, 1970), especially plants. A genome-scale phylogenomic study (Jiao et al., 2011) has provided new evidence for two ancient WGDs in plants, with potential importance for the evolution of seed and flowering plants. More precisely, the authors proposed a WGD in the common ancestor of all extant angiosperms and an even older one in the common ancestor of all extant seed plants (Fig. 9.3).

Jiao et al. constructed thousands of phylogenetic trees for gene families built from a collection of genes from sequenced genomes and ESTs from gymnosperms and basal angiosperms, finding overwhelming support for the presence of two groups of duplications: one in the common ancestor of all extant angiosperms and the other in the common ancestor of all seed plants (Jiao et al., 2011). The distribution of inferred duplication times of nodes on gene family trees was bimodal, with peaks at 192 ± 2 (95% CI) and 319 ± 3 mya. Dates were clustered in two relatively short time intervals, suggesting that these duplications were clearly not uniformly distributed (Fig. 9.4). Therefore, both molecular dating and phylogenetic analyses support two ancient genome-wide duplications: one in the ancestor of all extant seed plants and the other one in the ancestor of all extant angiosperms.

Prior studies based on Ks distributions of duplicated genes in basal angiosperms used much smaller numbers of ESTs (Cui et al., 2006; Soltis et al., 2009). These analyses had detected evidence of an ancient WGD event in several basal angiosperms, but not in *Amborella*, the sister to all other extant angiosperms (Soltis, Soltis, & Chase, 1999). However, Ks analysis with the greatly expanded set of ESTs in Jiao et al. (2,592,984) indicates two significant ancient duplication peaks: 1.97 and 2.76. Furthermore, 43 *Amborella* unigene pairs were identified from transcriptome Ks analysis where both genes mapped to a phylogenetic tree. Gene pairs with small Ks values (<1.5) were duplicated after the divergence of *Amborella* and the rest of the angiosperms, while gene pairs with large Ks values were generated by

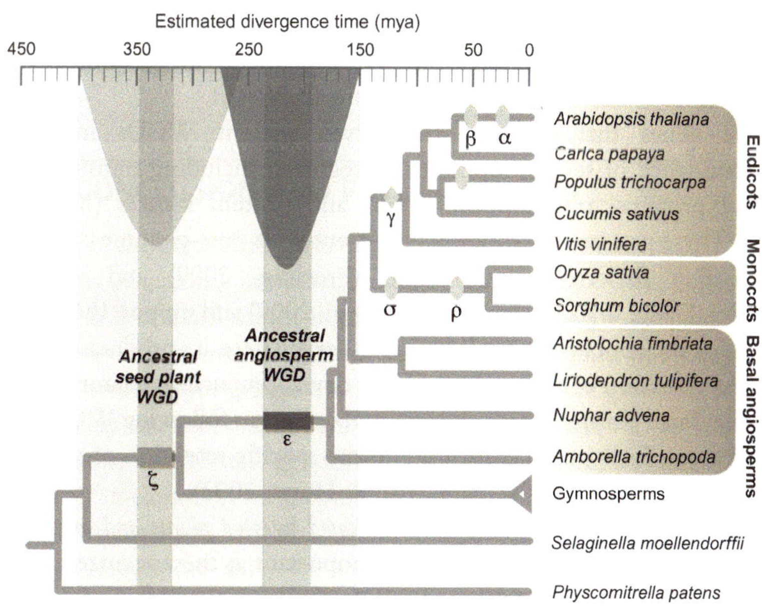

Figure 9.3 Ancestral polyploidy events identified in the common ancestors of extant seed plants and angiosperms. Ovals indicate the generally accepted WGDs identified in sequenced genomes. The diamond refers to the paleohexaploidization event thought to have been in a common ancestor of all core eudicots. *Adapted from Jiao et al. (2011).* (See colour plate.)

REFERENCES

Arabidopsis Genome Initiative. (2000). Analysis of the genome sequence of the flowering plant Arabidopsis thaliana. *Nature, 408*(6814), 796–815.

Archibald, J. M. (2009). The puzzle of plastid evolution. *Current Biology, 19*(2), R81–R88.

Banks, J. A., Nishiyama, T., Hasebe, M., Bowman, J. L., Gribskov, M., dePamphilis, C., et al. (2011). The Selaginella genome identifies genetic changes associated with the evolution of vascular plants. *Science, 332*(6032), 960–963.

Barker, M. S., Vogel, H., & Schranz, M. E. (2009). Paleopolyploidy in the Brassicales: Analyses of the *Cleome* transcriptome elucidate the history of genome duplications in *Arabidopsis* and other Brassicales. *Genome Biology and Evolution, 1*, 391–399.

Barker, M. S., & Wolf, P. G. (2010). Unfurling fern biology in the genomics age. *Bioscience, 60*(3), 177–185.

Bateman, R. M., Crane, P. R., DiMichele, W. A., Kenrick, P. R., Rowe, N. P., Speck, T., et al. (1998). Early evolution of land plants: Phylogeny, physiology, and ecology of the primary terrestrial radiation. *Annual Review of Ecology and Systematics, 29*, 263–292.

Bautista, R., Villalobos, D. P., Diaz-Moreno, S., Canton, F. R., Canovas, F. M., & Claros, M. G. (2007). Toward a Pinus pinaster bacterial artificial chromosome library. *Annals of Forest Science, 64*(8), 855–864.

Becker, B., & Marin, B. (2009). Streptophyte algae and the origin of embryophytes. *Annals of Botany, 103*(7), 999–1004.

Bell, C. D., Soltis, D. E., & Soltis, P. S. (2010). The age and diversification of the angiosperms re-revisited. *American Journal of Botany, 97*(8), 1296–1303.

Bottomley, W., & Bohnert, H. J. (1982). The biosynthesis of chloroplast proteins. In *Nucleic acids and proteins in plants II* (pp. 531–596). Berlin: Springer.

Bowe, L. M., Coat, G., & dePamphilis, C. W. (2000). Phylogeny of seed plants based on all three genomic compartments: Extant gymnosperms are monophyletic and Gnetales' closest relatives are conifers. *Proceedings of the National Academy of Sciences of the United States of America, 97*(8), 4092–4097.

Bowers, J. E., Chapman, B. A., Rong, J., & Paterson, A. H. (2003). Unravelling angiosperm genome evolution by phylogenetic analysis of chromosomal duplication events. *Nature, 422*(6930), 433–438.

Burleigh, J. G., Barbazuk, W. B., Davis, J. M., Morse, A. M., & Soltis, P. S. (2012). Exploring diversification and genome size evolution in extant gymnosperms through phylogenetic synthesis. *Journal of Botany, 2012*, 1–6.

Burleigh, J. G., & Mathews, S. (2004). Phylogenetic signal in nucleotide data from seed plants: Implications for resolving the seed plant tree of life. *American Journal of Botany, 91*(10), 1599–1613.

Chaw, S. M., Parkinson, C. L., Cheng, Y. C., Vincent, T. M., & Palmer, J. D. (2000). Seed plant phylogeny inferred from all three plant genomes: Monophyly of extant gymnosperms and origin of Gnetales from conifers. *Proceedings of the National Academy of Sciences of the United States of America, 97*(8), 4086–4091.

Crane, P. R. (1985). Phylogenetic analysis of seed plants and the origin of angiosperms. *Annals of the Missouri Botanical Garden, 72*(4), 716–793.

Cui, L., Wall, P. K., Leebens-Mack, J. H., Lindsay, B. G., Soltis, D. E., Doyle, J. J., et al. (2006). Widespread genome duplications throughout the history of flowering plants. *Genome Research, 16*(6), 738–749.

Dechaine, J. M., Gardner, G., & Weinig, C. (2009). Phytochromes differentially regulate seed germination responses to light quality and temperature cues during seed maturation. *Plant, Cell & Environment, 32*(10), 1297–1309.

Delevoryas, T. (1979). Polyploidy in gymnosperms. *Basic Life Sciences, 13*, 215–218.

Der, J. P., Barker, M. S., Wickett, N. J., dePamphilis, C. W., & Wolf, P. G. (2011). De novo characterization of the gametophyte transcriptome in bracken fern, Pteridium aquilinum. *BMC Genomics*, *12*, 99.

Devlin, P. F., Patel, S. R., & Whitelam, G. C. (1998). Phytochrome E influences internode elongation and flowering time in Arabidopsis. *Plant Cell*, *10*(9), 1479–1487.

Donoghue, M. J., & Doyle, J. A. (2000). Seed plant phylogeny: Demise of the anthophyte hypothesis? *Current Biology*, *10*(3), R106–R109.

Downie, S. R., & Palmer, J. D. (1992). *Use of chloroplast DNA rearrangements in reconstructing plant phylogeny*. Berlin: Springer.

Doyle, J. A. (2006). Seed ferns and the origin of angiosperms. *Journal of the Torrey Botanical Society*, *133*(1), 169–209.

Doyle, J. A., & Donoghue, M. J. (1986). Seed plant phylogeny and the origin of angiosperms—An experimental cladistic approach. *Botanical Review*, *52*(4), 321–431.

Edger, P. P., & Pires, J. C. (2009). Gene and genome duplications: The impact of dosage-sensitivity on the fate of nuclear genes. *Chromosome Research*, *17*(5), 699–717.

Felsenstein, J. (1978). Cases in which parsimony or compatibility methods will be positively misleading. *Systematic Zoology*, *27*(4), 401–410.

Freeling, M. (2009). Bias in plant gene content following different sorts of duplication: Tandem, whole-genome, segmental, or by transposition. *Annual Review of Plant Biology*, *60*, 433–453.

Freeling, M., & Thomas, B. C. (2006). Gene-balanced duplications, like tetraploidy, provide predictable drive to increase morphological complexity. *Genome Research*, *16*(7), 805–814.

Gott, J. M., & Emeson, R. B. (2000). Functions and mechanisms of RNA editing. *Annual Review of Genetics*, *34*, 499–531.

Graham, L. E., Cook, M. E., & Busse, J. S. (2000). The origin of plants: Body plan changes contributing to a major evolutionary radiation. *Proceedings of the National Academy of Sciences of the United States of America*, *97*(9), 4535–4540.

Gray, M. W., & Doolittle, W. F. (1982). Has the endosymbiont hypothesis been proven. *Microbiological Reviews*, *46*(1), 1–42.

Grotkopp, E., Rejmanek, M., Sanderson, M. J., & Rost, T. L. (2004). Evolution of genome size in pines (Pinus) and its life-history correlates: Supertree analyses. *Evolution*, *58*(8), 1705–1729.

Haas, B. J., Delcher, A. L., Wortman, J. R., & Salzberg, S. L. (2004). DAGchainer: A tool for mining segmental genome duplications and synteny. *Bioinformatics*, *20*(18), 3643–3646.

Hackett, J. D., Yoon, H. S., Li, S., Reyes-Prieto, A., Rummele, S. E., & Bhattacharya, D. (2007). Phylogenomic analysis supports the monophyly of cryptophytes and haptophytes and the association of rhizaria with chromalveolates. *Molecular Biology and Evolution*, *24*(8), 1702–1713.

Hajibabaei, M., Xia, J., & Drouin, G. (2006). Seed plant phylogeny: Gnetophytes are derived conifers and a sister group to Pinaceae. *Molecular Phylogenetics and Evolution*, *40*(1), 208–217.

Hall, S. E., Dvorak, W. S., Johnston, J. S., Price, H. J., & Williams, C. G. (2000). Flow cytometric analysis of DNA content for tropical and temperate New World pines. *Annals of Botany*, *86*(6), 1081–1086.

Hamberger, B., Hall, D., Yuen, M., Oddy, C., Hamberger, B., Keeling, C. I., et al. (2009). Targeted isolation, sequence assembly and characterization of two white spruce (Picea glauca) BAC clones for terpenoid synthase and cytochrome P450 genes involved in conifer defence reveal insights into a conifer genome. *BMC Plant Biology*, *9*, 106.

Harris, E. H., Burkhart, B. D., Gillham, N. W., & Boynton, J. E. (1989). Antibiotic resistance mutations in the chloroplast 16S and 23S rRNA genes of Chlamydomonas reinhardtii:

Correlation of genetic and physical maps of the chloroplast genome. *Genetics*, *123*(2), 281–292.

Hendy, M. D., & Penny, D. (1989). A framework for the quantitative study of evolutionary trees. *Systematic Zoology*, *38*(4), 297–309.

Hu, W., dePamphilis, C. W., & Ma, H. (2008). Phylogenetic analysis of the plant-specific zinc finger-homeobox and mini zinc finger gene families. *Journal of Integrative Plant Biology*, *50*(8), 1031–1045.

International Rice Genome Sequencing Project. (2005). The map-based sequence of the rice genome. *Nature*, *436*(7052), 793–800.

Jiao, Y., Wickett, N. J., Ayyampalayam, S., Chanderbali, A. S., Landherr, L., Ralph, P. E., et al. (2011). Ancestral polyploidy in seed plants and angiosperms. *Nature*, *473*(7345), 97–100.

Karol, K. G., McCourt, R. M., Cimino, M. T., & Delwiche, C. F. (2001). The closest living relatives of land plants. *Science*, *294*(5550), 2351–2353.

Kassahn, K. S., Dang, V. T., Wilkins, S. J., Perkins, A. C., & Ragan, M. A. (2009). Evolution of gene function and regulatory control after whole-genome duplication: Comparative analyses in vertebrates. *Genome Research*, *19*(8), 1404–1418.

Kenrick, P., & Crane, P. R. (1997). The origin and early evolution of plants on land. *Nature*, *389*(6646), 33–39.

Klekowski, E. J., Jr., & Baker, H. G. (1966). Evolutionary significance of polyploidy in the pteridophyta. *Science*, *153*(3733), 305–307.

Kovach, A., Wegrzyn, J. L., Parra, G., Holt, C., Bruening, G. E., Loopstra, C. A., et al. (2010). The Pinus taeda genome is characterized by diverse and highly diverged repetitive sequences. *BMC Genomics*, *11*, 420.

Kramer, E. M., Dorit, R. L., & Irish, V. F. (1998). Molecular evolution of genes controlling petal and stamen development: Duplication and divergence within the APETALA3 and PISTILLATA MADS-box gene lineages. *Genetics*, *149*(2), 765–783.

Kramer, E. M., Jaramillo, M. A., & Di Stilio, V. S. (2004). Patterns of gene duplication and functional evolution during the diversification of the AGAMOUS subfamily of MADS box genes in angiosperms. *Genetics*, *166*(2), 1011–1023.

Kugita, M., Kaneko, A., Yamamoto, Y., Takeya, Y., Matsumoto, T., & Yoshinaga, K. (2003). The complete nucleotide sequence of the hornwort (Anthoceros formosae) chloroplast genome: Insight into the earliest land plants. *Nucleic Acids Research*, *31*(2), 716–721.

Lemieux, C., Otis, C., & Turmel, M. (2000). Ancestral chloroplast genome in Mesostigma viride reveals an early branch of green plant evolution. *Nature*, *403*(6770), 649–652.

Lemieux, C., Turmel, M., Seligy, V. L., & Lee, R. W. (1985). The large subunit of ribulose-1,5-bisphosphate carboxylase-oxygenase is encoded in the inverted repeat sequence of the Chlamydomonas-eugametos chloroplast genome. *Current Genetics*, *9*(2), 139–145.

Lewis, L. A., & McCourt, R. M. (2004). Green algae and the origin of land plants. *American Journal of Botany*, *91*(10), 1535–1556.

Loiseaux-de Goer, S., Markowicz, Y., Dalmon, J., & Audren, H. (1988). Physical maps of the two circular plastid DNA molecules of the brown alga Pylaiella littoralis (L) Kjellm. *Current Genetics*, *14*(2), 8.

Lynch, M. (2007). *The origins of genome architecture*. Sunderland, MA: Sinauer Associates.

Magallon, S., & Castillo, A. (2009). Angiosperm diversification through time. *American Journal of Botany*, *96*(1), 349–365.

Magbanua, Z. V., Ozkan, S., Bartlett, B. D., Chouvarine, P., Saski, C. A., Liston, A., et al. (2011). Adventures in the enormous: A 1.8 million clone BAC library for the 21.7 Gb genome of loblolly pine. *PLoS One*, *6*(1), e16214.

Manton, I. (1950). *Problems of cytology and evolution in the Pteridophyta*. Cambridge: Cambridge University Press.

Mathews, S. (2009). Phylogenetic relationships among seed plants: Persistent questions and the limits of molecular data. *American Journal of Botany*, *96*(1), 228–236.
Mathews, S., Burleigh, J. G., & Donoghue, M. J. (2003). Adaptive evolution in the photosensory domain of phytochrome A in early angiosperms. *Molecular Biology and Evolution*, *20*(7), 1087–1097.
Merchant, S. S., Prochnik, S. E., Vallon, O., Harris, E. H., Karpowicz, S. J., Witman, G. B., et al. (2007). The Chlamydomonas genome reveals the evolution of key animal and plant functions. *Science*, *318*(5848), 245–250.
Miller, C. N. (1999). Implications of fossil conifers for the phylogenetic relationships of living families. *Botanical Review*, *65*(3), 239–277.
Morse, A. M., Peterson, D. G., Islam-Faridi, M. N., Smith, K. E., Magbanua, Z., Garcia, S. A., et al. (2009). Evolution of genome size and complexity in Pinus. *PLoS One*, *4*(2), e4332.
Murray, B., Leitch, I., & Bennett, M. (2012). Gymnosperm DNA C-values database (release 5.0, Dec. 2012). http://www.kew.org/cvalues/.
Nakazato, T., Jung, M. K., Housworth, E. A., Rieseberg, L. H., & Gastony, G. J. (2006). Genetic map-based analysis of genome structure in the homosporous fern Ceratopteris richardii. *Genetics*, *173*(3), 1585–1597.
Nickrent, D. L., Parkinson, C. L., Palmer, J. D., & Duff, R. J. (2000). Multigene phylogeny of land plants with special reference to bryophytes and the earliest land plants. *Molecular Biology and Evolution*, *17*(12), 1885–1895.
Nystedt, B., Street, N. R., Wetterbom, A., Zuccolo, A., Lin, Y. C., Scofield, D. G., et al. (2013). The Norway spruce genome sequence and conifer genome evolution. *Nature*, *497*(7451), 579–584.
Ohno, S. (1970). *Evolution by gene duplication.* New York: Springer-Verlag, ISBN: 0-04-575015-7.
Ohyama, K., Fukuzawa, H., Kohchi, T., Shirai, H., Sano, T., Sano, S., et al. (1986). Chloroplast gene organization deduced from complete sequence of liverwort marchantia-polymorpha chloroplast DNA. *Nature*, *322*(6079), 572–574.
Palmer, J. D. (1985). Comparative organization of chloroplast genomes. *Annual Review of Genetics*, *19*, 325–354.
Palmer, J. D., & Thompson, W. F. (1982). Chloroplast DNA rearrangements are more frequent when a large inverted repeat sequence is lost. *Cell*, *29*(2), 537–550.
Papp, B., Pal, C., & Hurst, L. D. (2003). Dosage sensitivity and the evolution of gene families in yeast. *Nature*, *424*(6945), 194–197.
Parry, G., Calderon-Villalobos, L. I., Prigge, M., Peret, B., Dharmasiri, S., Itoh, H., et al. (2009). Complex regulation of the *TIR1/AFB* family of auxin receptors. *Proceedings of the National Academy of Sciences of the United States of America*, *106*(52), 22540–22545.
Prigge, M. J., & Clark, S. E. (2006). Evolution of the class III HD-Zip gene family in land plants. *Evolution & Development*, *8*(4), 350–361.
Prochnik, S. E., Umen, J., Nedelcu, A. M., Hallmann, A., Miller, S. M., Nishii, I., et al. (2010). Genomic analysis of organismal complexity in the multicellular green alga Volvox carteri. *Science*, *329*(5988), 223–226.
Pryer, K. M., Schneider, H., Zimmer, E. A., & Banks, J. A. (2002). Deciding among green plants for whole genome studies. *Trends in Plant Science*, 7(12), 550–554.
Rensing, S. A., Ick, J., Fawcett, J. A., Lang, D., Zimmer, A., Van de Peer, Y., et al. (2007). An ancient genome duplication contributed to the abundance of metabolic genes in the moss Physcomitrella patens. *BMC Evolutionary Biology*, 7, 130.
Rensing, S. A., Lang, D., Zimmer, A. D., Terry, A., Salamov, A., Shapiro, H., et al. (2008). The Physcomitrella genome reveals evolutionary insights into the conquest of land by plants. *Science*, *319*(5859), 64–69.
Rothwell, G. W., & Serbet, R. (1994). Lignophyte phylogeny and the evolution of spermatophytes—A numerical cladistic-analysis. *Systematic Botany*, *19*(3), 443–482.

CHAPTER TEN

Decoding the Epigenomes of Herbaceous Plants

Robert J. Schmitz[*,1], **Xiaoyu Zhang**[†,1]
[*]Department of Genetics, University of Georgia, Athens, Georgia, USA
[†]Department of Plant Biology, University of Georgia, Athens, Georgia, USA
[1]Corresponding author: e-mail address: schmitz@uga.edu; xiaoyu@plantbio.uga.edu

Contents

Abstract

One of the most exciting findings in recent years is the discovery that the two main components of the chromatin, DNA and histones, can both be covalently modified by a myriad of pathways. The combination of high-throughput technologies and well-established biochemical assays has made it possible to analyse chromatin modifications at a genome-wide level with increasingly high resolution. As described in this chapter, results from these 'epigenomics' studies have significantly broadened our understanding of the complex and dynamic nature of plant genomes, as well as the diverse and important roles of chromatin modification in regulating gene expression and plant development.

Advances in Botanical Research, Volume 69
ISSN 0065-2296
http://dx.doi.org/10.1016/B978-0-12-417163-3.00010-X

1. INTRODUCTION

The first assembled plant genome was published in 2000 for *Arabidopsis thaliana*, a model system for plant genetics (AGI, 2000) (see Chapter 4). Since that milestone, an ever-growing number of high-quality plant genomes have been released for other model plant systems as well as major crop plants. With recent advancements in high-throughput sequencing technologies, even more genomes are expected in the coming years.

Although the sequences for some plant genomes have been available for over a decade, our understanding of how these sequences lead to development of different cell types and responses to the environment is still in its infancy. At the core of these developmental processes are finely tuned mechanisms that govern gene expression levels. These mechanisms include sequence-specific transcription factors (TFs), chromatin-modifying enzymes that alter posttranslational modifications to histone proteins or alter nucleosome positioning, histone variants, noncoding RNAs and even direct modification to DNA by methylation of cytosine bases. The pattern of these events as they occur in the genome of each cell and how they change in response to different endogenous or environmental cues to coordinate gene expression levels is referred to as the 'epigenome'.

A recent study on fruit ripening in *Solanum lycopersicum* beautifully demonstrates the power of using a comprehensive epigenomics approach to understanding a fundamental process in a crop species (Zhong et al., 2013). By combining the analysis of DNA methylomes, transcriptomes, small RNAs and TF binding over a developmental time course, Zhong et al. revealed evidence for reprogramming of the epigenome leading to fruit ripening. Furthermore, their ChIP-seq analysis of a master regulatory TF revealed that DNA hypomethylation occurs in regions that neighbour binding sites (Zhong et al., 2013). The result of this multifaceted data is a manageable number of genomic regions that can be targeted by genome and epigenome engineering techniques to control fruit ripening.

In this chapter, we outline the current understanding of plant epigenomes and how novel high-throughput sequencing methodologies (Schmitz & Zhang, 2011) are advancing our ability to rapidly map epigenomes of herbaceous land plants.

2. DNA METHYLATION

Herbaceous land plant genomes are extensively methylated at cytosine positions, which can significantly affect gene expression patterns depending on their context and location in the genome. Methylcytosines are found in the CG, CHG and CHH (where H=A, C or T) sequence contexts and, in general, there exist two major domains of DNA methylation in flowering plant genomes. The first domain strictly contains CG methylation and is preferentially enriched in the bodies of some genes, whereas the second domain contains methylation in all sequence contexts (CG, CHG and CHH) and is typically associated with repeat and transposon sequences and genes in rarer cases. Small RNAs in the size range of 23–24 nucleotides (nt) distinguish these two domains of DNA methylation, being abundant at regions of the genome methylated in all cytosine sequence contexts and absent in regions that contain CG-gene body DNA methylation. The small RNAs are an integral component of the RNA-directed DNA methylation (RdDM) pathway (for review, see Law & Jacobsen, 2010), which forms a positive-feedback loop between the generation of small RNAs and DNA methylation. In *A. thaliana*, the small RNAs guide the *de novo* DNA methyltransferase DOMAINS REARRANGED METHYLASE 2 to the target region where it can methylate cytosines in any sequence context (Cao & Jacobsen, 2002). Targeting by RdDM reduces the number of transcripts that can form from the target sequences, which functions to protect the genome from rapid expansion of mobile elements. The positive-feedback loop that exists between small RNAs and DNA methylation is one mechanism that enables stable inheritance of DNA methylation patterns through cell divisions. Alternatively, methylated CG dinucleotides are maintained by a maintenance methyltransferase through a semiconservative replication process (Bostick et al., 2007). In *A. thaliana*, this enzyme is referred to as MET1 (Finnegan, Peacock, & Dennis, 1996; Ronemus, Galbiati, Ticknor, Chen, & Dellaporta, 1996). Mutants in this enzyme are essentially depleted of all CG methylation whether the methylated CG is in a gene body or in a region targeted by RdDM (Cokus et al., 2008; Lister et al., 2008).

It is often stated that DNA methylation is negatively correlated with gene expression levels, but in flowering plants, this statement is not entirely accurate. Although it is true that sequences targeted by RdDM are transcriptionally silenced, genes that contain DNA methylation strictly at CG dinucleotides are expressed at moderately high levels (Zhang et al., 2006; Zilberman, Gehring, Tran, Ballinger, & Henikoff, 2006). Therefore, when studying DNA methylation in plant genomes, it cannot be emphasised enough the importance of considering contextual information, such as type and location of methylcytosines.

DNA methylation occurred at both CG-gene body methylation and regions targeted by RdDM (Hsieh et al., 2009; Ibarra et al., 2012; Zemach, Kim, et al., 2010). Lastly, analysis of DNA methylomes from the vegetative nucleus (VN) and sperm cells, which are contained within pollen grains, revealed evidence for both local and global alterations to patterns of DNA methylation (Calarco et al., 2012; Ibarra et al., 2012). The local changes revealed loss of DNA methylation near transposons, at recurrent epialleles and at some imprinted loci in the VN, whereas a global increase in CHH methylation was observed throughout the VN DNA methylome (Calarco et al., 2012; Ibarra et al., 2012). Therefore, although it is clear that, in some instances, both local and global patterns of DNA methylation are altered in a developmental context, it is unclear how widespread such changes are in other plant developmental contexts. Future studies that combine cell-type isolation methods (Birnbaum et al., 2005; Deal & Henikoff, 2010) with WGBS will be necessary to understand the contribution of dynamic DNA methylation to plant developmental processes.

2.4. Population epigenomics

In *A. thaliana* mutants, *met1* (Cokus et al., 2008; Lister et al., 2008; Zhang et al., 2006) and *ddm1* (*decreased DNA methylation 1*; Vongs, Kakutani, Martienssen, & Richards, 1993), a global reduction in DNA methylation is observed, which results in a wide array of developmental abnormalities (Finnegan, Brettell, & Dennis, 1993; Jacobsen, Sakai, Finnegan, Cao, & Meyerowitz, 2000; Kakutani, Jeddeloh, Flowers, Munakata, & Richards, 1996; Ronemus et al., 1996). It should be noted that not all of these phenotypes are directly linked to a methylation variant, as some families of transposons are activated in the absence of DNA methylation, which sometimes can cause functional knockouts by insertion in or near genes (Miura et al., 2001; Tsukahara et al., 2009).

The ability to perturb global patterns of DNA methylation is a useful approach to determine the contribution of DNA methylation to phenotypic variation. However, the effects on DNA methylation observed in either *met1* or *ddm1* are usually too drastic to find causal relationships, especially as mutant phenotypes are exacerbated over generational time (Kakutani et al., 1996; Mathieu, Reinders, Caikovski, Smathajitt, & Paszkowski, 2007). To break down the proportion of the DNA methylome that is altered in either *met1* or *ddm1*, an elegant scheme was devised, which resulted in the creation of two different populations of epigenetic recombinant inbred lines (epiRILs) (Johannes et al., 2009; Reinders et al., 2009). RIL populations are commonly created to identify the causal basis for phenotypic variation that exists between two genotypically distinct parental lines, but in the epiRIL populations, the genotypes of the epiRILs are essentially identical. Instead, the DNA methylome landscape is shuffled in each epiRIL line, revealing heritable phenotypic variation could be attributable to variation in DNA methylation. In fact, both epiRIL populations display a broad range of phenotypes including variation in disease susceptibility, time to flowering, biomass, plant height and fruit size to name a few (Johannes et al., 2009; Reinders et al., 2009; Roux et al., 2011). Future efforts to identify causal relationships among the methylation variants and these phenotypes will be of great interest.

The increasing throughput and the decreasing cost to obtain sequence data are enabling the first population-based studies of DNA methylomes. Recently, another population, referred to as 'mutation accumulation lines', was created to understand the effects of spontaneous mutations of reproductive traits by growing descendants of a single founder plant from *A. thaliana* by single seed descent for 30 generations (Shaw, Byers, & Darmo, 2000). Genome sequencing of a subset of the 31st generation lines revealed a spontaneous mutation rate of about one mutation per generation per line (Ossowski et al., 2010). This population was ideally situated for studying natural epigenetic variation because there existed multiple independent lineages and genetic variation was reduced to the lowest possible level (the spontaneous mutation rate). WGBS of this population showed that the single methylation polymorphism rate is about four orders of magnitude greater than the observed single-nucleotide polymorphism rate, whereas larger regions of differential DNA methylation were rare and similar in frequency to the spontaneous mutation rate (Becker et al., 2011; Schmitz et al., 2011). Fascinatingly, examples of recurrent variation in DNA methylation were identified over a single generation (Becker et al., 2011). It was proposed that one potential mechanism for the instability of these methylation states could be due to a failure to correctly transfer information from the parent to the germ line during sexual reproduction (Schmitz et al., 2011). The ability to assess natural epigenetic variation using this population provides a baseline with which to explore the potential for DNA methylomes to respond to ever-changing environments.

The identification of pure natural epigenetic variants, which are completely unlinked from their genotype, is not restricted to *A. thaliana*, as examples of variants that have also been identified using epigenomics approaches in *Z. mays* (Eichten et al., 2012). However, these types of methylation variants were rare in comparison to the frequency of methylation variants found associated with their genetic background, suggesting that a greater proportion of methylation variants are linked to a genetic variant. In fact, a pioneering epigenomics study between two different accessions of *A. thaliana* revealed evidence for extensive variation in DNA methylation (Vaughn et al., 2007), which was later supported by a similar study in *Z. mays* (Eichten et al., 2012). Therefore, plant epigenomes contain abundant variation in DNA methylation between genotypes, but identifying the causal basis for each methylation variant is a daunting task (Schmitz & Ecker, 2012).

The sequencing revolution has led to large-scale resequencing projects for entire germplasm collections for organisms that have a high-quality reference genome (Cao et al., 2011; Gan et al., 2011; Lam et al., 2010; Schmitz, Schultz, et al., 2013). There are numerous motivations behind these initiatives ranging from identifying the spectrum of DNA mutations to understanding the population genetics of an organism (Weigel & Mott, 2009). Arguably, one of the most important goals of these efforts is to enable the community to rapidly identify causal variants for traits of interest using genome-wide association (GWA) studies (Nordborg & Weigel, 2008). These approaches have already been proven useful in a wide range of plant species revealing novel variants associated with phenotypic variation (Atwell et al., 2010; Baxter et al., 2010; Buckler et al., 2009; Chan, Rowe, Corwin, Joseph, & Kliebenstein, 2011; Huang et al., 2010; Kump et al., 2011; Mandel et al., 2013; McMullen et al., 2009; Tian et al., 2011). Association mapping techniques can also be applied to molecular phenotypes such as expression, metabolite and even DNA methylation levels (Chan, Rowe, Hansen, & Kliebenstein, 2010; Gan et al., 2011; Johannes, Colot, & Jansen, 2008). Recently, a population-based survey of natural epigenomic variation in over 150 *A. thaliana* accessions showed extensive natural variation in DNA methylation within a plant species (Schmitz, Schultz, et al., 2013). Using a GWA approach, widespread evidence for linkage between genetic variants and methylation variants was identified and referred to as methylQTL (Schmitz, Schultz, et al., 2013). The greatest enrichment of methylQTL was found within 30 kb of the methylation variant indicating that there is likely a local genetic variant that is causal for the methylation variation. However, there was also strong evidence for both distant methylQTL (causal variants located at unlinked sites) and multiple methylQTL associated with a single methylation variant (causal variants located at multiple genomic regions) (Schmitz, Schultz, et al., 2013).

Identifying causal variants for each of these methylQTL will require large efforts to assemble high-quality genomes for this population, but in doing so, this will likely yield a wide range of potential mechanisms that can lead to variation in DNA methylation within plant genomes.

It was recently discovered that genes and transposons that are targeted by RdDM become expressed in developing pollen grains and seeds (Schmitz, Schultz, et al., 2013; Slotkin et al., 2009). The activation of gene and transposon expression in reproductive tissues was at first perplexing, as these tissues harbour the germ line and the specific activation of transposons could be deleterious. It was revealed that the transcriptional activation of transposons was compartmentalised, at least in the VN, and this provided information in the form of small RNAs to guide reprogramming of the nearby germ line epigenome (Calarco et al., 2012; Ibarra et al., 2012; Slotkin et al., 2009). Although it is becoming clear that reprogramming of the male germ line epigenome is likely important for successful reproduction in flowering plant species, it is unclear why genes targeted by RdDM are also transcriptionally activated at the same time. Most plant epigenomic surveys in natural populations usually focus on understanding phenotypic variation regardless of whether the trait is morphological or molecular in nature. However, a recent exploration of the invariably RdDM-targeted regions throughout a population of approximately 150 *A. thaliana* accessions revealed that not only these classes of genes are expressed in pollen grains but they are also enriched for gene ontology terms important for pollen germination and pollen tip growth (Schmitz, Schultz, et al., 2013). Therefore, it seems that in some plant species, certain genes important for pollen development are under control of a silencing system that typically targets transposons. This system is impaired in the VN, which allows information flow about specific transposons to the germ line epigenome and allows genes that are also targeted by RdDM to become activated, which ensures proper pollen development. Whether this particular system is restricted to *A. thaliana* or is widespread throughout herbaceous land plants is yet to be determined, but future population-epigenomics studies will address this key question.

3. NUCLEOSOME POSITIONING AND OCCUPANCY

Plant nuclear DNA is packaged into the chromatin by interacting with positively charged histone proteins. The basic repeating unit of chromatin is the nucleosome, composed of approximately 147 base pairs of DNA wrapped around a core histone octamer (two copies each of H2A, H2B,

in the small nuclei of male gametes (Ingouff & Berger, 2010). The genome-wide distribution patterns of two histone variants, H2A.Z and H3.3, have been analysed in detail, both of which appear to be tightly coupled with gene expression.

4.1. H2A.Z

The histone variant H2A.Z from different eukaryotes shares approximately 90% sequence identities (~60% between H2A.Z and H2A), indicating that H2A.Z diverged from H2A very early during evolution (Suto, Clarkson, Tremethick, & Luger, 2000). The SWRL1 protein complex responsible for replacing H2A with H2A.Z also appears to be shared by all eukaryotes (Deal, Topp, McKinney, & Meagher, 2007; Noh & Amasino, 2003). That both the primary sequence and the deposition mechanism of H2A.Z are highly conserved suggests that this variant may play an important function. Despite the sequence divergence, the crystal structures of H2A- and H2A.Z-containing nucleosomes with regard to protein–protein and protein–DNA interactions are very similar (Suto et al., 2000). There are, however, two interesting differences. First, the interaction between the H2A–H2B dimer and the H3–H4 tetramer is weakened in H2A.Z-containing nucleosomes because several residues important for H2A–H3 interactions are replaced in H2A.Z. Second, the incorporation of H2A.Z generates an uninterrupted acidic patch on the nucleosome surface. Such an acidic patch could interact with additional protein factors through an ion molecule or negatively affect the interaction between an H2A.Z-containing nucleosome and the N-terminal tails of histone H4 in a neighbouring nucleosome (particularly when H4 is acetylated) (Suto et al., 2000).

In *A. thaliana*, H2A.Z is predominantly located within genes, resulting in an apparent enrichment in the euchromatin and depletion in the heterochromatin (Zilberman, Coleman-Derr, Ballinger, & Henikoff, 2008). Interestingly, the local distribution of H2A.Z in genes is tightly correlated with the variability of gene expression (Coleman-Derr & Zilberman, 2012). That is, H2A.Z is highly enriched in the first two to three nucleosomes in genes that are constitutively expressed (housekeeping genes) but depleted from the remainder of nucleosomes in the same genes. In contrast, H2A.Z is present at a moderate and uniform level in genes that are only expressed under specific conditions (hypervariable genes) (Coleman-Derr & Zilberman, 2012). This observation raised the interesting possibility that H2A.Z might facilitate chromatin configuration changes during transcriptional activation.

Consistent with this notion, loss of H2A.Z in an *A. thaliana* mutant leads to numerous developmental abnormalities as well as the preferential misregulation of hypervariable genes.

4.2. H3.3

In both plants and animals, the histone variant H3.3 differs from the canonical H3 (also called 'H3.1') by only four amino acid substitutions, three of which occur at the same positions. However, the sequences of these substitutions are different in plants and animals, indicating that H3.3 may have evolved from H3.1 independently in the two kingdoms (curiously, H3.1 appears to have been lost from fungal genomes) (Waterborg, 2011). A second significant difference between H3.1 and H3.3 in animals lies in the timing and the mechanisms of their synthesis and deposition. H3.1 is only synthesised during the S phase and incorporated into the chromatin by the CAF1 complex during DNA replication. In contrast, H3.3 is synthesised throughout the cell cycle and deposited by the HIRA and the DAXX complexes in a replication-independent but transcription-coupled manner (Ahmad & Henikoff, 2002; Tagami, Ray-Gallet, Almouzni, & Nakatani, 2004). For this reason, H3.3 is also referred to as the 'replacement' H3. The HIRA complex is conserved between plants and animals, but direct evidence for the cell cycle-independent synthesis and HIRA-mediated deposition of H3.3 has yet to be provided.

The distributions of H3.1 and H3.3 in the *A. thaliana* genome are largely mutually exclusive (Stroud, Otero, et al., 2012; Wollmann et al., 2012). H3.3 is located in actively transcribed genes where it is highly enriched toward the 3′ half but depleted in the 5′ half. In contrast, repressed genes and transposable elements contain high and uniform levels of H3.1 and very little H3.3. Because a viable mutant lacking H3.3 or defective in its deposition has not been described in plants, it is not yet clear whether H3.3 performs any other function beyond packaging DNA when H3.1 is evicted during transcription. However, circumstantial evidence seems to indicate that H3.3 might indeed play a role in active transcription. For example, most histone modifications involved in transcriptional regulation (detailed later) occur in the N-terminal tail of H3, and mass spectrometry analyses have shown that H3.1 and H3.3 contain drastically different sets of modifications. As an example, the repressive modification H3K27me3 is exclusively found in H3.1, whereas H3.3 contains a high level of the active modification H3K36me3 (Johnson et al., 2004). These results raised the interesting possibility that H3.3-containing chromatin may adapt a more permissive configuration and thereby facilitate active transcription.

5. HISTONE MODIFICATIONS

In addition to the positioning and composition, individual nucleosomes can also differ from each other by posttranslational histone modifications. Mass spectrometry studies in plants have identified over two-dozen residues containing four major covalent modifications, namely, methylation, acetylation, phosphorylation and ubiquitination (Zhang, Sridhar, Zhu, Kapoor, & Zhu, 2007). In the cases of methylation, an arginine residue can be mono- or dimethylated (either symmetrically or asymmetrically), and a lysine residue can be mono-, di- or trimethylated. In theory, the number of possible modification combinations is therefore so large that each nucleosome in a eukaryotic genome can have its own unique set of modifications, thereby providing an additional layer of information superimposed on the underlying DNA sequence. However, genome-wide empirical analyses indicate that most histone modifications occur in a limited number of combinations that are tightly correlated with the activities of various chromatin-based molecular processes (Zhang, 2008). These results suggest that distinct histone modification pathways may function in concert at specific loci to regulate important processes such as transcription, replication, DNA repair and recombination. The mechanisms, distribution patterns and functions of histone modifications are described later.

5.1. Lysine methylation and demethylation

In *A. thaliana*, four lysine residues on H3 can be methylated: H3K4, H3K9, H3K27 and H3K36. Lysine methylation is catalysed by a group of evolutionarily conserved histone methyltransferases (HMTases) containing the SET domain, referred to as 'SET domain genes' (SDGs) in plants. Phylogenetic analysis defined five major classes of SDGs (I–V), each with different activities and substrate specificities (Springer et al., 2003). Two distinct types of lysine demethylases (KDMs) mediate lysine demethylation. The lysine-specific demethylase 1 (LSD1) type of KDMs contains the polyamine oxidase domain and removes methyl groups by amine oxidation, whereas the Jumonji C (JmjC) domain-containing KDMs catalyse demethylation through hydroxylation (Shi et al., 2004; Tsukada et al., 2006). As described later, the combined actions of HMTases and KDMs set up four types of methylation profiles that are correlated with actively transcribed genes, developmentally repressed genes, permanently silenced transposable element and intergenic regions with virtually no transcription activity.

5.1.1 H3K4 methylation

Class III SDGs are homologous to the H3K4 HMTases in yeast (SET1) and animals (Trithorax-like). Two *A. thaliana* Class III SDGs, SDG2/ATXR3 and SDG27/ATX1, specifically methylate H3K4 *in vitro*, whereas the loss of SDG2 and SDG27 resulted in decreased di- and trimethylation of H3K4 (H3K4me2/3) *in vivo* (Alvarez-Venegas et al., 2003; Guo, Yu, Law, & Zhang, 2010). The SDG30/ATX2 and SDG25/ATXR7 proteins in Class III are also required for H3K4me2/3 in a locus-specific manner (Berr et al., 2009; Saleh et al., 2008). Available biochemical and genetic evidence suggests that H3K4 demethylation is catalysed by both the LSD1-type and the KDM5/JARID1 group of JmjC-type KDMs (Jiang, Yang, He, & Amasino, 2007; Liu et al., 2007; Lu, Cui, Zhang, Liu, & Cao, 2010).

H3K4 methylation is highly enriched in actively transcribed genes and depleted in all other genomic regions, including repressed genes, transposable elements or intergenic regions (Zhang, Bernatavichute, Cokus, Pellegrini, & Jacobsen, 2009). Specifically, H3K4me3 is enriched in the ~1 kb region at the 5′-end of genes, with the highest level found at ~400 bp downstream of the TSS. H3K4me2 largely overlaps with H3K4me3, with H3K4me2 being slightly downstream of H3K4me3 by roughly one nucleosome. The remainder of the transcribed region is enriched for H3K4me1. The specific localization of H3K4 methylation in virtually all expressed genes with H3K4me3/2/1 arranged in a 5′–3′ gradient suggests that these modifications may play general and important roles in active transcription. Consistent with this notion, abnormal levels of H3K4 methylation in HMTase and KDM mutants are associated with the misregulation of thousands of genes as well as severe developmental phenotypes.

5.1.2 H3K9 methylation

Class V SDGs are homologous to the SUPPRESSOR OF VARIEGATION 3–9 [Su(var)3–9] group of animal H3K9 HMTases. Several members of this family have been shown to methylate H3K9 *in vivo*, including SDG33/SUVH4/KYP, SDG9/SUVH5, SDG23/SUVH6 and SDG3/SUVH2 (Ebbs, Bartee, & Bender, 2005; Ebbs & Bender, 2006; Jackson, Lindroth, Cao, & Jacobsen, 2002; Malagnac, Bartee, & Bender, 2002; Naumann et al., 2005). H3K9 demethylation is catalysed by the

2008). Curiously, homologues of the KDM2/JHDM1 group of animal H3K36 KDMs appear to be missing in plant genomes (Lu et al., 2008).

Similar to H3K4 methylation, H3K36 methylation is also present in actively transcribed genes (Oh, Park, & van Nocker, 2008; Roudier et al., 2011). H3K36me3 largely overlaps with H3K4me2 or H3K4me1 and H3K36me2 is located just upstream of the 3′ ends. H3K36me1 is the most downstream modification in genes, covering the 3′ ends, as well as a short region downstream of the polyadenylation site. The distribution of H3K36me3/2/1 along transcribed regions in active genes (but not in promoters) suggests that these modifications might function in transcription elongation.

5.2. Lysine acetylation and deacetylation

A number of lysine residues can be acetylated in *A. thaliana*, including H2AK4/K144, H2BK6/K11/K27/K32, H3K9/K14/K18/K23 and H4K5/K8/K12/K16/K20. Histone lysine acetylation and deacetylation are catalysed by histone acetyltransferases (HATs) and histone deacetylases (HDACs), respectively. The *A. thaliana* genome encodes 12 HATs (five GNAT/MYST-like, five CBP-like and two TAFII250-like) and 16 HDACs (10 RPD3/HDA1-like, four HD2-like and two SIR2-like) (Pandey et al., 2002). The specificities of several HATs have been characterised by *in vitro* assays, whereas other HATs appear to be more promiscuous and can act on multiple lysines (Earley, Shook, Brower-Toland, Hicks, & Pikaard, 2007).

To date, the distribution profiles of several acetylated lysines on H3 and H4 have been determined in *A. thaliana*. Interestingly, all acetylated lysines examined so far share the same distribution pattern. They are highly enriched in the first few nucleosomes in actively transcribed genes, the same region where H3K4me3 is located. These results suggest that hyperacetylation of histones in the first few nucleosomes may be required for active transcription, perhaps by introducing extra negative charges to weaken DNA–histone interactions.

5.3. Other histone modifications

The genome-wide distribution patterns of several other types of histone modifications have not been determined. This is largely due to technical difficulties such as the inaccessibility of modified residues by antibodies or the transient appearance of these modifications during the cell cycle.

These modifications nevertheless play critically important roles in many processes and will be briefly described later.

5.3.1 Lysine ubiquitination and deubiquitination

Polyubiquitination is a critical step to targeting proteins to degradation pathways. In contrast, monoubiquitination of histones may function in transcriptional activation (e.g. by disrupting DNA–histone interactions) rather than targeting histones for degradation. The only ubiquitinated residue detected in plant histones to date is lysine 143 on histone H2B (H2BK143). H2BK143ubq depends on the E3 ligases HISTONE MONOUBIQUITINATION1 (HUB1) and HUB2 (Cao, Dai, Cui, & Ma, 2008), whereas H2BK143 deubiquitination is catalysed by the ubiquitin-specific proteases (or deubiquitinases) UBP26/SUP32 and OTLD1 (Krichevsky, Zaltsman, Lacroix, & Citovsky, 2011; Sridhar et al., 2007). Loss of H2BK143ubq in the *hub1* and *hub2* mutants causes several developmental phenotypes such as pale green leaves, dwarfism and early flowering. Importantly, the loss of H2BK143ubq from the flowering regulator gene *FLC* is associated with downregulation of *FLC* expression and the removal of H2BK143ubq is required for the transcriptional repression of *FLC* (Krichevsky et al., 2011; Schmitz, Tamada, Doyle, Zhang, & Amasino, 2009). These results indicate that H2BK143ubq is critically important for transcriptional activation.

5.3.2 Arginine methylation

A growing list of arginine residues in core histones has been found to be methylated. The *A. thaliana* genome encodes nine protein arginine methyltransferases (PRMTs) (Niu, Lu, Pei, Liu, & Cao, 2007; Niu, Zhang, Pei, Liu, & Cao, 2008). The activities of several PRMTs have been characterised. For example, AtPRMT4a and AtPRMT4b are capable of dimethylating H3R2/R17/R26, whereas AtPRMT5/SKB1 and AtPRMT10 can dimethylate H4R3 and H2A. Abnormal levels of arginine methylation in some PRMT mutants (e.g. *atprmt5*, *atprmt10* and *atprmt4a atprmt4b*) are associated with various developmental abnormalities, indicating that histone arginine methylation may play important roles in regulating gene expression. Interestingly, some PRMTs are also involved in other important processes by methylating arginines in nonhistone proteins (Deng et al., 2012). For example, AtPRMT5 can methylate many proteins involved in RNA processing and metabolism, and the loss of AtPRMT5 causes splicing defects at hundreds of genes. These results uncovered a novel link between histone modifications and cotranscriptional RNA splicing, two important process that occur in close proximity of each other.

5.3.3 Serine and threonine phosphorylation

Histone serine and threonine phosphorylation functions in cell cycle progression and DNA damage repair in animals. Mass spectrometry studies have shown that several serine and threonine residues on histone H2A, H2B and H3 are phosphorylated in *A. thaliana*. Two families of kinases are responsible for H3 phosphorylation. The Aurora kinases phosphorylate H3S10 and H3S28, whereas AtHaspin phosphorylates H3T3 and H3T11. The kinases involved in H2A and H2B phosphorylation have not been identified (Ashtiyani et al., 2011; Demidov, Van Damme, Geelen, Blattner, & Houben, 2005; Kawabe et al., 2005; Kurihara, Matsunaga, Omura, Higashiyama, & Fukui, 2011).

H3 phosphorylation is tightly regulated during the cell cycle. H3S10p and H3S28p accumulate in the pericentromeric regions at metaphase in mitosis at metaphase II in meiosis and throughout the chromatin at metaphase I in meiosis (Gernand, Demidov, & Houben, 2003; Kaszas & Cande, 2000; Manzanero, Arana, Puertas, & Houben, 2000). Inhibition of Aurora kinase activity in tobacco cells frequently leads to lagging chromosomes at anaphase, indicating that H3S10p and H3S28p may be required for the dissociation of chromatid cohesion (Kurihara et al., 2006). In contrast, H3T3p and H3T11p are present over the entire condensed chromosomes in both mitosis and meiosis and decreased levels of H3T3p and H3T11p in *AtHaspin* RNAi plants are associated with reduced chromatin condensation (Ashtiyani et al., 2011; Houben, Demidov, Rutten, & Scheidtmann, 2005).

6. CONCLUDING REMARKS AND FUTURE DIRECTIONS

During the past decade, major breakthroughs in high-throughput technologies have made it possible to study genome-wide patterns of DNA and histone modifications at very high resolutions. Results from these epigenomics studies have significantly broadened our understanding of the structure, organisation, composition and function of plant genomes. We now know that a myriad of chromatin-modifying pathways are at work during cell cycles, throughout development and in response to environmental cues. With regard to gene expression and plant development, it is important to note that the four major chromatin modification profiles correspond to the four types of transcription activities. (1) Stably silenced transposable elements are associated with CG, CHG and CHH DNA methylation, siRNAs and H3K9me2. (2) Developmentally repressed genes are associated with H3.1 and H3K27me3. (3) Actively transcribed genes are associated with low nucleosome occupancy in the promoter, H3K4me2/3 and H3ac/H4ac at the 5′ end; H3.3, H3K36me3 and H4K4me1 in the remainder of transcribed regions; and H3K36me1/2 at the 3′ end. (4) Intergenic regions with no transcription activity are virtually devoid of any modification. These tight correlations as well as the larger number of genes that become misregulated in individual mutants all support a critically important role for chromatin modification in transcriptional regulation. The important questions of how chromatin-modifying machineries are targeted to specific genes and how the resulting modifications may interact with the transcription apparatus should be addressed by future studies.

It should also be noted that most plant epigenome studies rely on nucleic acids isolated from populations of diverse cell types. Therefore, future efforts to improve methods for isolating specific cell types (Birnbaum et al., 2005; Deal & Henikoff, 2011) and to improve methods for constructing low-input sequencing libraries will be of great importance to advancing our understanding of plant epigenomes.

It is clear that perturbation to DNA methylation profiles in some plant species can have profound effects of phenotypic variation, but our ability to systematically link methylation variants to these phenotypes is still limited. One area currently unexplored in plants is the potential for epigenome-wide association studies (Liu et al., 2013; Rakyan, Down, Balding, & Beck, 2011) to link causal methylation variants to morphological variation. The epiRILs described earlier are an ideal test case for these types of studies as genetic variation is essentially eliminated allowing an easier path to identifying causal methylation variants that are linked to phenotypes. Epigenome-wide association studies could even be useful for identifying valuable phenotypes that appear naturally in the field under native environments, noting that a rapidly occurring phenotype in multiple independent plants under the same conditions would not likely have a genetic basis. Our knowledge of plant epigenomes is still in its infancy, but as the numbers of known causal epigenomic variants linked to phenotypic variation increase, our attention will turn to a growing need to begin engineering epigenomes. Such engineering efforts may be accelerated by progress in efforts currently under way to engineer genomes using TALs, CRSPERs or some yet undiscovered method (Boch et al., 2009; Mali et al., 2013; Moscou & Bogdanove, 2009; Qi et al., 2013).

Molecular and Cellular Biology, *25*(23), 10507–10515. http://dx.doi.org/10.1128/MCB.25.23.10507-10515.2005, [pii] 25/23/10507.

Ebbs, M. L., & Bender, J. (2006). Locus-specific control of DNA methylation by the Arabidopsis SUVH5 histone methyltransferase. *Plant Cell*, *18*(5), 1166–1176.

Eichten, S. R., Ellis, N. A., Makarevitch, I., Yeh, C. T., Gent, J. I., & Guo, L. (2012). Spreading of heterochromatin is limited to specific families of maize retrotransposons. *PLoS Genetics*, *8*(12), e1003127. http://dx.doi.org/10.1371/journal.pgen.1003127, [pii] PGENETICS-D-12-01022.

Feng, S., Cokus, S. J., Zhang, X., Chen, P. Y., Bostick, M., & Goll, M. G. (2010). Conservation and divergence of methylation patterning in plants and animals. *Proceedings of the National Academy of Sciences of the United States of America*, *107*(19), 8689–8694. http://dx.doi.org/10.1073/pnas.1002720107, [pii] 1002720107.

Finnegan, E. J., Brettell, R. I., & Dennis, E. S. (1993). The role of DNA methylation in the regulation of plant gene expression. *EXS*, *64*, 218–261.

Finnegan, E. J., Peacock, W. J., & Dennis, E. S. (1996). Reduced DNA methylation in Arabidopsis thaliana results in abnormal plant development. *Proceedings of the National Academy of Sciences of the United States of America*, *93*(16), 8449–8454.

Frommer, M., McDonald, L. E., Millar, D. S., Collis, C. M., Watt, F., & Grigg, G. W. (1992). A genomic sequencing protocol that yields a positive display of 5-methylcytosine residues in individual DNA strands. *Proceedings of the National Academy of Sciences of the United States of America*, *89*(5), 1827–1831.

Gan, X., Stegle, O., Behr, J., Steffen, J. G., Drewe, P., & Hildebrand, K. L. (2011). Multiple reference genomes and transcriptomes for Arabidopsis thaliana. *Nature*, *477*(7365), 419–423. http://dx.doi.org/10.1038/nature10414, [pii] nature10414.

Gent, J. I., Ellis, N. A., Guo, L., Harkess, A. E., Yao, Y., & Zhang, X. (2013). CHH islands: De novo DNA methylation in near-gene chromatin regulation in maize. *Genome Research*, *23*(4), 628–637. http://dx.doi.org/10.1101/gr.146985.112, [pii] gr.146985.112.

Gernand, D., Demidov, D., & Houben, A. (2003). The temporal and spatial pattern of histone H3 phosphorylation at serine 28 and serine 10 is similar in plants but differs between mono- and polycentric chromosomes. *Cytogenetic and Genome Research*, *101*(2), 172–176. http://dx.doi.org/10.1159/000074175, [pii] 74175.

Guo, L., Yu, Y., Law, J. A., & Zhang, X. (2010). SET DOMAIN GROUP2 is the major histone H3 lysine [corrected] 4 trimethyltransferase in Arabidopsis. *Proceedings of the National Academy of Sciences of the United States of America*, *107*(43), 18557–18562. http://dx.doi.org/10.1073/pnas.1010478107, [pii] 1010478107.

Hayatsu, H. (1976). Bisulfite modification of nucleic acids and their constituents. *Progress in Nucleic Acid Research and Molecular Biology*, *16*, 75–124.

Hayatsu, H., Wataya, Y., Kai, K., & Iida, S. (1970). Reaction of sodium bisulfite with uracil, cytosine, and their derivatives. *Biochemistry*, *9*(14), 2858–2865.

Houben, A., Demidov, D., Rutten, T., & Scheidtmann, K. H. (2005). Novel phosphorylation of histone H3 at threonine 11 that temporally correlates with condensation of mitotic and meiotic chromosomes in plant cells. *Cytogenetic and Genome Research*, *109*(1–3), 148–155. http://dx.doi.org/10.1159/000082394, [pii] 82394.

Hsieh, T. F., Ibarra, C. A., Silva, P., Zemach, A., Eshed-Williams, L., & Fischer, R. L. (2009). Genome-wide demethylation of Arabidopsis endosperm. *Science*, *324*(5933), 1451–1454. http://dx.doi.org/10.1126/science.1172417, [pii] 324/5933/1451.

Huang, X., Wei, X., Sang, T., Zhao, Q., Feng, Q., & Zhao, Y. (2010). Genome-wide association studies of 14 agronomic traits in rice landraces. *Nature Genetics*, *42*(11), 961–967. http://dx.doi.org/10.1038/ng.695, [pii] ng.695.

Ibarra, C. A., Feng, X., Schoft, V. K., Hsieh, T. F., Uzawa, R., & Rodrigues, J. A. (2012). Active DNA demethylation in plant companion cells reinforces transposon methylation in gametes. *Science, 337*(6100), 1360–1364. http://dx.doi.org/10.1126/science.1224839, [pii] 337/6100/1360.

Ingouff, M., & Berger, F. (2010). Histone3 variants in plants. *Chromosoma, 119*(1), 27–33. http://dx.doi.org/10.1007/s00412-009-0237-1.

Jackson, J. P., Lindroth, A. M., Cao, X., & Jacobsen, S. E. (2002). Control of CpNpG DNA methylation by the KRYPTONITE histone H3 methyltransferase. *Nature, 416*(6880), 556–560.

Jacob, Y., Feng, S., LeBlanc, C. A., Bernatavichute, Y. V., Stroud, H., & Cokus, S. (2009). ATXR5 and ATXR6 are H3K27 monomethyltransferases required for chromatin structure and gene silencing. *Nature Structural and Molecular Biology, 16*(7), 763–768. http://dx.doi.org/10.1038/nsmb.1611, [pii] nsmb.1611.

Jacob, Y., Stroud, H., Leblanc, C., Feng, S., Zhuo, L., & Caro, E. (2010). Regulation of heterochromatic DNA replication by histone H3 lysine 27 methyltransferases. *Nature, 466*(7309), 987–991. http://dx.doi.org/10.1038/nature09290, [pii] nature09290.

Jacobsen, S. E., Sakai, H., Finnegan, E. J., Cao, X., & Meyerowitz, E. M. (2000). Ectopic hypermethylation of flower-specific genes in Arabidopsis. *Current Biology, 10*(4), 179–186.

Jiang, D., Yang, W., He, Y., & Amasino, R. M. (2007). Arabidopsis relatives of the human lysine-specific Demethylase1 repress the expression of FWA and FLOWERING LOCUS C and thus promote the floral transition. *Plant Cell, 19*(10), 2975–2987. http://dx.doi.org/10.1105/tpc.107.052373, [pii] tpc.107.052373.

Johannes, F., Colot, V., & Jansen, R. C. (2008). Epigenome dynamics: A quantitative genetics perspective. *Nature Reviews Genetics, 9*(11), 883–890. http://dx.doi.org/10.1038/nrg2467, [pii] nrg2467.

Johannes, F., Porcher, E., Teixeira, F. K., Saliba-Colombani, V., Simon, M., & Agier, N. (2009). Assessing the impact of transgenerational epigenetic variation on complex traits. *PLoS Genetics, 5*(6), e1000530. http://dx.doi.org/10.1371/journal.pgen.1000530.

Johnson, L. M., Bostick, M., Zhang, X., Kraft, E., Henderson, I., & Callis, J. (2007). The SRA methyl-cytosine-binding domain links DNA and histone methylation. *Current Biology, 17*(4), 379–384. http://dx.doi.org/10.1016/j.cub.2007.01.009, [pii] S0960-9822(07)00810-X.

Johnson, L., Mollah, S., Garcia, B. A., Muratore, T. L., Shabanowitz, J., & Hunt, D. F. (2004). Mass spectrometry analysis of Arabidopsis histone H3 reveals distinct combinations of post-translational modifications. *Nucleic Acids Research, 32*(22), 6511–6518.

Kakutani, T., Jeddeloh, J. A., Flowers, S. K., Munakata, K., & Richards, E. J. (1996). Developmental abnormalities and epimutations associated with DNA hypomethylation mutations. *Proceedings of the National Academy of Sciences of the United States of America, 93*(22), 12406–12411.

Kaszas, E., & Cande, W. Z. (2000). Phosphorylation of histone H3 is correlated with changes in the maintenance of sister chromatid cohesion during meiosis in maize, rather than the condensation of the chromatin. *Journal of Cell Science, 113*(Pt. 18), 3217–3226.

Kawabe, A., Matsunaga, S., Nakagawa, K., Kurihara, D., Yoneda, A., & Hasezawa, S. (2005). Characterization of plant Aurora kinases during mitosis. *Plant Molecular Biology, 58*(1), 1–13. http://dx.doi.org/10.1007/s11103-005-3454-x.

Kim, S. Y., He, Y., Jacob, Y., Noh, Y. S., Michaels, S., & Amasino, R. (2005). Establishment of the vernalization-responsive, winter-annual habit in Arabidopsis requires a putative histone H3 methyl transferase. *Plant Cell, 17*(12), 3301–3310.

Krichevsky, A., Zaltsman, A., Lacroix, B., & Citovsky, V. (2011). Involvement of KDM1C histone demethylase-OTLD1 otubain-like histone deubiquitinase complexes in plant gene repression. *Proceedings of the National Academy of Sciences of the United States of America, 108*(27), 11157–11162. http://dx.doi.org/10.1073/pnas.1014030108, [pii] 1014030108.

Kump, K. L., Bradbury, P. J., Wisser, R. J., Buckler, E. S., Belcher, A. R., & Oropeza-Rosas, M. A. (2011). Genome-wide association study of quantitative resistance to southern leaf blight in the maize nested association mapping population. *Nature Genetics*, *43*(2), 163–168. http://dx.doi.org/10.1038/ng.747, [pii] ng.747.

Kurihara, D., Matsunaga, S., Kawabe, A., Fujimoto, S., Noda, M., & Uchiyama, S. (2006). Aurora kinase is required for chromosome segregation in tobacco BY-2 cells. *Plant Journal*, *48*(4), 572–580. http://dx.doi.org/10.1111/j.1365-313X.2006.02893.x, [pii] TPJ2893.

Kurihara, D., Matsunaga, S., Omura, T., Higashiyama, T., & Fukui, K. (2011). Identification and characterization of plant Haspin kinase as a histone H3 threonine kinase. *BMC Plant Biology*, *11*, 73. http://dx.doi.org/10.1186/1471-2229-11-73, [pii] 1471-2229-11-73.

Lafos, M., Kroll, P., Hohenstatt, M. L., Thorpe, F. L., Clarenz, O., & Schubert, D. (2011). Dynamic regulation of H3K27 trimethylation during Arabidopsis differentiation. *PLoS Genetics*, 7(4), e1002040. http://dx.doi.org/10.1371/journal.pgen.1002040.

Lam, H. M., Xu, X., Liu, X., Chen, W., Yang, G., & Wong, F. L. (2010). Resequencing of 31 wild and cultivated soybean genomes identifies patterns of genetic diversity and selection. *Nature Genetics*, *42*(12), 1053–1059. http://dx.doi.org/10.1038/ng.715, [pii] ng.715.

Law, J. A., & Jacobsen, S. E. (2010). Establishing, maintaining and modifying DNA methylation patterns in plants and animals. *Nature Reviews Genetics*, *11*(3), 204–220. http://dx.doi.org/10.1038/nrg2719, [pii] nrg2719.

Lippman, Z., Gendrel, A. V., Black, M., Vaughn, M. W., Dedhia, N., & McCombie, W. R. (2004). Role of transposable elements in heterochromatin and epigenetic control. *Nature*, *430*(6998), 471–476.

Lister, R., O'Malley, R. C., Tonti-Filippini, J., Gregory, B. D., Berry, C. C., & Millar, A. H. (2008). Highly integrated single-base resolution maps of the epigenome in Arabidopsis. *Cell*, *133*(3), 523–536. http://dx.doi.org/10.1016/j.cell.2008.03.029, [pii] S0092-8674(08)00448-0.

Liu, Y., Aryee, M. J., Padyukov, L., Fallin, M. D., Hesselberg, E., & Runarsson, A. (2013). Epigenome-wide association data implicate DNA methylation as an intermediary of genetic risk in rheumatoid arthritis. *Nature Biotechnology*, *31*(2), 142–147. http://dx.doi.org/10.1038/nbt.2487, [pii] nbt.2487.

Liu, F., Quesada, V., Crevillen, P., Baurle, I., Swiezewski, S., & Dean, C. (2007). The Arabidopsis RNA-binding protein FCA requires a lysine-specific demethylase 1 homolog to downregulate FLC. *Molecular Cell*, *28*(3), 398–407. http://dx.doi.org/10.1016/j.molcel.2007.10.018, [pii] S1097-2765(07)00700-9.

Lu, F., Cui, X., Zhang, S., Jenuwein, T., & Cao, X. (2011). Arabidopsis REF6 is a histone H3 lysine 27 demethylase. *Nature Genetics*, *43*(7), 715–719. http://dx.doi.org/10.1038/ng.854, [pii] ng.854.

Lu, F., Cui, X., Zhang, S., Liu, C., & Cao, X. (2010). JMJ14 is an H3K4 demethylase regulating flowering time in Arabidopsis. *Cell Research*, *20*(3), 387–390. http://dx.doi.org/10.1038/cr.2010.27, [pii] cr201027.

Lu, F., Li, G., Cui, X., Liu, C., Wang, X. J., & Cao, X. (2008). Comparative analysis of JmjC domain-containing proteins reveals the potential histone demethylases in Arabidopsis and rice. *Journal of Integrative Plant Biology*, *50*(7), 886–896. http://dx.doi.org/10.1111/j.1744-7909.2008.00692.x, [pii] JIPB692.

Malagnac, F., Bartee, L., & Bender, J. (2002). An Arabidopsis SET domain protein required for maintenance but not establishment of DNA methylation. *EMBO Journal*, *21*(24), 6842–6852.

Mali, P., Yang, L., Esvelt, K. M., Aach, J., Guell, M., & DiCarlo, J. E. (2013). RNA-guided human genome engineering via Cas9. *Science*, *339*(6121), 823–826. http://dx.doi.org/10.1126/science.1232033, [pii] 1232033.

Mandel, J. R., Nambeesan, S., Bowers, J. E., Marek, L. F., Ebert, D., & Rieseberg, L. H. (2013). Association mapping and the genomic consequences of selection in sunflower. *PLoS Genetics*, *9*(3), e1003378. http://dx.doi.org/10.1371/journal.pgen.1003378, [pii] PGENETICS-D-12-01937.

Manzanero, S., Arana, P., Puertas, M. J., & Houben, A. (2000). The chromosomal distribution of phosphorylated histone H3 differs between plants and animals at meiosis. *Chromosoma*, *109*(5), 308–317.

Mathieu, O., Reinders, J., Caikovski, M., Smathajitt, C., & Paszkowski, J. (2007). Transgenerational stability of the Arabidopsis epigenome is coordinated by CG methylation. *Cell*, *130*(5), 851–862. http://dx.doi.org/10.1016/j.cell.2007.07.007.

McMullen, M. D., Kresovich, S., Villeda, H. S., Bradbury, P., Li, H., & Sun, Q. (2009). Genetic properties of the maize nested association mapping population. *Science*, *325*(5941), 737–740. http://dx.doi.org/10.1126/science.1174320, [pii] 325/5941/737.

Miura, A., Yonebayashi, S., Watanabe, K., Toyama, T., Shimada, H., & Kakutani, T. (2001). Mobilization of transposons by a mutation abolishing full DNA methylation in Arabidopsis. *Nature*, *411*(6834), 212–214.

Moscou, M. J., & Bogdanove, A. J. (2009). A simple cipher governs DNA recognition by TAL effectors. *Science*, *326*(5959), 1501. http://dx.doi.org/10.1126/science.1178817, [pii] 1178817.

Naumann, K., Fischer, A., Hofmann, I., Krauss, V., Phalke, S., & Irmler, K. (2005). Pivotal role of AtSUVH2 in heterochromatic histone methylation and gene silencing in Arabidopsis. *EMBO Journal*, *24*(7), 1418–1429. http://dx.doi.org/10.1038/sj.emboj.7600604, [pii] 7600604.

Niu, L., Lu, F., Pei, Y., Liu, C., & Cao, X. (2007). Regulation of flowering time by the protein arginine methyltransferase AtPRMT10. *EMBO Reports*, *8*(12), 1190–1195. http://dx.doi.org/10.1038/sj.embor.7401111, [pii] 7401111.

Niu, L., Zhang, Y., Pei, Y., Liu, C., & Cao, X. (2008). Redundant requirement for a pair of PROTEIN ARGININE METHYLTRANSFERASE4 homologs for the proper regulation of Arabidopsis flowering time. *Plant Physiology*, *148*(1), 490–503. http://dx.doi.org/10.1104/pp.108.124727, [pii] pp.108.124727.

Noh, Y. S., & Amasino, R. M. (2003). PIE1, an ISWI family gene, is required for FLC activation and floral repression in Arabidopsis. *Plant Cell*, *15*(7), 1671–1682.

Nordborg, M., & Weigel, D. (2008). Next-generation genetics in plants. *Nature*, *456*(7223), 720–723. http://dx.doi.org/10.1038/nature07629, [pii] nature07629.

Oh, S., Park, S., & van Nocker, S. (2008). Genic and global functions for Paf1C in chromatin modification and gene expression in Arabidopsis. *PLoS Genetics*, *4*(8), e1000077. http://dx.doi.org/10.1371/journal.pgen.1000077.

Ossowski, S., Schneeberger, K., Lucas-Lledo, J. I., Warthmann, N., Clark, R. M., & Shaw, R. G. (2010). The rate and molecular spectrum of spontaneous mutations in Arabidopsis thaliana. *Science*, *327*(5961), 92–94. http://dx.doi.org/10.1126/science.1180677, [pii] 327/5961/92.

Pandey, R., Muller, A., Napoli, C. A., Selinger, D. A., Pikaard, C. S., & Richards, E. J. (2002). Analysis of histone acetyltransferase and histone deacetylase families of Arabidopsis thaliana suggests functional diversification of chromatin modification among multicellular eukaryotes. *Nucleic Acids Research*, *30*(23), 5036–5055.

Pien, S., & Grossniklaus, U. (2007). Polycomb group and trithorax group proteins in Arabidopsis. *Biochimica et Biophysica Acta*, *1769*(5–6), 375–382. http://dx.doi.org/10.1016/j.bbaexp.2007.01.010, [pii] S0167-4781(07)00030-9.

Qi, L. S., Larson, M. H., Gilbert, L. A., Doudna, J. A., Weissman, J. S., & Arkin, A. P. (2013). Repurposing CRISPR as an RNA-guided platform for sequence-specific control of gene expression. *Cell*, *152*(5), 1173–1183. http://dx.doi.org/10.1016/j.cell.2013.02.022.

Takuno, S., & Gaut, B. S. (2013). Gene body methylation is conserved between plant orthologs and is of evolutionary consequence. *Proceedings of the National Academy of Sciences of the United States of America*, *110*(5), 1797–1802. http://dx.doi.org/10.1073/pnas.1215380110, [pii] 1215380110.

Thorstensen, T., Grini, P. E., Mercy, I. S., Alm, V., Erdal, S., & Aasland, R. (2008). The Arabidopsis SET-domain protein ASHR3 is involved in stamen development and interacts with the bHLH transcription factor ABORTED MICROSPORES (AMS). *Plant Molecular Biology*, *66*(1–2), 47–59. http://dx.doi.org/10.1007/s11103-007-9251-y.

Tian, F., Bradbury, P. J., Brown, P. J., Hung, H., Sun, Q., & Flint-Garcia, S. (2011). Genome-wide association study of leaf architecture in the maize nested association mapping population. *Nature Genetics*, *43*(2), 159–162. http://dx.doi.org/10.1038/ng.746, [pii] ng.746.

Tsukada, Y., Fang, J., Erdjument-Bromage, H., Warren, M. E., Borchers, C. H., & Tempst, P. (2006). Histone demethylation by a family of JmjC domain-containing proteins. *Nature*, *439*(7078), 811–816. http://dx.doi.org/10.1038/nature04433, [pii] nature04433.

Tsukahara, S., Kobayashi, A., Kawabe, A., Mathieu, O., Miura, A., & Kakutani, T. (2009). Bursts of retrotransposition reproduced in Arabidopsis. *Nature*, *461*(7262), 423–426. http://dx.doi.org/10.1038/nature08351, [pii] nature08351.

Valouev, A., Johnson, S. M., Boyd, S. D., Smith, C. L., Fire, A. Z., & Sidow, A. (2011). Determinants of nucleosome organization in primary human cells. *Nature*, *474*(7352), 516–520. http://dx.doi.org/10.1038/nature10002, [pii] nature10002.

Vaughn, M. W., Tanurd Ić, M., Lippman, Z., Jiang, H., Carrasquillo, R., & Rabinowicz, P. D. (2007). Epigenetic natural variation in Arabidopsis thaliana. *PLoS Biology*, *5*(7), e174. http://dx.doi.org/10.1371/journal.pbio.0050174.

Vongs, A., Kakutani, T., Martienssen, R. A., & Richards, E. J. (1993). Arabidopsis thaliana DNA methylation mutants. *Science*, *260*(5116), 1926–1928.

Wang, R. Y., Gehrke, C. W., & Ehrlich, M. (1980). Comparison of bisulfite modification of 5-methyldeoxycytidine and deoxycytidine residues. *Nucleic Acids Research*, *8*(20), 4777–4790.

Waterborg, J. H. (2011). Evolution of histone H3: Emergence of variants and conservation of post-translational modification sites. *Biochemistry and Cell Biology*, *90*(1), 79–95. http://dx.doi.org/10.1139/o11-036.

Weigel, D., & Mott, R. (2009). The 1001 genomes project for Arabidopsis thaliana. *Genome Biology*, *10*(5), 107. http://dx.doi.org/10.1186/gb-2009-10-5-107, [pii] gb-2009-10-5-107.

Wollmann, H., Holec, S., Alden, K., Clarke, N. D., Jacques, P. E., & Berger, F. (2012). Dynamic deposition of histone variant H3.3 accompanies developmental remodeling of the Arabidopsis transcriptome. *PLoS Genetics*, *8*(5), e1002658. http://dx.doi.org/10.1371/journal.pgen.1002658, [pii] PGENETICS-D-11-02569.

Xu, L., Zhao, Z., Dong, A., Soubigou-Taconnat, L., Renou, J. P., & Steinmetz, A. (2008). Di- and tri- but not monomethylation on histone H3 lysine 36 marks active transcription of genes involved in flowering time regulation and other processes in Arabidopsis thaliana. *Molecular and Cellular Biology*, *28*(4), 1348–1360. http://dx.doi.org/10.1128/MCB.01607-07, [pii] MCB.01607-07.

Yen, K., Vinayachandran, V., Batta, K., Koerber, R. T., & Pugh, B. F. (2012). Genome-wide nucleosome specificity and directionality of chromatin remodelers. *Cell*, *149*(7), 1461–1473. http://dx.doi.org/10.1016/j.cell.2012.04.036, [pii] S0092-8674(12)00592-2.

Zemach, A., Kim, M. Y., Silva, P., Rodrigues, J. A., Dotson, B., & Brooks, M. D. (2010). Local DNA hypomethylation activates genes in rice endosperm. *Proceedings of the National*

Academy of Sciences of the United States of America, *107*(43), 18729–18734. http://dx.doi.org/10.1073/pnas.1009695107, [pii] 1009695107.

Zemach, A., McDaniel, I. E., Silva, P., & Zilberman, D. (2010). Genome-wide evolutionary analysis of eukaryotic DNA methylation. *Science*, *328*(5980), 916–919. http://dx.doi.org/10.1126/science.1186366, [pii] science.1186366.

Zhang, X. (2008). The epigenetic landscape of plants. *Science*, *320*(5875), 489–492. http://dx.doi.org/10.1126/science.1153996, [pii] 320/5875/489.

Zhang, X., Bernatavichute, Y. V., Cokus, S., Pellegrini, M., & Jacobsen, S. E. (2009). Genome-wide analysis of mono-, di- and trimethylation of histone H3 lysine 4 in Arabidopsis thaliana. *Genome Biology*, *10*(6), R62. http://dx.doi.org/10.1186/gb-2009-10-6-r62, [pii] gb-2009-10-6-r62.

Zhang, X., Clarenz, O., Cokus, S., Bernatavichute, Y. V., Pellegrini, M., & Goodrich, J. (2007). Whole-genome analysis of histone H3 lysine 27 trimethylation in Arabidopsis. *PLoS Biology*, *5*(5), e129. http://dx.doi.org/10.1371/journal.pbio.0050129, [pii] 06-PLBI-RA-2480R2.

Zhang, K., Sridhar, V. V., Zhu, J., Kapoor, A., & Zhu, J. K. (2007). Distinctive core histone post-translational modification patterns in Arabidopsis thaliana. *PLoS ONE*, *2*(11), e1210. http://dx.doi.org/10.1371/journal.pone.0001210.

Zhang, Z., Wippo, C. J., Wal, M., Ward, E., Korber, P., & Pugh, B. F. (2011). A packing mechanism for nucleosome organization reconstituted across a eukaryotic genome. *Science*, *332*(6032), 977–980. http://dx.doi.org/10.1126/science.1200508, [pii] 332/6032/977.

Zhang, X., Yazaki, J., Sundaresan, A., Cokus, S., Chan, S. W., & Chen, H. (2006). Genome-wide high-resolution mapping and functional analysis of DNA methylation in Arabidopsis. *Cell*, *126*(6), 1189–1201. http://dx.doi.org/10.1016/j.cell.2006.08.003.

Zhao, Z., Yu, Y., Meyer, D., Wu, C., & Shen, W. H. (2005). Prevention of early flowering by expression of FLOWERING LOCUS C requires methylation of histone H3 K36. *Nature Cell Biology*, 7(12), 1256–1260.

Zhong, S., Fei, Z., Chen, Y. R., Zheng, Y., Huang, M., & Vrebalov, J. (2013). Single-base resolution methylomes of tomato fruit development reveal epigenome modifications associated with ripening. *Nature Biotechnology*, *31*(2), 154–159. http://dx.doi.org/10.1038/nbt.2462, [pii] nbt.2462.

Zilberman, D., Coleman-Derr, D., Ballinger, T., & Henikoff, S. (2008). Histone H2A.Z and DNA methylation are mutually antagonistic chromatin marks. *Nature*, *456*(7218), 125–129. http://dx.doi.org/10.1038/nature07324, [pii] nature07324.

Zilberman, D., Gehring, M., Tran, R. K., Ballinger, T., & Henikoff, S. (2006). Genome-wide analysis of Arabidopsis thaliana DNA methylation uncovers an interdependence between methylation and transcription. *Nature Genetics*, *39*(1), 61 69. http://dx.doi.org/10.1038/ng1929.

CHAPTER ELEVEN

The Plant Microbiome

Marnie E. Rout[1]
Department of Forensics and Investigative Genetics, University of North Texas Health Science Center, Fort Worth, Texas, USA
[1]Corresponding author: e-mail address: marnie.rout@unthsc.edu

Contents

Abstract

How do we define the 'plant microbiome' and what is its significance to the plant genome? Before addressing what the microbiome is in relation to plants, it is important to first understand the concept of the microbiome; what this means in relation to the host becomes an extension of this working concept. Conceptualizing the microbiome requires a fusion of microbial ecology and bioinformatics, integrated with an understanding of both host biology and ecology. The analysis of microbiome structure and function was pioneered in studies of human hosts and has become widely recognized as essential to understanding genetic and functional capacity otherwise attributed to the host, including important aspects of metabolism and physiology. Plants are teeming with microbial organisms, including those that colonize internal tissues, in addition to those that adhere to external surfaces. Combined with the vast diversity of microorganisms in the soil rhizosphere, these plant–soil-associated microbes comprise the plant microbiome. The microbiome is intricately involved in plant health and serves as a reservoir of additional genes that plants can access when needed. Understanding the regulation of plant trait expression, hence plant performance and how this in turn impacts ecosystem function, requires that we study the impacts of the plant microbiome. Herein, the importance of the plant microbiome to plant

Advances in Botanical Research, Volume 69
ISSN 0065-2296
http://dx.doi.org/10.1016/B978-0-12-417163-3.00011-1

genomics is addressed by defining the plant microbiome in relation to the ecology of the system with emphasis on habitats occurring belowground at the plant–soil interface, where focus is on the role of exudates as currency in this system.

1. BACKGROUND

The concept of the microbiome and the relevance it has to host health, diseases state, and immune function have been the focus of research over the past decade that has led to significant advances in our understanding of the enormous power of the small unseen majority—the microbes. Coinvolvement of the microbiome, the environment, and host genetics and immunity is now recognized to produce several human disease states. Each individual human has a 'personalized' microbiome when examined at fine scales (genus level and below), but at the group and phylum level, there are characteristic trends in abundance that have value in defining a 'healthy-state' microbiome from disease states such as obesity, irritable bowel (Greenblum, Turnbaugh, & Borenstein, 2012), and atherosclerosis (Koeth et al., 2013). The Human Genome Project has referred to this concept as the 'supraorganism'. Turnbaugh et al. (2007) stated it in this way: 'If humans are thought of as a composite of microbial and human cells, the human genetic landscape as an aggregate of the genes in the human genome and the microbiome, and human metabolic features as a blend of human and microbial traits, then the picture that emerges is one of a human "supraorganism"'.

To then borrow from those that pioneered the human microbiome project (Turnbaugh et al., 2007), the concepts of their definition can be expanded to centralize upon plants rather than humans. Here then, the core plant microbiome comprises the set of genes present in a given habitat associated with a given plant (which can be further scaled based on plant phylogeny, phenology, etc.). This begs the question of defining the habitat, which is scaled over a range—from the whole organism (plant as an individual) to specific regions of the macroorganism (e.g. roots, leaves, shoots, flowers, and seeds), including out to zones of interaction between roots and the adhering/surrounding soil, the rhizosphere. The rhizosphere refers to the zone of influence created by the roots through their exudates and by the exudates of the microorganisms within the soil matrix (discussed in detail in the succeeding text) and is not merely the soil in contact with the roots (or other belowground structures of the plant); exudates effectively extend the functional boundary of the belowground plant–microbe interface. The plant microbiome can be further compartmentalized and subdivided to zoom in on and observe these interactions as information highways (Bais, Park, Weir, Callaway, & Vivanco, 2004) at the interface where plants and microbes exchange information. These habitats of the microbiome include the rhizosphere, as well as the plant organs (below- to aboveground structures) where microbes can interact in an associated, adherent (epiphytic), or internal (endophytic) manner.

Thinking about the soil–microbe–plant interface is not new—however, the concept of the plant microbiome perceives this interaction more along the lines of the 'microbe–soil–microbe–plant–microbe interface' rather than the 'soil–microbe–plant interface'. What is the difference? The difference is understanding that the microbiome is comprised of genomes vastly more complex than that of the plant alone, and by the nature of microbial interactions, these genomes serve as an extension of the plant's genetic compendium—effectively coined the plant's 'second genome' (Bernedsen, Pieterse, & Bakker, 2012).

Focusing on the belowground habitats within the plant microbiome, the subsections of bulk soil, rhizosphere, rhizoplane, epiphyte, and endophyte (Fig. 11.1) are addressed. Bulk soil technically refers to areas of soil not penetrated by roots. This region is beyond the plant's zone of influence through root exudation. A large body of research on plant root exudates has documented higher concentrations of organic compounds consistently reported in the rhizospheres and not found in bulk soils (reviewed by Jones, 1998). The rhizosphere is a key habitat documented to contain vast microbial diversity (Egamberdiyeva et al., 2008; Mendes et al., 2011), where soil functions as the medium in which complex signalling occurs among microbes and plants, accomplished by exudates, creating zones of interaction between roots and the adhering/surrounding soil. Influenced by climatic factors, the rhizosphere in turn impacts the plants and microbiota that utilize this habitat as an information highway (Bais et al., 2004). Moving closer into proximity with the plant, the next habitat is the rhizoplane, which refers to the surface of the plant tissues in contact with the soil (i.e. roots and rhizomes). Microbes that can exist in an adherent form to the plant tissues are termed epiphytes. Endophytes refer to the microbial genomes located inside plant tissues (Bulgarelli, Schlaeppi, Spaepen, Ver Loren van Themaat, & Schulze-Lefert, 2013). It is important to understand that microbial lifestyles are complex and many microorganisms are not restricted in their interactive potential, thus enabling them to exist as facultative epiphytes and endophytes, as dictated by other biotic and abiotic factors. The interrelation

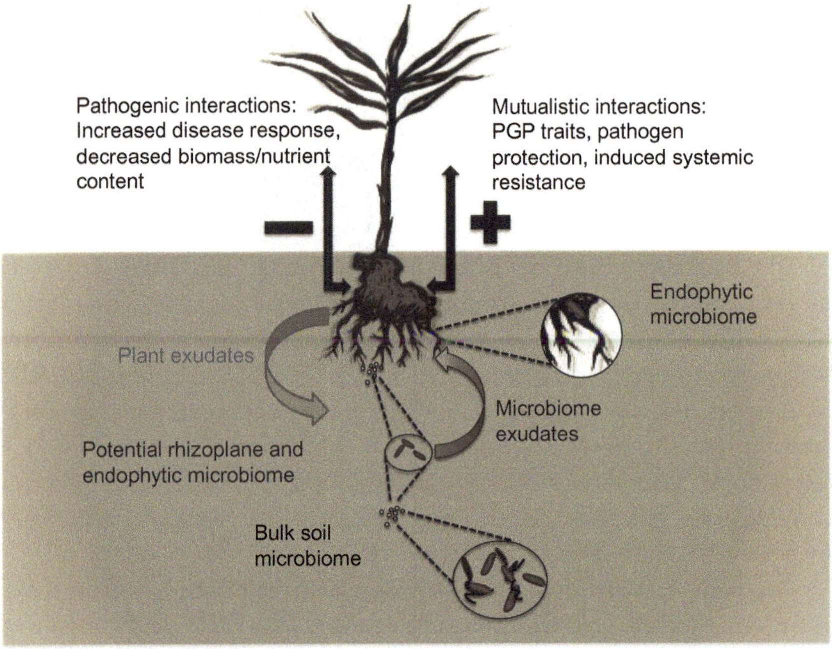

Figure 11.1 Model of the plant microbiome, with emphasis on the belowground habitats. The belowground regions of the plant microbiome include microorganisms inhabiting areas surrounding the plant roots at the root–soil interface or rhizosphere, those that are adherent to the root surface referred to as the rhizoplane, and those that colonize the internal root tissues that are known as endophytes. The microbial community residing in the bulk soil is primarily under the influence of environmental factors and is not under the direct influence of the plant (roots and exudates). (See colour plate.)

and importance of epiphytic and endophytic lifestyles to the plant microbiome is critical and thus is explored in detail herein.

Due to this interplay, the rhizosphere–rhizoplane is a dynamic environment. Microbiome structure is both influenced by and has an influence on the rhizosphere, contributing to some of the major differences between rhizosphere and bulk soil. Immobile cations, such as phosphorus, potassium, and ammonium, will be quickly depleted in the rhizosphere, while more mobile ions can be restored (Neumann and Romheld, 2002). The pH of the rhizosphere can differ up to 2–3 units from bulk soil as a direct result of biological activity, which can likewise impact the relative solubilities of essential nutrients—for example, phosphorus occurs in soils most abundantly in insoluble inorganic forms that can be solubilized through the actions of plants and microbes (Neumann and Romheld, 2002).

Extrapolating commonalities from current measurements of rhizosphere biota using structural assessments has been difficult to apply with any success across biogeographical contexts or over spatial and temporal fluxes. This suggests that local variations and adaptations of the rhizosphere biota are occurring, and evidence of this effect is demonstrated in invaded ecosystems where nutrient cycles are altered in a consistent but seemingly paradoxical way (Rout and Callaway, 2009, 2012). The implications of this are that the plant microbiome can influence ecosystem functions, increasing plant-available forms and soil stocks of carbon and nitrogen.

Ecosystem services are intricately linked to plant functional traits, of which several are likely mediated by microbes including soil formation, decomposition of organic matter, nutrient mineralization, and primary productivity (de Bello, Lavorel, Diaz, Harrington, & Cornelissen, 2010). The impact of the rhizosphere microbiome on plant productivity has not escaped those that are familiar with crops, where nodulating soybean (*Glycine max*) cultivars have been historically manipulated to enhance yield through alterations to their interactions with various *Rhizobium* microbial partners (Harris, Pacovsky, & Paul, 1985; Heath and Tiffin, 2009; Kiers and Denison, 2008). There are many examples of microbially mediated plant-growth-promoting (PGP) activity in the literature. The PGP activities that many rhizosphere-dwelling prokaryotes provide to plants include nitrogen (N_2) fixation (James, 2000; Martinez-Romero, 2006), phosphate solubilization, and production of plant-growth hormones (reviewed by Hardoim, van Overbeek, & van Elsas, 2008).

Clearly, any contribution from the microbiome to plant hormone signalling pathways has huge implications for plant genomics. Changes in the plant microbiome ecosystem that confer novel functionality may be closely associated with macroevolutionary events in the host, such as genome duplication (polyploidization). Whether from a macro (plant-centric) or a micro (microbiome-centric) perspective, the plant microbiome can exert influences on plant trait expression through top-down and bottom-up interactions. To understand the dynamics within ecosystem function and the regulation of plant trait expression accounting for plant performance, we must study the impacts of the microbiome. Herein, the plant microbiome is defined and discussed in relation to the currency of the complex communication pathways that occur within it, the ecology of the habitats of the plant microbiome with emphasis on those occurring belowground, and the importance of the plant microbiome to plant genomics.

2. CURRENCY OF THE MICROBIOME: EXUDATES

Exudates in the soil rhizosphere are defined here as those chemicals released from plants (primarily root cap cells) or from microbes. Through exudates, plants are able to communicate with microorganisms to elicit assistance in environmental acclimation to alleviate stresses, such as pathogen attack, drought-limiting nutrient acquisition, and metal toxicity to name a few. Microbes benefit from plant exudates by using them as a resource, in many cases carbon but also other nutrients. Through the exudate currency, the plant microbiome serves as an extension of the plant genome. A discussion of plant and microbial exudate uptake/release and impacts on plant and microbial functions are addressed.

2.1. Plant uptake and release

Plant exudate structure and constituents can be quite varied and often plant-species-specific and include high-molecular-weight molecules (e.g. sugars, lipids, and proteins) and low-molecular-weight signalling molecules (e.g. organic and amino acids) (Badri and Vivanco, 2009). Sugars and amino acids comprise the majority of plant root exudates (Jaeger, Lindow, Miller, Clark, & Firestone, 1999), and they serve a variety of functions including as antimicrobials, allelopathic molecules, and pathogen/herbivore defences. Many of these exudates also serve as an energy source for the microbiome; prokaryotes can utilize many different sources of carbon, including many of the plant exudates. For example, in the invasive grass *Sorghum halepense*, the exudate sorgoleone is the dominant chemical exuded from root hairs (Czarnota, Paul, Dayan, Nimbal, & Weston, 2001; Kagan, Rimando, & Dyan, 2003) and has well-known allelopathic properties (Czarnota, Rimando, & Weston, 2003; Nielsen and Moller, 1999). Further work using mineralization kinetics confirmed that rhizosphere microbes in invaded soils were capable of utilizing this allelopathic molecule as a carbon source (Gimsing et al., 2009). The multiple functions of this exudate reiterate the importance it serves as multiple currencies of the plant microbiome. Studies of the feedbacks between the influence of the microbiome and plant production of this and other allelochemical exudates are important, yet currently unknown, aspects of the functional contribution of the plant microbiome.

Roots function to uptake nutrients and chemical signalling molecules from the rhizosphere while simultaneously depositing nutrients and chemical signalling molecules into this same space. These exudates serve as a chemical currency to the plant, in turn providing a wide range of services to other components of the plant microbiome (e.g. as a nutrient substrate, a signalling molecule eliciting protective response, or other microbiome ecosystem service). The chemical components released by plants often are unique to a genus or family (Lesuffleur, Paynel, Bataille, & Cliquet, 2007) and can range in concentration and composition due to a wide variety of factors (Carvalhais et al., 2011; Matilla et al., 2010). Terpenoids, flavonoids, and isoflavonoids comprise many of the plant's antimicrobial defences (Hardoim et al., 2008). Isoprenoids are the most diverse group of plant metabolitic compounds both structurally and functionally. As primary metabolites, they regulate cellular processes. For example, the biological molecules that are the end result of isoprenoid metabolites are essential aspects of photosynthesis (as phytopigments) and seed emergence (as gibberellic and abscisic acids). As secondary metabolites, they provide a range of benefits to plants in the form of defences from pathogens and niche expansion through allelopathic molecules.

Environmental factors influencing root exudate composition and quantity can include elevated levels of CO_2, drought, and nutrient deprivation (particularly nitrogen and phosphorus). All of these factors can have profound impacts on the phytochemistry released as exudates. For example, increased carbon allocation to roots has been documented in CO_2-fertilization experiments, resulting in shifts in exudate composition and concentration that varied according to plant species (Cheng and Gershenson, 2007; Phillips, Fow, & Six, 2006). These species-specific impacts can result in increased yield, in no net productivity increase, or can be detrimental to plant growth and production. An example of this was shown in positive biomass responses in rye and clover to CO_2 fertilization, while maize showed no net biomass benefit (Phillips et al., 2006). However, maize demonstrated increased exudation of several amino acids under CO_2 fertilization. These findings are not surprising considering that the C4 photosynthetic pathway facilitates growth under high levels of CO_2; however, the impacts of increased release of amino acids into the rhizosphere by the C4 grass (maize) might play a role in a multitude of feedbacks between other plants and microbes (Bever, 1994; Klironomos, 2002). The extent of this latter aspect is largely unknown and is a major reason why many scientists are studying the plant microbiome.

Not all environmental stressors elicit the same responses from plant roots, and nutrient deprivation has been shown to have an inhibitory effect on root

Increased root development, and hence increased nutrient uptake, has been documented after inoculation of wheat with rhizosphere bacteria expressing ACC deaminase activity (Shaharoona, Naveed, Arshad, & Zahir, 2008). The mode of action of this enzyme is suppression of ethylene production in growing roots by catalysing the cleavage of ACC, the immediate precursor of ethylene in plants, to α-ketobutyrate and ammonia (Honma and Shimomura, 1978). The bacteria capable of producing ACC deaminase essentially shunt plant ACC away from synthesis in the form of ethylene, thereby minimizing impacts of various environmental stresses, which typically trigger increased ethylene production (Glick et al., 2007; Hardoim et al., 2008). The interplay in hormone signalling between the plant and the microbiome will be further explored in the following sections where the ecology of the microbiome is dissected.

Microbial exudates comprise much of the basis of modern antibiotics and antifungals. The excretion of antimicrobial substances is a common feature found in prokaryotic and eukaryotic microorganisms, and only a miniscule fraction of these organisms are cultivable (Piel, 2011), making the development of metatranscriptomics and metabolomics research more imperative. It might be that the complexity of these antimicrobial interactions within the microbiome actually maintains the astounding diversity previously documented in soils (Buée et al., 2009; Curtis, Sloan, & Scannell, 2002; Torsvik, Goksøyr, & Daae, 1990; Torsvik, Øvreås, & Thingstad, 2002). The complexity of these antimicrobial interactions was demonstrated by a research that found that nearly 35% of *E. coli* strains produce the antimicrobial compound known as colcin. Findings indicated that most tested strains were sensitive to at least one form of colcin, multiple resistance was found to be common among strains, and nearly one-fourth of the strains showed total resistance to all compounds tested (Achtman et al., 1983; Riley, 2011; Riley and Gordon, 1992). These findings support the hypothesis that antimicrobial interactions within microbial communities serve to maintain diversity; this idea was developed using simulation models (Czaran, Hoekstra, & Pagie, 2002). In addition to exuding antimicrobials that assist in plant immunity, microbes also exude low-molecular-weight effector molecules detectible to plants through what are known as pathogen- or microbe-associated molecular patterns in plants that trigger the immune response (Boller and He, 2009). Thus, microbial exudates in the rhizosphere can act directly on other microbes within the microbiome in suppressive mechanisms or can act directly on the plant to stimulate immune responses, often triggering plant exudation.

As our ability to detect and quantify these microbial exudates in the soil matrix improves, we are discovering multiple functions of the same microbiological substances. This should not be surprising considering that functional redundancy is a hallmark of soil microbial communities, one that again is demonstrated by the nitrogen cycle in soils (e.g. the range of bacterial taxa that perform N_2 fixation and denitrification in rhizospheres). The multiplicity of function is a complementary aspect of functional redundancy in microbial communities.

2.3. Impact on plant functions

The microbiome of plant roots was best summarized with the analogous reference to the microbiome of the human gut; both provide similar functional roles to their respective hosts (Bernedsen et al., 2012). These roles include nutrient uptake (Van der Heijden, Bardgett, & van Straalen, 2008), protection from pathogens (Doornbos et al., 2012), and regulation of host immunity (Neal, Ahmad, Gordon-Weeks, & Ton, 2012; Neal and Ton, 2013; Van der Ent et al., 2009). There are also costs to the plant to participate in nutrient exchange and exudate communication, and these logically vary depending upon the molecule and the energy required for the plant (alone or through help from the microbiome) to acquire and/or release exudate currency. High-molecular-weight compounds like those documented in many plant phenolic exudates require active transport mechanisms to cross the plasma membrane and gain release into the rhizosphere (Badri and Vivanco, 2009). One of the primary active transport systems, the ATP-binding cassette transporter, seems to play an important role in root exudation composition and concentration (Badri et al., 2009). Low-molecular-weight molecules can be released through membrane diffusion as in the case of amino acids or through protein channels (Badri and Vivanco, 2009; Jones and Dangle, 2006). There are many factors influencing plant exudation; thus, estimating the costs to the plant becomes both impossible and irrelevant outside of the context of a given system. Regardless, it is clear that plants can benefit from interactions with the microbiome through the direct effects of diminished pathogenic and/or enhanced mutualistic interactions and indirect effects manifest through alterations of ecological and ecosystem processes (i.e. enhanced nutrient availability).

Plants are often the beneficiaries of antimicrobial exudates provided from the microbiome. Recent research has indicated that plants can actively construct the rhizosphere microbiome and that this community is, at least to some degree, regulated or recruited by the plant to serve as protection from pathogens (Bernedsen et al., 2012; Friesen et al., 2011). An example of this is found in the bacterial production of enzymes that degrade *N*-acyl-L-homoserine lactones (AHLs), commonly found among several prokaryotic genera, which inhibit quorum sensing. Research has shown that plants were capable of recruiting beneficial bacteria expressing high levels of AHL-degrading enzymes when exposed to a pathogen, thus suppressing virulence gene expression (Reading and Sperandio, 2006). Plants also recruit activity within the microbiome specifically to stimulate AHL degradation (Teplitski, Robinson, & Bauer, 2000). Another example is demonstrated by fluorescent pseudomonads, many of which are capable of producing the antimicrobials 2,4-diacetylphloroglucinol (2,4-DAPG) and derivatives of phenazine (Phz). These bacteria are common to the rhizosphere of a diverse array of plant species (Mavrodi, Mavrodi, Parejko, Weller, & Thomashow, 2011). Both antimicrobials are broad spectrum and provide protection against a wide range of plant pathogens, many of which are fungal (Raaijmakers, Paulitz, Steinberg, Alabouvette, & Moenne-Loccoz, 2009). Pathogen success is a function of the soil microbial community, and in some systems, pathogen suppression can persists for extended lengths of time. This suppressive activity can be induced through recruitment of microbes that secrete valuable antimicrobials. The build-up of 2,4-DAPG in the rhizosphere has been observed in many ecosystems and is correlated with the suppression of the disease known as take-all in wheat (Mendes et al., 2011; Raaijmakers et al., 2009; Weller, Raaijmakers, Gardener, & Thomashow, 2002). Interestingly, recent research suggests that 2,4-DAPG and Phz derivatives also serve to reduce mineral content in the rhizosphere, perform plant regulatory and signalling functions that include alterations to exudate profiles, and play a role in the induction of plant systemic resistance (Doornbos et al., 2012; Matilla, Espinosa-Urgel, Rodriguez-Herva, Ramos, & Ramos-Gonzalez, 2007; Mavrodi et al., 2011).

In addition to suppression of pathogens, plant trait expression can also be a function of the interplay with the microbiome. Microbes mediate plant functional traits by providing novel biochemical capabilities and through altering existing plant pathways. Compared to the plant hosts, microbes have a far more diverse metabolic library, enabling them to synthesize biologically active chemicals that can mimic those produced by plants (e.g. hormones), or are totally novel to plants (e.g. specific antimicrobials). All plant hormones currently known can be produced by microbes (Friesen et al., 2011). This ability allows microbial communities to alter plant physiological pathways by producing or manipulating phytohormones. For example, production of IAA has been reported in nearly 80% of bacterial taxa in plant rhizospheres (Loper and Schroth, 1986). Low concentrations of IAA promote root growth in many plants (Glick, 1999; Patton and Glick, 1996); high concentrations of IAA inhibit plant growth and cause developmental perturbations typical of plant pathogenic bacteria (Sarwar and Kremer, 1995). A support for this variation in IAA secretion has been documented for members of the microbiome associated with the invasive grass, *S. halepense*, where prokaryotes most closely related to known plant pathogens secreted high concentrations of IAA *in situ* compared to lower levels found in prokaryotes not closely related to known pathogens (Rout, Chrzanowski, DeLuca, et al., 2013). Fluctuations in the plant's need for increased or decreased levels of this hormone could also be driven by climate stresses and plant phenology. Selection pressures favour those PGP members of the microbiome that can influence plant traits in ways that increase abiotic stress tolerance and/or plant performance to ensure plant persistence, and empirical evidence supports this (as reviewed by Friesen et al., 2011; Kaplan et al., 2013). PGPB that contain and express multiple genes important for maintaining the plant–microbial association have been called 'competent' (Hardoim et al., 2008). This idea is similar to the 'dual trait' phenomenon where PGPB were more effective at promoting plant growth when they were capable of dual growth-promoting traits (i.e. phosphate solubilization and ACC deaminase production, Baig, Arshad, Shaharoona, Khalid, & Ahmed, 2012). Here again is an example of multiplicity of microbial function as a complementary aspect of functional redundancy.

2.4. Impact on bacterial functions

The preference for organic and amino acids (demonstrated as chemotaxis) is a common phenomenon among rhizosphere microbes (Nelson, 2004). Root exudates are known to induce a range of phenotypic expression in rhizosphere-dwelling bacteria and include chemotaxis and stress tolerance (Amador, Canosa, Govantes, & Santero, 2010), polychlorinated biphenyl degradation (Toussaint, Pham, Barriault, & Sylvestre, 2012), modulation of genes involved in competence and sporulation (Mader et al., 2002), and biofilm formation on plant roots (Rudrappa, Czymmek, Pare, & Bais, 2008). Root exudates (specifically, amino acids) are intimately involved in biofilm formation and disassembly of rhizosphere-dwelling bacteria (Kolodkin-Gal et al., 2010). The surface of roots and admission into the internal structures of plants are protected from many microbial inhabitants by plant exudates, such as terpenoids, flavonoids, and isoflavonoids

(Hardoim et al., 2008). Whether living adherent to roots (epiphytes) or internal to cellular structures (endophytes), these microbiome symbionts are able to persist for some period of their life cycle in these intimate relationships and include not only successful pathogens but also those that can typically promote plant growth through a variety of mechanisms (including hormone signalling).

3. ECOLOGY OF THE MICROBIOME

The aspects of the plant microbiome covered in this overview are restricted to the rhizosphere, rhizoplane (epiphytes), and internal endosymbionts (endophytes) of the belowground organs of the plant, to address the importance of the soil-associated plant microbiome to plant trait expression and ecosystem functions. However, to do so admittedly ignores the impacts of the rhizosphere microbiome on the aboveground interactions, including protection from herbivory, pollination, and seed predation, as well as pathogen attack from aboveground structures (see the review by Friesen et al., 2011). Clearly, all of these aboveground interactions exert an influence on the plant genome, and a complete integration of the 'microbe–soil–microbe–plant–microbe' microbiome will certainly need to be expanded to include the various pollinators (and potentially their microbiomes) for the complete biochemical network map. However, since the 'black box' concept of soils still persists, a key to dissecting the interactions in the rhizosphere will be through a deeper understanding of the plant microbiome that is in contact with this region of soil. Many factors driving plant microbiome structure are a function of the currency in this system—exudates. Three factors influence the currency; not surprisingly, these are three components of the system: soils, microbes, and plants. However, separating these three as independent aspects is both incomplete and unlikely to discover the underlying organization or functional hierarchy within the plant microbiome.

3.1. Rhizosphere and rhizoplane

Defining the microbial community of the rhizosphere in any form of a static or consistent description is impossible. While we are struggling to increase our understanding of the complexity of the rhizosphere microbiome through the use of 'next-generation' and now 'third-generation' technologies, we have known for quite some time that the rhizosphere influences plant health (Hiltner, 1904). The mechanisms enabling this interaction are only now beginning to unfold, thanks to modern advances in molecular science permitting mining of environmental DNA for functional genes of interest. The magnitude of the differences in rhizosphere soils compared to bulk soils is largely a function of abiotic and biotic stresses influenced by climate. Thus, as stresses/disturbances increase at the macroscale (e.g. drought, plant invasion, pathogenic attack, and metal toxicities), these differences are manifested at the microscale and are demonstrated in the magnitude of difference in microbial community composition between these zones. This phenomenon has been repeatedly demonstrated in arid environments, where low microbial abundance and diversity are typical of bulk soils and rhizosphere soils contain abundantly diverse microbial communities (Aguirre-Garrido et al., 2012; Ben-David, Zaady, Sher, & Nejidat, 2011; Kaplan et al., 2013; Yu, Grishkan, & Steinbrener, 2012). In general, rhizosphere soils have higher water-holding capacity, increased nutrient availability, and greater microbial biomass compared to bulk soils (Schade and Hobbie, 2005). Spatiotemporal shifts have been documented in the rhizosphere microbiome (Houlden, Timms-Wilson, Day, & Bailey, 2008; Kaplan et al., 2013), but how much of this is driven by direct effects of abiotic impacts (seasonal stochasticity) and indirect effects of abiotic stressors as exhibited by biotic interactions (e.g. plant life stage and plant community structure) remains tightly entangled.

Similar to the function of the gut, the rhizosphere region of the plant microbiome is informative about plant health, at both the individual and community level (particularly in monoculture crops), where a healthy versus diseased state of the plant community can be reflected in the composition of the rhizosphere microbiome (Burdon and Thrall, 2009). The rhizosphere and rhizoplane microbiomes are being tapped for a wide range of services (e.g. crops, bioremediation, and energy) to maximize quality, productivity, and sustainability. It is important to remember the context in which these microbes are interacting with the plant and surrounding members of the microbiome and that this interaction is where ecosystem functions become expressed, rather than being an intrinsic property of a specific, or suite of, microbe(s). For example, some members of the microbiome (particularly those classified as PGPB) stimulate induced systemic resistance (ISR) in plants, marked by priming of jasmonic acid-inducible genes in leaves (Van Wees, Van der Ent, & Pieterse, 2008), effectively stimulating immunity conferring resistance to a broad range of pathogens (Pineda, Zheng, van Loon, Pieterse, & Dicke, 2010; Van der Ent et al., 2009; Van Oosten et al., 2008). In turn, plants with activated ISR display increased or enhanced defence signalling aboveground (Ahmad et al., 2011) and belowground (Neal et al., 2012; Neal and Ton, 2013). A network of communication begins to develop in the rhizosphere microbiome, and three-way interactions between multiple microbial partners and plants are as simple as we can hope for in the microbiome. As the largest reservoir of biological diversity currently known (Bernedsen et al., 2012; Curtis et al., 2002), the rhizosphere microbiome should pique the interest of anyone interested in venturing into the unknown to solve problems relevant to a changing climate, including bioremediation, maximization of crop yields (increased efficiency and sustainability), enhanced soil fertility of marginal lands, and discovery of novel antimicrobials, to name but a few.

3.2. Epiphytes and endophytes

The root epiphyte and endophyte communities can differ widely from those in the rhizosphere, which supports the concept of coevolution among host plant and microbial symbionts (Boller and He, 2009; Compant, Clement, & Sessitsch, 2010; Compant, Duffy, Nowak, Clement, & Barka, 2005), in which the plant engages in recognition and selection of microbiomes that promote a homeostatic relationship with the plant. The ability of the microbial members of the plant microbiome to colonize the plant tissues, whether in an adherent (epiphytic) or internal (endophytic) manner, is another complexity that we are beginning to catch glimpses of with molecular methods like metabolomics and metatranscriptomics.

The colonization process that occurs between microbes and plant hosts involves exudate communication through microbe–microbe signalling interactions (Deakin and Broughton, 2009; Elasri et al., 2001) in addition to those between microbes and plants (Boller and He, 2009; Deakin and Broughton, 2009; Friesen et al., 2011; Hardoim et al., 2008). The ability of a microbial partner to colonize the external root surfaces appears to be, at least to some degree, a coevolved communication exchange. For example, secondary metabolite root exudates in maize that were released due to an ISR response were responsible for the recruitment of the PGPB *P. putida*, based on chemotaxis preferences (Neal et al., 2012). Supporting this, genes for chemotaxis have been identified in a wide range of microbes inhabiting the rhizosphere, and expression of these genes is an essential step towards adherence and colonization of the root surface (Hardoim et al., 2008). Here again, a plant-mediated response is apparent as demonstrated by different plant chemical exudates used to recruit beneficial PGPB; in tomato, organic acids are the major chemotactic agent (de Weert et al., 2002), while amino acids serve this function in rice (Bacilio-Jimenez et al., 2003). Several bacterial genes enabling establishment on plant roots have been identified and include those that code for the type IV pilus and twitching motility (Bohm, Hurek, & Reinhold-Hurek, 2007), isoflavonoid efflux pump (Palumbo, Kado, & Phillips, 1998), and DNA rearrangements affecting colony aggregation (Dekkers, Phoelich, van der Fits, & Lugtenberg, 1998).

One critical question at the crux of the communication pathways of the plant microbiome is how the plant detects a microbial mutualist from pathogen. This involves complex signalling between plants and microbes that trigger plant immune responses, the subject of several reviews (see Deakin and Broughton, 2009; Hardoim et al., 2008; or the May 2009 special issue in *Science* for starting material). Once adherent to the roots, biofilm and microcolony development proceeds (Compant et al., 2008; James et al., 2002), which potentially serve as gateways into internal colonization of the plant in an endophytic life stage (Hardoim et al., 2008).

The microbiome located inside plant tissues reflects the microorganisms living in an endophytic lifestyle (Bulgarelli et al., 2013). Many plant species have been shown to harbour endophytes, and these include domesticated crops (Compant et al., 2008; Hallmann, Quadt-Hallmann, Mahaffee, & Kloepper, 1997; James et al., 2002), and wild cultivars, including invasive species (Rout and Chrzanowski, 2009). Beyond a few well-known endophytic microbe–plant interactions (e.g. arbuscular mycorrhizal fungi and nodule-forming N_2-fixing bacteria in legumes), the importance of these endosymbionts to the plant microbiome is largely unknown. It should be noted that the importance of N_2-fixing bacterial endophytes in crop grasses of sugar cane and sorghum has been investigated for decades (Baldani et al., 1996; James, Olivares, Baldani, & Dobreiner, 1997; Kirchof et al., 2001).

Microbiome components demonstrating detectible plant-growth-promoting genes are of considerable interest to biotechnology and agriculture, where the motivating hypothesis is that multiple genes involved in establishment, persistence, and thriving of plant-beneficial microbes will be in higher abundance within these organisms. Some support for this is observed in phosphate-solubilizing *Bacillus* strains, where their effectiveness at plant-growth promotion was additive when the strain had the additional ability to produce the enzyme ACC deaminase (Baig et al., 2012). Other phosphate-solubilizing bacteria act in concert with mycorrhizal fungi to enhance plant-growth promotion (Zaidi and Khan, 2005), and while not a dual trait from an individual organismic level, this does function as a dual trait from the microbiome perspective.

In general, putative epiphytic and endophytic microbes have to communicate with the plant for a series of interactions with the plant hosts that enable their establishment, persistence, and thriving. At the rhizosphere and bulk soil level, this is largely driven by soil factors, expanding to macroscale environmental inputs, such as climatic factors. On a microscale, this involves currency exchanges among plant and microbial exudates. Persistence is a function of timing of colonization. Therefore, it is not surprising that plant phenology is correlated with endophyte microbiome composition shifts (van Overbeek and van Elsas, 2008), and achieving persistence on or in the plant heavily relies upon colonization and compatibility (Hardoim et al., 2008). Shifts in endophyte communities as a function of plant developmental stage is a likely factor driving recruitment of PGPB that demonstrate high ACC deaminase activity to respond to high levels of ethylene in the plant, whether that be due to aspects of plant phenology (e.g. seedling emergence or fruit ripening) or as a response to abiotic stresses like high salinity (Cheng and Glick, 2007) and heavy metal toxicity (Zang et al., 2011). These findings support the viewpoint of others that the plant–endophyte relationship is coevolutionary (Boller and He, 2009; Compant et al., 2010, 2005) and suggest that cross-communication of plant developmental stage is conveyed through exudates. Potentially, these interactions achieve homeostasis. Simply stated, this is when the interaction leans more towards mutualism on the spectrum of symbiosis rather than towards parasitism. One confounding influence that epiphytes and endophytes can express on the plant genome is manifested through horizontal and vertical transmission of the symbionts.

The majority of known plant endophytes and epiphytes are horizontally transmitted (Friesen et al., 2011). This creates a potential fitness conflict where antagonistic coevolution of functional trait expression likely arises between the plant and microbial symbionts, where one downstream impact on functionality might be expressed as an altered ecosystem service. Horizontal transmission enables host-to-host transfer of endosymbionts without involvement of plant sexual reproduction. In turn, this often, but certainly not always, involves pathogenic interactions, and plants are challenged to coevolve mechanisms facilitating beneficial trait expression from these endosymbionts (Bever, Richardson, Lawrence, Holmes, & Watson, 2009; Kiers and Denison, 2008). This brings to mind the evolutionary shifts along the mutualism–parasitism trajectory, where horizontal transmission of endophytes might be a coevolutionary mechanism contributing to this range of interactions.

Vertical transmission of endophytes relies upon host fitness and is characterized in fungi as those demonstrating an asymptomatic lifestyle in the plant host (Clay and Schardl, 2002). Since this strategy involves the transfer of endosymbionts through sexual reproduction of the host, an increased likelihood for coevolutionary mechanisms between these partners exists. Some have suggested that vertically transmitted endophytes confer increased host benefits over those horizontally transmitted, and there is some evidence supporting this idea (Clay and Schardl, 2002; Sachs, Mueller, Wilcox, & Bull, 2004). However, it is unlikely that a generalizable rule can be extrapolated to the mode of transmission that will be applicable in a broad ecological sense, as a wide range of factors are likely to influence endophyte transmission, including host evolutionary relationship. For example, the invasive grass *S. halepense* is a relatively newly evolved polyploid hybrid (Feltus et al., 2004) that can transmit at least some of the N_2-fixing bacterial endophytes within the rhizomes both vertically and horizontally (Rout, Chrzanowski, DeLuca, et al., 2013; Rout, Chrzanowski, Smith, et al., 2013), indicating that this successful invader might be capable of regulating potential pathogenic responses expressed by these endosymbionts. Ecological theory suggests that horizontally transmitted organisms achieve maximal transmission when available hosts are within close proximity. Support for this theory has been shown, where horizontally transmitted endophytes were positively correlated with plant density dependence, while vertically transmitted endophytes did not show this trend (Rudgers et al., 2009). A hallmark of plant invasions is increased density, which often approaches near monocultures (Tilman, 2000). Thus, it is not surprising that endophytes of the *S. halepense* microbiome can be horizontally transmitted through plant rhizomes. This strategy will likely be a common one utilized by other invasive plants, particularly ones that clonally reproduce through rhizomes, as these organs have been shown to harbour endophytes in a wide range of plants (Baldani et al., 1996; James et al., 1997; Kirchof et al., 2001; Rout and Chrzanowski, 2009). A comprehensive and holistic consideration of the plant microbiome interface will be required to determine plant and microbial trait expression, genetics, and outcomes of their impacts on ecological functions.

4. IMPORTANCE OF THE MICROBIOME TO PLANT GENOMICS

A main goal of plant genomics is to uncover the mechanisms that regulate plant trait expression, ultimately to enhance plant performance.

Impacts of the plant microbiome community structure are clearly manifest in plant trait expression. For example, the plant microbiome influences plant root architecture (Morris and Djordjevic, 2006; Spaepen, Dobbelaere, Croonenborghs, & Vanderleyden, 2008), which should not be surprising given the repertoire of microbial phytohormones that the plant can elicit from the microbiome. Alterations to root architecture have important implications for overall plant health. Take, for an example, border cells in roots; increased numbers of border cells conferred increased resistance to fungal pathogen infection (Chen, Chen, & Wu, 2012), likely due to the function of border cells in extracellular trapping of microbes that has been shown to provide a defence for the root tip (Curlango-Rivera et al., 2013). Influencing plant architecture through enhancing root growth is a primary motivation behind mining the microbiome for this PGP trait, which is of interest to agribusiness. Enhanced root (and rhizome) growth is also of great ecological significance, since many successful invasive plants express increased belowground biomass in their nonnative ranges, and this serves as a primary means of niche expansion. Thus, it is possible that this ability confers an increased probability of invasiveness. Some evidence supports this notion, linking the microbiome to belowground growth in one of the world's worst invaders (Rout, Chrzanowski, DeLuca, et al., 2013; Rout, Chrzanowski, Smith, & Gough, 2013). Suppression of the invasive grass microbiome with antibiotics reduced belowground (rhizome) growth (over fivefold declines) compared to growth with an unsuppressed microbiome. The precise role(s) of the plant and the microbial contributions to the microbiome will be revealed in this system, and others like it, with the use of advances in molecular techniques, particularly the use of RNA-Seq allowing for separation of prokaryotic from eukaryotic mRNA in metagenomic samples. To what extent the microbiome drives traits associated with invasive plants and what generalities might be extrapolated across broad phylogenetic or biogeographic ranges remain to be determined. Nevertheless, the plant microbiome has been implicated in the mediation of over a dozen plant functional traits including root, vegetative, regenerative, stem, and leaf traits (reviewed by Friesen et al., 2011), so it follows that extremes in plant trait expression observed in the majority of invasive plants, including traits of physiology, leaf-area allocation, shoot allocation, growth rate, fitness, and biomass (detailed in the meta-analysis by van Kleunen, Weber, & Fischer, 2010), likely involve help from the microbiome.

The microbial members within the plant microbiome cause fundamental differences in expression of plant phenotypes, due to their influence on plant functional traits. In turn, the plant microbiome also influences ecosystem functioning through an indirect mechanism, the plant as a community. The influence on the expression of individual functional traits impacts population dynamics, particularly for sessile organisms such as plants, best demonstrated through pathogen spread documented in wheat and rye (Burdon and Thrall, 2009). Besides assisting plants in niche expansion (or invasion spread) through enhanced root and rhizome production, the microbiome can impact plant ecology through influences exerted on plant–plant competition (Klironomos, 2002), pollination (Cahill, Elle, Smith, & Shore, 2008), herbivory, and defence (Friesen et al., 2011). Given that plant primary productivity is impacted by the availability of limiting nutrients and possessing the traits to enhance resource acquisition (Lambers, Raven, Shaver, & Smith, 2008), the role of the microbiome in soil biogeochemical cycles is a fundamental factor influencing plant productivity. Primary productivity is considered an aspect, or measurable outcome from a land management perspective, of ecosystem services.

It is important to keep in mind that broader ecological aspects must also be considered when trying to disentangle the interactions within the plant microbiome. Ecosystems where we can measure shifts in microbiome community structure and function are the precise areas of attention that need focus. Ecology will dictate the adaptive traits required of the PGP microbiomes that enable plant persistence. Plant-invaded communities will likely provide many exciting insights about the plant microbiome, including the influence of the microbiome on invasive plant traits and their role in profound alterations to nutrient cycling documented in many invaded ecosystems and (Liao et al., 2008). Extreme environments, such as deserts, are another area where plant microbiome research efforts have correlated mechanisms of the microbiome permitting the plant to tolerate drought stress (Kaplan et al., 2013). To what extent plant communities regulate and are regulated by their microbiomes is an area that needs to be explored and raises questions regarding important mechanisms for acclimating and adapting to disturbances or tolerating stress.

Perception of the microbiome as a component of the plant genome is certainly evident in an obligate endosymbiotic relationship, such as the most well-known and evolutionarily significant one between plants and chloroplasts. This relationship occurred, presumably, over long periods of time when the bacterial organism switched between expressions of facultative–obligate endophyte lifestyles. When the parasitism–mutualism continuum is viewed in this way, it becomes clearer how the microbiome can impact the plant genome without having to become a permanent component of the genome. We know that microbes of the rhizosphere can influence nearly every plant functional trait, resulting in impacts on plant trait variation of similar magnitude to those of the genotype (Friesen et al., 2011). Given the complexity of chemical communication capabilities within the microbiome, including microbial characteristics of multiple metabolic pathways for carbon utilization, the ability to utilize multiple terminal electron acceptors in addition to oxygen, and production of mimic plant phytohormones, the link between microbially mediated ecosystem process of nutrient cycling and plant productivity lies embedded in the rhizosphere matrix. Disentangling the plant–microbe communications in this matrix will enable us to determine the drivers of a vast array of desirable plant traits that will, potentially, be harnessed to address some of the many challenges currently facing our world, such as degradation of soil resources (Montgomery, 2007), depletion of water supplies (Rosegrant, Cai, & Cline, 2002), and losses of biodiversity (Pimentel, Lach, Zuniga, & Morrison, 2000) as examples. The human global population places many demands on the Earth beyond the need for increased food production, for example, an increased need to harvest more solar energy for fuel production (e.g. biofuel crops). This combination of events makes it likely that increased conversion of lands for agricultural uses will compete with human urbanization, a fact that we can already observe in most large U.S. cities. Balancing these two demands will inevitably require harnessing resources from fragile ecosystems or marginal landscapes (Tilman et al., 2009) while simultaneously maximizing plant productivity. The plant microbiome will likely play a prominent role towards achieving this requirement.

Clearly, the microbiome plays a huge role in plant function and performance. While microbiome analysis has only recently been widely recognized, it is now viewed as essential to understanding genetic and functional capacity otherwise attributed to host organisms, including important aspects of metabolism and physiology. High-throughput technologies of next- and third-generation sequencing will enable library construction of trait loci involved in plant phenotypic expression. Quantitative trait locus maps have been used in several plant species to identify genomic regions where genes responsible for a given trait of interest exist (e.g. ;Hu et al., 2003 Jang, Lemke, Tang, Bowers, & Paterson, 2008; Paterson, Schertz, Lin, Liu, & Chang, 1995). By knowing plant genes involved in phenotypic variation of a trait, the hormonal signalling pathways linked to trait expression can be tested for microbiome contributions to signalling cascades. The roles of microbiome in the regulation of plant hormone signalling cascades will also be influenced by abiotic and biotic stresses such as drought, nutrient limitation, and pathogen attack. More complex still is deciphering the extent to which the microbiome is under selection by the host plant. Specific plant attributes and environmental factors that increase the plant's ability to select a microbiome are currently lacking from most plant ecology studies, but we know that plants can select their microbiome communities through the exudate currency shared among the plant and its colonizers (Doornbos et al., 2012).

5. CONCLUSIONS

Strong evidence illustrates the importance of understanding the multitude of plant microbiome associations that contribute to plant plasticity in a given environment (Friesen et al., 2011). Recognition of the plant microbiome as an integrated aspect of the plant genome expands on the ecological concept of 'feedbacks' (Bever, 1994). Feedbacks were certainly an important concept that forged the path ahead for integrating the soil–plant–microbe matrix. The reciprocity of the effects that soil biota and plants exert over this interaction varies over time as a function of attenuation of microbiome members that span the spectrum from parasitic to mutualistic in their interactions with plants (reviewed by Callaway and Rout, 2011). Disproportional accumulation of microbiome parasites (expressed as pathogenic effects) leads to negative feedbacks, while the disproportional accumulation of microbiome mutualists leads to positive feedbacks (Klironomos, 2002). Increased knowledge about these interactions and how shifts in biodiversity impact functions in the context of ecosystem services (plant productivity, biogeochemical pools, and fluxes) will be a critical factor for elucidating plant microbiome growth and gene expression patterns. Many of these patterns likely exhibit species-specific or other phylogenetically based distributions among plants. Rapid microbial generation times and the prevalence of horizontal gene transfer provide potential mechanisms for the development of regional genetic differences, or ecotypes, to arise in response to the effects of local plant species and communities (Rout and Callaway, 2012). As the integration of the plant microbiome unfolds, a new approach is emerging that includes aspects of microbial ecology, microbiomes, and transcriptomes into plant genetics. This is certainly motivated by the vast diversity documented in the rhizosphere microbiome (Bernedsen et al., 2012; Curtis et al., 2002) and correlated with the functional redundancy of genes responsible for essential nutrient transformations, like those involved in N_2 fixation (Zehr et al., 2003), previously discussed in earlier sections. Selection favours the plants that can motivate/manipulate their microbiome in ways that favour plant persistence, particularly under a variety of stochastic disturbances (de Bello et al., 2010). Recent findings from many different plants in a wide range of ecosystems support this, as demonstrated by the ability of the plant to control the composition of the microbiome (reviewed by Bernedsen et al., 2012).

Understanding the regulation of such complex communication pathways within the plant microbiome involves detecting and quantifying the multiple functions of microbial and plant exudates and their impacts on gene transcription and translation. The holistic approach to understanding any organismic function and structure is to understand the organism in its entirety; the microbiome and its functional contribution are certainly integral for all higher organisms on the planet. This should not be a surprise. How the microbiome is influencing or being influenced by the plant will likely vary among species, as well as by environmental and genetic factors. Studies of the plant microbiome need to document microbial community structural and functional diversity and shifts in these metrics as a function of spatiotemporal changes associated with ecological habitats. Further development of functional screening that utilizes metagenomics and metatranscriptomics will lead us to predicting plant traits based on knowledge of the microbiome, in addition to knowing how and when this 'second genome' functions as an organ system of the plant.

REFERENCES

Achtman, M., Mercer, A., Kusecek, B., Pohl, A., Heuzenroeder, M., Aaronson, W., et al. (1983). Six widespread bacterial clones among *Escherichia coli* K1 isolates. *Infectious Immunology, 39*, 315–335.

Aguirre-Garrido, J. F., Montiel-Lugo, D., Hernandez-Rodriguez, F., Torres-Cortes, G., Millan, V., Toro, N., et al. (2012). Bacterial community structure in the rhizosphere of three cactus species from semi-arid highlands in central Mexico. *Antonie van Leeuwenhoek, 101*, 891–904.

Ahmad, S., Veyrat, N., Gordon-Weeks, R., Zhang, Y., Martin, J., Smart, L., et al. (2011). Benzoxazinoid metabolites regulate innate immunity against aphids and fungi in maize. *Plant Physiology, 157*, 317–327.

Amador, C. I., Canosa, I., Govantes, F., & Santero, G. (2010). Lack of CbrB in *Pseudomonas putida* affects not only amino acids metabolism but also different stress responses and biofilm development. *Environmental Microbiology, 12*, 1748–1761.

Bacilio-Jimenez, M., Aguilar-Flores, S., Ventura-Zapata, E., Perez-Campos, E., Bouquelet, S., & Zenteno, E. (2003). Chemical characterization of root exudates from rice (Oryza sativa) and their effects on the chemotactic response of endophytic bacteria. *Plant and Soil, 249*, 271–277.

Badri, D. V., Quintana, N., El Kassis, E. G., Kim, K., Choi, Y. H., et al. (2009). An ABC transporter mutation alters root exudation of phytochemicals that provoke an overhaul of natural soil microbiota. *Plant Physiology, 151*, 2006–2017.
Badri, D. V., & Vivanco, J. M. (2009). Regulation and function of root exudates. *Plant, Cell & Environment, 32*, 666–681.
Baig, K. S., Arshad, M., Shaharoona, B., Khalid, A., & Ahmed, I. (2012). Comparative effectiveness of *Bacillus* spp. possessing either dual or single growth-promoting traits for improving phosphorus uptake, growth and yield of wheat (*Triticum aestivum* L.). *Annals of Microbiology, 62*, 1109–1119.
Bais, H. P., Park, S.-W., Weir, T. L., Callaway, R. M., & Vivanco, J. M. (2004). How plants communicate using the underground information superhighway. *Trends in Plant Science, 9*, 26–32.
Baldani, J. I., Pot, B., Kirchof, G., Falsen, E., Baldani, V. L. D., Olivares, F. L., et al. (1996). Emended description of herbaspirillum; inclusion of [pseudomonas] rubrisubalbicans, a mild plant pathogen, as herbaspirillum rubrisubalbicans comb. Nov. And classification of a group of clinical isolates (ef group 1) as herbaspirillum species 3. *International Journal of Systematic Bacteriology, 46*, 802–810.
Ben-David, E. A., Zaady, E., Sher, Y., & Nejidat, A. (2011). Assessment of the spatial distribution of soil microbial communities in patchy arid and semi-arid landscapes of the Negev Desert using combined PLFA and DGGE analyses. *FEMS Microbiology Ecology, 76*, 492–503.
Bernedsen, R. L., Pieterse, C. M. J., & Bakker, A. H. M. (2012). The rhizosphere microbiome and plant health. *Trends in Plant Science, 17*, 478–486.
Bever, J. D. (1994). Feedback between plants and their soil communities in an old field community. *Ecology, 75*, 1965–1977.
Bever, J. D., Richardson, S. C., Lawrence, B. M., Holmes, J., & Watson, M. (2009). Preferential allocation to beneficial symbiont with spatial structure maintains mycorrhizal mutualism. *Ecology Letters, 12*, 13–21.
Bohm, M., Hurek, T., & Reinhold-Hurek, B. (2007). Twitching motility is essential for endophytic rice colonization by the N2-fixing endophyte Azoarcus sp. strain BH72. *Molecular Plant Microbe Interactions, 20*, 526–533.
Boller, T., & He, S. Y. (2009). Innate immunity in plants: An arms race between pattern recognition receptors in plants and effectors in microbial pathogens. *Science, 324*, 742–744.
Buée, M., Reich, M., Murat, C., Morin, E., Nilsson, R. H., Uroz, S., et al. (2009). 454 Pyrosequencing analyses of forest soils reveal an unexpectedly high fungal diversity. *New Phytologist, 184*, 844–856.
Bulgarelli, D., Schlaeppi, K., Spaepen, S., Ver Loren van Themaat, E., & Schulze-Lefert, P. (2013). Structure and functions of the bacterial microbiota of plants. *Annual Review of Plant Biology, 64*, 9.1–9.32.
Burdon, J. J., & Thrall, P. H. (2009). Coevolution of plants and their pathogens in natural habitats. *Science, 324*, 755–756.
Cahill, J. F., Elle, E., Smith, G. R., & Shore, B. H. (2008). Disruption of a belowground mutualism alters interactions between plants and their floral visitors. *Ecology, 89*, 791–801.
Callaway, R. M., & Rout, M. E. (2011). Soil biota and plant invasions: Biogeographic effects on plant-microbe interactions. In D. H. Richardson (Ed.), *Fifty years of invasion ecology: the legacy of Charles Elton* (pp. 131–142). West Sussex: Wiley-Blackwell.
Carvalhais, L. C., Dennis, P. G., Fedoseyenko, D., Hajirezaei, M. R., Borriss, R., & von Wiren, N. (2011). Root exudation of sugars, amino acids, and organic acids by maize as affected by nitrogen, phosphorus, potassium, and iron deficiency. *Journal of Plant Nutrition and Soil Science, 174*, 3–11.

James, E. K., Gyaneshwar, P., Mathan, N., Barraquio, W. L., Reddy, P. M., Iannetta, P. P., et al. (2002). Infection and colonization of rice seedlings by the plant growth-promoting bacterium Herbaspirillum seropedicae Z67. *Molecular Plant Microbe Interactions, 15*, 894–906.

James, E. K., Olivares, F. L., Baldani, J. I., & Dobreiner, J. (1997). Herbaspirillum, an endophytic diazotroph colonizing vascular tissue in leaves of sorghum bicolor. *Journal of Experimental Botany, 48*, 785–797.

Jang, C. S., Lemke, C., Tang, H., Bowers, J. E., & Paterson, A. H. (2008). Evolutionary fate of rhizome specific genes in a non-rhizomatous Sorghum genotype. *Heredity, 102*, 266–273.

Jones, D. L. (1998). Organic acids in the rhizosphere—A critical review. *Plant and Soil, 1*, 25–44.

Jones, J. D. G., & Dangle, J. L. (2006). The plant immune system. *Nature, 444*, 323–329.

Kagan, I. A., Rimando, A. M., & Dyan, F. E. (2003). Elucidation of the biosynthetic pathway of the allelochemical sorgoleone using retro biosynthetic NMR analysis. *Journal of Biological Chemistry, 278*, 28607–28611.

Kaplan, D., Maymon, M., Agapakis, C. M., Lee, A., Wang, A., Prigge, B. A., et al. (2013). A survey of the microbial community in the rhizosphere of two dominant shrubs of the Negev Desert highlands, Zygophyllum dumosum Boiss. and Atriplex halimus, using cultivation-dependent and -independent methods. *American Journal of Botany, 100*, 1713–1725.

Kielland, K. (1994). Amino acid absorption by Arctic plants: Implications for plant nutrition and nitrogen cycling. *Ecology, 75*, 2373–2383.

Kiers, E. T., & Denison, R. F. (2008). Sanctions, cooperation, and the stability of plant-rhizosphere mutualisms. *Annual Reviews of Ecology, Evolution & Systematics, 39*, 215–236.

Kirchof, G., Eckert, B., Stoffels, M., Baldani, J. I., Reis, V., & Hartman, A. (2001). Herbaspirillum frisingense sp. Nov. A new nitrogen-fixing bacterial species that occurs in c4-fibre plants. *International Journal of Systematic and Evolutionary Microbiology, 51*, 157–168.

Klironomos, J. N. (2002). Feedback with soil biota contributes to plant rarity and invasiveness in communities. *Nature, 417*, 67–70.

Koeth, R. A., Wang, Z., Levison, B. S., Buffa, J. A., Org, E., Sheehy, B. T., et al. (2013). Intestinal microbiota metabolism of L-carnitine, a nutrient in red meat, promotes atherosclerosis. *Nature Medicine, 19*, 576–585. http://dx.doi.org/10.1038/nm.3145.

Kolodkin-Gal, I., Romero, D., Cao, S., Clardy, J., Kolter, R., & Losick, R. (2010). D-Amino acids trigger biofilm disassembly. *Science, 328*, 627–629.

Lambers, H., Raven, J. A., Shaver, G. R., & Smith, S. E. (2008). Plant nutrient-acquisition strategies change with soil age. *Trends Ecology & Evolution, 23*, 95–103.

Lesuffleur, F., Paynel, F., Bataille, M. P., & Cliquet, J. B. (2007). Root amino acid exudation: Measurement of high efflux rates of glycine and serine from six different plant species. *Plant and Soil, 294*, 235–246.

Liao, E., Peng, R., Luo, Y., Zhou, X., Wu, X., Fang, C., et al. (2008). Altered ecosystem carbon and nitrogen cycles by plant invasion: A metaanalysis. *New Phytologist, 177*, 706–714.

Loper, J. E., & Schroth, M. N. (1986). Influence of bacterial source of indole-3-acetic acid on root elongation of sugar beet. *Phytopathology, 76*, 386–389.

Mader, U., Antelmann, H., Buder, T., Dahl, M. K., Hecker, M., & Homuth, G. (2002). Bacillus subtilis functional genomics: Genome-wide analysis of the DegS-DegU regulon by transcriptomics and proteomics. *Molecular Genetics Genomics, 268*, 455–467.

Martinez-Romero, E. (2006). Dinitrogen-fixing prokaryotes. In M. Dworkin, S. Falkow, E. Rosenberg, K. H. Schleifer, & E. Stackebrandt (Eds.), *The Prokaryotes* (pp. 793–817). New York: Springer.

Matilla, M. A., Espinosa-Urgel, M., Rodriguez-Herva, J. J., Ramos, J. L., & Ramos-Gonzalez, M. I. (2007). Genomic analysis reveals the major driving forces of bacterial life in the rhizosphere. *Genome Biology*, *8*, 13.

Matilla, M. A., Ramos, J. L., Bakker, P., Doornbos, R., Badri, D. V., & Vivanco, J. M. (2010). Pseudomonas putida KT2440 caused induced systemic resistance and changes in Arabidopsis root exudation. *Environmental Microbiology Reports*, *2*, 381–388.

Mavrodi, D. V., Mavrodi, O. V., Parejko, J. A., Weller, D. M., & Thomashow, L. S. (2011). The role of 2,4-diacetylphloroglucinol- and phenazine-1-carboxylic acid-producing Pseudomonas spp. In D. K. Maheshwari (Ed.), *Natural protection of wheat from soilborne pathogens. Bacteria in agrobiology: Plant nutrient management* (pp. 60–63). Germany: Springer.

Mendes, R., Kruijt, M., de Bruijn, I., Dekkers, E., van der Voort, M., Schneider, J. H. M., et al. (2011). Deciphering the rhizosphere microbiome for disease-suppressive bacteria. *Science*, *332*, 1097–1100.

Montgomery, D. R. (2007). Is agriculture eroding civilization's foundation? *Geological Society of America Today*, *17*, 4–9.

Moreno, R., Martines-Gomariz, M., Yuste, L., Gil, C., & Rojo, F. (2009). The Pseudomonas putida Crc global regulator controls the hierarchical assimilation of amino acids in a complete medium: Evidence from proteomic and genomic analyses. *Proteomics*, *9*, 2910–2928.

Morris, A. C., & Djordjevic, M. A. (2006). The *rhizobium leguminosarum* biovar *trifolii* ANU794 includes novel developmental responses on the subterranean clover cultivar Woogenellup. *Molecular Plant Microbe Interactions*, *19*, 471–479.

Neal, A. L., Ahmad, S., Gordon-Weeks, R., & Ton, J. (2012). Benzoxazinoids in root exudates of maize attract Pseudomonas putida to the rhizosphere. *PLoS One*, *7*, e35498. http://dx.doi.org/10.1371/journal.pone.0035498.

Neal, A. L., & Ton, J. (2013). *Systemic defense priming by Pseudomonas putida KT2440 in maize depends on benzoxazinoid exudate from roots. Plant Signaling & Behavior*, *8*, e22655. http://dx.doi.org/10.4161/psb.22655.

Nelson, E. B. (2004). Microbial dynamics and interactions in the spermosphere. *Annual Review of Phytopathology*, *42*, 271–309.

Neumann, G., & Romheld, V. (2002). Root-induced changes in the availability of nutrients in the rhizosphere. In Y. Waisel, E. Eshram, & T. Beeckman (Eds.), *Plant roots: the hidden half* (pp. 617–649). New York: Marcel Dekker, Inc.

Nielsen, J. S., & Moller, B. L. (1999). Biosynthesis of cyanogenic glucosides in *Triglochin maritima* and the involvement of cytochrome P450 enzymes. *Archives Biochemistry Biophysiology*, *368*, 121–130.

Owen, A. G., & Jones, D. L. (2001). Competition for amino acids between wheat roots and rhizosphere microorganisms and the role of amino acids in plant N acquisition. *Soil Biology and Biochemistry*, *33*, 651–657.

Palumbo, J. D., Kado, C. I., & Phillips, D. A. (1998). An isoflavonoid-inducible efflux pump in Agrobacterium tumefaciens is involved in competitive colonization of roots. *Journal of Bacteriology*, *180*, 3107–3113.

Paterson, A. H., Schertz, K. F., Lin, Y. R., Liu, S. C., & Chang, Y. L. (1995). The weediness of wild plants: Molecular analysis of genes influencing dispersal and persistence in johnsongrass, Sorghum halepense (L.) Pers. *Proceedings of the National Academy of Sciences of the United States of America*, *92*, 6127–6131.

Patton, C. L., & Glick, B. R. (1996). Bacterial biosynthesis of indole-3-acetic acid. *Canadian Journal of Microbiology*, *42*, 207–220.

Phillips, D. A., Fow, T. C., & Six, J. (2006). Root exudation (net efflux of amino acids) may increase rhizodeposition under elevated CO_2. *Global Change Biology*, *12*, 561–567.

Piel, J. (2011). Approaches to capturing and designing biologically active small molecules produced by uncultured microbes. *Annual Reviews in Microbiology, 65*, 431–453.
Pimentel, D., Lach, L., Zuniga, R., & Morrison, D. (2000). Environmental and economic costs of nonindigenous species in the United States. *Bioscience, 50*, 53–65.
Pineda, A., Zheng, S., van Loon, J. J. A., Pieterse, C. M. J., & Dicke, M. (2010). Helping plants to deal with insects: The role of beneficial soil-borne microbes. *Trends in Plant Science, 15*, 507–514.
Raaijmakers, J. M., Paulitz, T. C., Steinberg, C., Alabouvette, C., & Moenne-Loccoz, Y. (2009). The rhizosphere: A playground and battlefield for soilborne pathogens and beneficial microorganisms. *Plant and Soil, 321*, 341–361.
Reading, N. C., & Sperandio, V. (2006). Quorum sensing: The many languages of bacteria. *FEMS Microbial Letters, 254*, 1–11.
Riley, M. A. (2011). Bacteriocin-mediated competitive interactions of bacterial populations and communities. In D. Drider & S. Rebuffat (Eds.), *Prokaryotic antimicrobial peptides: from genes to applications* (pp. 13–26). Springer: New York.
Riley, M. A., & Gordon, D. M. (1992). A survey of Col plasmids in natural isolates of Escherichia coli and an investigation into the stability of Col-plasmid lineages. *Journal of General Microbiology, 138*, 1345–1352.
Rosegrant, M. W., Cai, X., & Cline, S. A. (2002). *World water and food 2025: Dealing with scarcity*. Washington, DC: International Food Policy Research Institute and the International Water Management Institute.
Rout, M. E., & Callaway, R. M. (2009). An invasive plant paradox. *Science, 324*, 724–725.
Rout, M. E., & Callaway, R. M. (2012). Interactions between exotic invasive plants and soil microbes in the rhizosphere suggest 'everything is not everywhere'. *Annals of Botany, 110*, 213–222.
Rout, M. E., & Chrzanowski, T. H. (2009). The invasive sorghum halepense harbors endophytic N2-fixing bacteria and alters soil biogeochemistry. *Plant and Soil, 315*, 163–172.
Rout, M. E., Chrzanowski, T. H., DeLuca, T. H., Westlie, T. K., Callaway, R. M., & Holben, W. E. (2013). Bacterial endophytes enhance invasive plant competition. *American Journal of Botany, 100*, 1726–1737.
Rout, M. E., Chrzanowski, T. H., Smith, W. K., & Gough, L. (2013). Ecological impacts of the invasive grass sorghum halepense on native tallgrass prairie. *Biological Invasions, 15*, 327–339.
Rudgers, J. A., Afkhami, M. A., Rua, M. A., Davitt, S. H., Hammer, S., & Huguet, V. M. (2009). A fungus among us: Broad pattern of endophyte distribution in the grasses. *Ecology, 90*, 1531–1539.
Rudrappa, T., Czymmek, K. J., Pare, P. W., & Bais, H. P. (2008). Root-secreted malic acid recruits beneficial soil bacteria. *Plant Physiology, 148*, 1547–1556.
Sachs, J. L., Mueller, U. G., Wilcox, T. P., & Bull, J. J. (2004). The evolution of cooperation. *Quarterly Reviews in Biology, 79*, 135–160.
Sarwar, M., & Kremer, R. J. (1995). Enhanced suppression of plant growth through production of L-tryptophan-derived compounds by deleterious rhizobacteria. *Plant and Soil, 172*, 261–269.
Schade, J., & Hobbie, S. E. (2005). Spatial and temporal variation in the islands of fertility in the Sonoran Desert. *Biogeochemistry, 73*, 541–553.
Shaharoona, B., Naveed, M., Arshad, M., & Zahir, Z. A. (2008). Fertilizer-dependent efficiency of pseudomonads for improving growth, yield and nutrient use efficiency of wheat (*Triticum aestivum* L.). *Applied Microbiology and Biotechnology, 79*, 147–155.
Spaepen, S., Dobbelaere, S., Croonenborghs, A., & Vanderleyden, J. (2008). Effects of *Azospirillum brasilense* indole-3-acetic acid production on inoculated wheat plants. *Plant and Soil, 312*, 15–23.
Stevenson, F. J., & Cole, M. A. (1999). *Cycles of soil: Carbon, nitrogen phosphorus, sulphur and micronutrients* (2nd ed.). New York: John Wiley & Sons, Inc.

Teplitski, M., Robinson, J. B., & Bauer, W. D. (2000). Plants secrete substances that mimic bacterial N-acyl homoserine lactone signal activities and affect population density-dependent behaviors in associated bacteria. *Molecular Plant-Microbe Interactions*, *13*, 637–648.

Tilman, D. (2000). Global environmental impacts of agricultural expansions: The need for sustainable and efficient practices. *Proceedings of the National Academy of Sciences of the United States of America*, *96*, 5995–6000.

Tilman, D., Socolow, R., Foley, J. A., Hill, J., Larson, E., Lynd, L., et al. (2009). Beneficial biofuels—The food, energy, and environment trilemma. *Science*, *325*, 270–271.

Torsvik, V., Goksøyr, J., & Daae, F. L. (1990). Comparison of phenotypic diversity and DNA heterogeneity in a population of soil bacteria. *Applied Environmental Microbiology*, *56*, 776–781.

Torsvik, V., Øvreås, L., & Thingstad, T. F. (2002). Prokaryotic diversity—Magnitude, dynamics, and controlling factors. *Science*, *296*, 1064–1066.

Toussaint, J. P., Pham, T. T. M., Barriault, D., & Sylvestre, M. (2012). Plant exudates promote PCB degradation by a rhodococcal rhizobacteria. *Applied Microbiology and Biotechnology*, *95*, 1589–1603.

Turnbaugh, P. J., Ley, R. E., Hamady, M., Fraiser-Liggett, C. M., Knight, R., & Gordon, J. I. (2007). The human microbiome project: A strategy to understand the microbial components of the human genetic and metabolic landscape and how they contribute to normal physiology and predisposition to disease. *Nature*, *449*, 804–810.

Van der Ent, S., Van Hulten, M., Pozo, M. J., Czechowski, T., Udvardi, M. K., Pieterse, C. M. J., et al. (2009). Priming of plant innate immunity by rhizobacteria and b-aminobutyric acid: Differences and similarities in regulation. *New Phytologist*, *183*, 419–431.

Van der Heijden, M. G. A., Bardgett, R. D., & van Straalen, N. M. (2008). The unseen majority: Soil microbes as drivers of plant diversity and productivity in terrestrial ecosystems. *Ecology Letters*, *11*, 296–310.

van Hees, P. A. W., Jones, D. L., Nyberg, L., Holstrom, S. J. M., Godbold, D. L., & Lundstrom, U. S. (2005). Modeling low molecular weight organic acid dynamics inforest soils. *Soil Biology and Biochemistry*, *37*, 517–531.

van Kleunen, M., Weber, E., & Fischer, M. (2010). A meta-analysis of trait differences between invasive and non-invasive plant species. *Ecology Letters*, *13*, 235–245.

Van Oosten, V. R., Bodenhausen, N., Reymond, P., Van Pelt, J. A., Van Loon, L. C., Dicke, M., et al. (2008). Differential effectiveness against herbivorous insects in Arabidopsis. *Molecular Plant Microbe Interactions*, *21*, 919–930.

van Overbeek, L., & van Elsas, J. D. (2008). Effects of plant genotype and growth stage on the structure of bacterial communities associated with potato (*Solanum tuberosum* L.). *FEMS Microbiology Ecology*, *64*, 283–296.

Van Wees, S. C. M., Van der Ent, S., & Pieterse, C. M. J. (2008). Plant immune responses triggered by beneficial microbes. *Current Opinions in Plant Biology*, *11*, 443–448.

Weller, D. M., Raaijmakers, J. M., Gardener, B. B. M., & Thomashow, L. S. (2002). Microbial populations responsible for specific soil suppressiveness to plant pathogens. *Annual Reviews in Phytopathology*, *40*, 309–348.

Yu, J., Grishkan, I., & Steinbrener, Y. (2012). Microfungal community diversity in Zygophyllum dumosum and Hammada scoparia root zones in the northern Negev Desert. *Journal of Basic Microbiology*, *52*, 1–12.

Zaidi, A., & Khan, M. S. (2005). Interactive effect of rhizospheric microorganisms on growth, yield and nutrient uptake of wheat. *Journal of Plant Nutrition*, *28*, 2079–2092.

Zang, Y. F., He, L. Y., Chen, Z. J., Wang, Q. Y., Qian, M., & Sheng, X. F. (2011). Characterization of ACC deaminase-producing endophytic bacteria isolated from copper-tolerant plants and their potential in promoting the growth and copper accumulation of Brassica napus. *Chemosphere*, *83*, 57–62.

Zehr, J. P., Jenkins, B. D., Short, S. M., & Steward, G. F. (2003). Nitrogenase gene diversity and microbial community structure: A cross-system comparison. *Environmental Microbiology*, *5*, 539–554.

CHAPTER TWELVE

Population Genomics of Plant Species

Khaled M. Hazzouri[*], **Michael D. Purugganan**[*,†], **Jonathan M. Flowers**[*,†,1]
[*]Center for Genomics and Systems Biology, NYU Abu Dhabi Institute, New York University Abu Dhabi, Abu Dhabi, United Arab Emirates
[†]Department of Biology, Center for Genomics and Systems Biology, New York University, New York, USA
[1]Corresponding author: e-mail address: jmf11@nyu.edu

Contents

Abstract

This chapter addresses fundamental questions in population genomics and molecular evolution of plants and their domesticated relatives. We review the evolutionary forces that shape genetic diversity within and among species and highlight recent exceptional examples from the literature to illustrate how whole genome sequencing of entire populations of plant genomes has accelerated the pace of discovery of the mechanisms of plant genome evolution. Throughout the chapter, we emphasize the role of mating system and population size in plant genome evolution and discuss why plant genomes frequently show limited evidence of adaptation and display features that are inconsistent with simple models of linked selection. We close by highlighting recent studies that illustrate a range of discoveries enabled by sequencing of populations of plant genomes.

Advances in Botanical Research, Volume 69
ISSN 0065-2296
http://dx.doi.org/10.1016/B978-0-12-417163-3.00012-3

1. INTRODUCTION

High-throughput sequencing of whole genomes has forever altered all of biology, and plant biology is no exception. Plant molecular and cell biology, plant genomics, plant ecology, crop genetics, plant development, botany, and related fields have fundamentally changed as a result of the ability to acquire whole genome sequences. The field of plant population genomics—the sequencing and analysis of populations of individual genomes—has emerged as an exciting extension of population genetics. The emergence of this field offers the ability not only to obtain and analyze comprehensive DNA sequence and functional information from large numbers of genotypes but also to extend analyses formerly restricted to a few model and crop organisms to a broad array of taxonomically diverse plant species.

Why are studies of populations of genomes essential to our understanding of plant biology? The reasons are many. For the plant breeder, populations of genomes provide a means for genomic selection and the ability to create detailed genomic maps of breeding populations, which provide a means to identify the locations of key genes. For the plant molecular biologist, populations of genomes have facilitated entirely new fields of exploration such as population epigenomics, which characterizes the molecular mechanisms of epiallele formation and how natural selection may operate on inherited epigenetic variation to shape patterns of phenotypic diversity. For the plant pathologist, sequences from entire populations of cells offer the prospect of advancing our understanding of senescence and the molecular basis of plant disease.

In the field of plant population genomics, population-level analyses of genomes have illuminated in detail the genetic differences among individuals and facilitated a description of the patterns and mechanisms of evolution in various plant species. From the sequences of many closely related genotypes and their evolutionary relatives, we gain insight into many evolutionary phenomena, including the frequency and intensity of natural selection, the relative importance of adaptive and neutral processes in plant evolution, the genomic targets of selection, the historical movements of individuals and dynamics of plant populations, and the elusive connections between genes and the traits they control. Many of these subjects have been core areas of research in plant evolutionary and population genetics for decades, but questions once beyond reach are tractable in ways that they never have been before. In particular, comparative studies of populations of genomes between closely related individuals and species with known phylogeny are responsible for some of the most significant advances in our understanding of the factors at work in plant evolution (Flowers & Purugganan, 2008).

The field of plant population genomics benefits from integration of a wide array of experimental approaches. For example, field studies of plant fitness in their native environments are essential to bridging the divide between molecular population genetics and evolution in the wild (reviewed in Anderson, Willis, & Mitchell-Olds, 2011; Mitchell-Olds & Schmitt, 2006). Functional studies of plant populations including comparative transcriptomics, proteomics, and epigenomics (Schmitz et al., 2013) together with genetic mapping of ecologically significant traits provide a means to characterize the molecular mechanisms responsible for plant adaptation (reviewed in Bergelson & Roux, 2010). Integrating these approaches is essential to a comprehensive understanding of plant evolution.

In this chapter, we describe insights into the mechanisms of plant evolution gained from the study of genetic diversity and the operation of selection in natural populations of plants, including domesticated species. We begin by summarizing what is known about the factors that contribute to differences in nucleotide variation observed across plant species, describe how estimates of nucleotide diversity and the mutation rate provide an essential framework to understanding the mechanisms of plant evolution, and highlight questions of historical interest to plant population geneticists. We then identify ways in which whole genome sequences from populations of plant genomes are providing new insight into the molecular basis of adaptation and, in some cases, maladaptation in plant species. Throughout the chapter, we highlight recent advances that have shed light on mechanisms of evolution in plant genomes with emphasis on ways in which the mating system and effective size of populations influence their evolution.

2. GENETIC DIVERSITY IN PLANT POPULATIONS

A primary focus of population genetic studies of plant species has been to identify the neutral and selective mechanisms that determine the amount of genetic variation within and among populations. Central to this problem is the concept of the effective population size (N_e)—the size of an idealized population that experiences the same rate of genetic drift as the study population.

Why are population geneticists interested in N_e? First, under the neutral theory of molecular evolution, nucleotide diversity reflects a balance between the rate of spontaneous mutation and the loss of variation due to genetic drift. In a population at equilibrium between these processes, nucleotide diversity is proportional to the product of the effective population size and the per generation rate of mutation ($\theta = 4N_e\mu$). Thus, the effective population size determines the amount of genetic diversity in a population (reviewed in Charlesworth, 2009) and forms the basis for comparative studies that aim to characterize the effects of life history variation and other factors on the evolution of plant species.

Second, the efficacy of selection is a function of the effective population size and the distribution of fitness effects of mutations. Mutations behave neutrally and can escape selection if their effect on fitness is on the order of 1/Ne or less, while those with stronger effects are subject to selection. Thus, the effective population size provides insight into the relative importance of genetic drift and natural selection and the efficiency of selection at either removing harmful variants from a population or increasing the frequency of beneficial ones.

What factors determine the level of genetic diversity in a population? As can be anticipated from the equation in the preceding text, under neutral theory, the amount of genetic variation is determined by the effective population size and the mutation rate. Yet, there are many factors that contribute to the effective population size that must be distinguished. Moreover, neutral mechanisms are unlikely to be the only factors contributing to genomic patterns of variation despite the conspicuous absence of clear evidence of adaptation in most plant protein-coding genes (see succeeding text). Nevertheless, interest in this long-standing question has recently been rekindled in a sweeping survey of plant and animal species (Leffler et al., 2012), which reported a narrower than expected range of nucleotide diversity across species given the range of census population sizes. This observation may be explained by a variety of factors including a correlation between the effective size and the mutation rate (Sung, Ackerman, Miller, Doak, & Lynch, 2012), pervasive selection (see the following text), or other factors, which we describe in some detail in the succeeding text.

3. MUTATION RATE VARIATION

In plants, the mutation rate has been inferred by calibrating the rate of mutation using known dates of divergence from the fossil record (i.e. the 'molecular clock', Gaut, Morton, McCaig, & Clegg, 1996). A significant recent advance, however, is the use of whole genome sequencing of mutation accumulation lines as a means to assess the per generation rate of mutation in a land plant (*Arabidopsis thaliana*, 7.1×10^{-9} mutations per base pair per generation; Ossowski et al., 2010) and a chlorophyte (*Chlamydomonas reinhardtii*, 2.08×10^{-10} mutations per base pair per generation, Ness, Morgan, Colegrave, & Keightley, 2012; 6.76×10^{-11}, Sung et al., 2012). These are at present the only direct estimates of mutation rate and the spectrum of spontaneous mutations in plants and their close relatives. As a consequence, direct estimates of a key parameter necessary for estimating the effective population size (from the relationship $\theta = 4Ne\mu$) are unavailable for the vast majority of plant species. While estimates based on the molecular clock suggest that mutation rate variation may not be a primary factor contributing to differences in genetic diversity among species (Muse, 2000; Muse & Gaut, 1997), a rigorous test of this hypothesis awaits a more comprehensive understanding of mutation rate variation among species and the factors contributing to it. An intriguing observation is that across the tree of life, the mutation rate and the effective population size are negatively correlated, which could indicate that the mutation rate itself is subject to selection (Sung et al., 2012).

4. MATING-SYSTEM EFFECTS ON GENETIC DIVERSITY

A number of factors influence the rate of genetic drift and the long-term effective population size of plant species. These include variance in reproductive success, population expansion and contraction, distortion in the sex ratio, and the mating system. The effect of mating system has been of special interest in plant population genomics owing to a rich theory that predicts that components of life history, such as the outcrossing rate and seed dispersal mechanism, should have pronounced effects on the rate of genetic drift and the efficiency of selection (Glémin, 2007). For example, self-fertilization reduces the effective population size by up to twofold relative to an obligate outcrossing mode of reproduction owing to the effect of inbreeding (Charlesworth, Lyons, & Litchfield, 1993; Nordborg, 2000; Pollak, 1987). Selfing populations are also characterized by lower effective rates of recombination, which should increase losses of diversity due to linked selection (see succeeding text), and also frequently experience cycles of local extinction and recolonization, recurrent population bottlenecks, and reduced rates of migration—factors that further enhance the rate of genetic drift within, and the degree of isolation between, populations.

rates of recombination likely constrain the probability of fixation of beneficial mutations (Wright & Andolfatto, 2008). This area of research will benefit greatly from additional surveys of variation from related species with different effective population sizes and mating systems and expansion of comparative analyses to noncoding DNA.

Limited evidence for adaptive substitution using the approaches adopted in the studies in the preceding text does not preclude the significance of molecular adaptation in plants, including those species with small effective sizes. In fact, it is in situations where molecular evidence of adaptive evolution is rare that application of genome wide approaches may be particularly effective. For example, widespread plant species, that is, those with geographic distributions that span a wide range of climatic conditions, are excellent models for exploring the effect of climate on local adaptation. In a genome-wide survey of SNPs genotyped in 948 accessions of *A. thaliana*, Hancock et al. (2011) found that SNPs whose population frequencies showed the strongest correlations with climatic variables were significantly enriched for nonsynonymous SNPs relative to their overall proportions in the genome, while noncoding SNPs were not. This suggests that amino acid-altering variants improve plant fitness under local environmental conditions and provide strong support for climatic adaptation. In another example, Lee and Mitchell-Olds (2012) reported that after accounting for the effects of genomic factors, a relatively large percentage of the variation in nonsynonymous polymorphism is explained by the environment. Their analysis supports a model where spatially varying selection maintains polymorphism at environmentally responsive genes through the action of antagonistic pleiotropy. While examples such as this are at present limited to a few model plant species, the prospect of large samples of genotypes from different environments should provide a fruitful means to detect genes responding to selection.

A question of great interest in molecular population genetics is the importance of balancing selection in molecular evolution (Charlesworth, 2006; Wright & Andolfatto, 2008). This form of selection, which may include, but is not limited to, heterozygote advantage (i.e. overdominance), can lead to maintenance of variation for periods far longer than expected for neutral alleles. Well-known examples in plants include maintenance of conditionally adaptive alleles at self-incompatibility loci (Castric, Bechsgaard, Schierup, & Vekemans, 2008; Castric & Vekemans, 2007; Roux et al., 2013; Schierup, Mable, Awadalla, & Charlesworth, 2001; Wright, 1939) and disease resistance alleles over periods spanning millions of years, which in extreme cases can lead to maintenance of highly diverged alleles and sharing of polymorphisms among species (Gos, Slotte, & Wright, 2012; Liu et al., 2010; Meyer, Nelson, Clement, & Ramakrishnan, 2010; Rose, Michelmore, & Langley, 2007; Stahl, Dwyer, Mauricio, Kreitman, & Bergelson, 1999; Tian, Traw, Chen, Kreitman, & Bergelson, 2003). The population genetic signature of such 'balanced' polymorphisms—increases in coalescence times at sites tightly linked to the selected locus—has been detected in genomic scans of polymorphism in plants most prominently in regions enriched for disease resistance genes (Clark et al., 2007). Such whole genome resequencing studies facilitate a description of outlier regions with high levels of polymorphism and provide a means for identifying genes subject to long-term balancing selection.

While examples of long-term balancing selection are interesting cases, details concerning the mechanism (e.g. overdominance, frequency-dependent selection, or spatially varying selection) are frequently obscure. Moreover, there are likely to be many cases where various forms of diversity-maintaining selection do not operate for long enough to leave a diagnostic signature of balancing selection. In order to elucidate both the target(s) and mechanisms of selection in such cases, it is essential to use a combination of experimental approaches including genomic scans for selection, association studies, functional analyses, and field experiments. For example, using 39 years of aphid abundance data, Züst et al. (2012) correlated allele frequencies at *GS-ELONG*, a gene controlling the defence response in *A. thaliana,* with the abundance of two specialist aphid species. In addition to reporting a strong correlation between herbivore intensity and allele frequency, they carried out a multigenerational selection experiment and demonstrated a causal link between aphid abundance and variation at the defence locus. In another example, Fournier-Level et al. (2011) conducted common garden experiments at four field sites across Europe with *A. thaliana* ecotypes with different geographic origins. Using genome-wide association mapping (GWAS; see the succeeding text), they mapped components of lifetime fitness in the field and found that high-fitness alleles identified in their study were distributed closer to the site where they increase fitness compared with genomic controls. These and related studies offer clear insight into the mechanisms of plant evolution and help to close the gap between inferences drawn purely from population genetic patterns and the operation of selection in the wild.

6. DELETERIOUS VARIATION IN PLANT POPULATIONS

Small population sizes are relatively permissive of deleterious variation owing to increased rates of genetic drift and the effective neutrality of weakly selected alleles (Charlesworth, 2009; Gossmann et al., 2012; Pollak, 1987; Wright, Ness, Foxe, & Barrett, 2008). Therefore, species of plants with small effective population sizes, including many domesticated crops, are subject to the accumulation of deleterious variants whose selective effect is on the order of the reciprocal of the effective population size (i.e. the so-called effectively neutral polymorphisms). In *A. thaliana*, there is a large class of potentially deleterious amino acid polymorphisms segregating in natural populations (Wright & Andolfatto, 2008; Bustamante et al., 2002). A similar pattern is observed in the outcrossing relative *Arabidopsis lyrata*, suggesting that the excess of replacement polymorphisms in *A. thaliana* is not the result of reduced efficiency of selection in a selfing species (Foxe et al., 2008), although patterns of selection on codon usage are consistent with an effect of mating system (Qiu, Zeng, Slotte, Wright, & Charlesworth, 2011). These studies point to a large class of potentially harmful mutations segregating both in *A. thaliana* and *A. lyrata*, but the effect of mating system is only apparent for very weakly selected sites. By contrast, in a comparison between the primarily outcrossing *C. grandiflora* and its selfing relative *Capsella rubella*, there is a strong signature of relaxed purifying selection in the self-fertilizing species (Hazzouri et al., 2013; Slotte et al., 2013).

In other cases, population size contractions may be responsible for the release of formerly deleterious mutations into the effectively neutral class. This has received some empirical support in crop plants, where *tropical* and *temperate japonica* rice segregate proportionally more effectively neutral nonsynonymous mutations than their wild relative *Oryza rufipogon* (Gossmann et al., 2010), and there is evidence of genome-wide relaxation of purifying selection in *japonica* (Yang et al., 2012). The ability of effectively neutral mutations to escape the effect of purifying selection is expected to increase the fixation probability of deleterious mutations and may account for the higher rate of amino acid substitution in rice cultivars with smaller effective size (Lu et al., 2006; Yang et al., 2012). As illustrated by these examples, reductions in population size associated with life history transitions to self-fertilization or population bottlenecks may allow harmful alleles to accumulate in plant populations. In some cases, this may contribute to declines in mean population fitness (Kondrashov, 1984; Phillips, Otto, & Whitlock, 2000; Schultz & Lynch, 1997).

Differences in the efficiency of selection against harmful mutations may also have profound consequences for many aspects of plant genome evolution. For example, it has been postulated that variation in genome size across the web of life is the result of differences in the effective size among species

(Lynch & Conery, 2003). This hypothesis proposes that mutations that increase genome size (e.g. TE proliferation) have a weak fitness cost and natural selection will work more efficiently to constrain genome expansion in species with larger effective size, whereas mutations of the same effect in species with small effective size can drift to fixation. This hypothesis has received mixed support both overall and within plants. For example, genomes of self-fertilizing plant species are smaller than those of their outcrossing relatives (Wright et al., 2008) and species with restricted geographic ranges can have larger genome sizes compared with their more common relatives (Vinogradov, 2003), although this may not be a universal effect (Grotkopp, Rejmánek, Sanderson, & Rost, 2004). However, in a comparison across populations of *A. lyrata*, there is evidence that TEs are subject to purifying selection in large populations, whereas they evolve neutrally in smaller populations (Lockton, Ross-Ibarra, & Gaut, 2008). A study of 10 species in the genus *Oryza* found no relationship between the effective population size and genome size using lineage-specific mutation rates to estimate the effective size and phylogenetic independent contrasts to control for phylogenetic nonindependence (Ai, Wang, & Ge, 2012). Examination of 205 species of seed plants similarly found no support for the Lynch and Conery (2003) model when phylogenetic relationships were factored into the analysis (Whitney et al., 2010). Studies such as these, which consider genome size in the context of phylogeny, control for phylogenetic nonindependence, and incorporate lineage-specific mutation rates to estimate the effective population size, provide the best framework for addressing these and related questions concerning the accumulation of deleterious variation and the effect of population size on genome evolution (Charlesworth & Barton, 2004; Flowers & Purugganan, 2008).

7. THE EFFECTS OF LINKED SELECTION

In previous sections, we highlighted work on the incidence of positive and negative selection in plant genomes. A related approach is to study the effects of natural selection on patterns of variation at neutral sites across the genome. Analysis of linked neutral sites can greatly enrich our understanding of selection and genes targeted by selection can be identified. Moreover, studies of neutral variation can provide a possible answer to the riddle of why genetic diversity varies so little among species that differ so dramatically in census population size (Leffler et al., 2012).

The "hitchhiking" model is perhaps the best-known model of linked selection that describes the rise in frequency of a new beneficial mutation due to positive selection and the concurrent loss of genetic diversity at linked neutral sites (Maynard Smith and Haigh 1974). This model predicts that the loss of variation in regions subject to positive selection is proportional to the recombination rate. Under this model, fixation of beneficial mutations in regions of suppressed recombination (e.g. heterochromatic regions in many plants, Gaut, Wright, Rizzon, Dvorak, & Anderson, 2007) will lead to losses of diversity over greater physical distances than fixation of an equivalent allele in regions with normal rates of crossing over. Charlesworth et al. (1993) discovered that "background selection"—loss of genetic diversity at linked sites due to purifying selection—will similarly operate to eliminate variation to a greater extent in regions with reduced rates of crossing over. In principle, both genetic hitchhiking and background selection should yield a positive correlation between estimates of nucleotide diversity and the recombination rate provided that the rate of deleterious or beneficial mutations is roughly the same in different genomic regions.

This expectation is met in many organisms including Drosophila and Caenorhabditis, but studies of plant populations suggest that this pattern is absent in Arabidopsis, *O. rufipogon*, and *Oryza sativa* (Flowers et al., 2012; Nordborg et al., 2005) and is rare in plant species studied to date (reviewed in Cutter & Payseur, 2013). Why does polymorphism in plant species frequently not conform to expectations of simple models of linked selection? In *O. rufipogon*, significant negative associations are observed between polymorphism and gene density, but a correlation with recombination rate is not apparent. The probable cause of this pattern is the joint effect of an uneven distribution of deleterious mutations across the genome (i.e. enrichment of deleterious mutations in gene dense regions) and a reduced outcrossing rate (Flowers et al., 2012), although other factors including dominance of deleterious mutations and the distribution of fitness effects may contribute to the pattern in rice and other species (Cutter & Payseur, 2013). It remains to be seen what factors underlie the different patterns observed in different plant species, but it is clear that the physical organization of genes on recombining chromosomes, mating system, population structure, and mutation rate effects can generate patterns that differ significantly from the classic Drosophila-like association between polymorphism and the rate of crossing over (Begun & Aquadro, 1992).

Genome scans of nucleotide polymorphism also provide a means to identify regions targeted by selection. In plants, genome scans have been applied most frequently in crop species with the primary goals of identifying the number and function of genes subject to selection (reviewed in Alonso-Blanco et al., 2009; Olsen & Wendel, 2013) and quantifying the strength and frequency of selection during the course of domestication (reviewed in Olsen & Wendel, 2013). A recent study of maize domestication inferred that 7.6% of the genome showed evidence of linked selection during the "domestication" phase with large regions showing evidence of selective sweeps and relatively large estimates of the strength of selection ($s=0.015$). Regions subject to selection during the "improvement phase" were subject to less intense selection ($s=0.003$) and were smaller in size and contained fewer genes. While up to 1–2% of the genes in the genome may have been the direct targets of selection for domestication or improvement traits, another 3000 genes not directly targeted are genetically impoverished of variation due to linkage with genes subject to selection (Hufford et al., 2012) above and beyond the losses of variation associated with population bottlenecks. Similar whole genome resequencing studies of domesticated *indica* and *japonica* rice reported more than 700 10 kb regions containing on the order of a thousand genes that are severely depleted of variation in both groups of cultivars (Xu et al., 2011) and a genotyping study of sorghum landraces identified selective sweeps of up to several megabases (Morris et al., 2013). These studies suggest a mating-system effect on the breadth of selective sweeps with self-pollinating species of rice and sorghum (lower effective rate of recombination) having larger sweep regions on average than outcrossing maize (higher recombination rate; Morris et al., 2013).

Genomic regions depauperate of variation are frequently caused by "hard sweeps" that lead to losses of variation due to linkage with new advantageous mutations (Clark, Linton, Messing, & Doebley, 2004; Olsen et al., 2006; Wang et al., 2008). An interesting area of future research will be to examine whether the numbers of genes subject to selection have been underestimated because of current methodological limitations in detecting "soft sweeps"—incomplete or partial loss of variation at linked sites owing to selection on standing variation (Pritchard, Pickrell, & Coop, 2010). Nevertheless, studies such as these not only provide insight into the operation of selection under domestication but also suggest regions of the genome where trait mapping approaches will provide novel and powerful insights into the evolutionary history of crop plants (Chia et al., 2012; Hufford et al., 2012; Jiao et al., 2012; Mandel et al., 2013).

8. MAPPING OF QUANTITATIVE TRAIT LOCI

Traditionally, QTL mapping in plants was the method of choice to identify loci responsible for the phenotypic changes associated with crop evolution (Burke, Tang, Knapp, & Rieseberg, 2002; Doebley, Stec, Wendel, & Edwards, 1990; Xiong, Liu, Dai, Xu, & Zhang, 1999) and the evolution of mating systems (Fishman, Kelly, & Willis, 2002; Goodwillie, Ritland, & Ritland, 2006; Grillo et al., 2009; Lin & Ritland, 1997; Slotte, Hazzouri, Stern, Andolfatto, & Wright, 2012). The traditional method or single-family QTL approach is limited by small mapping population sizes and issues in model construction (Beavis, 1998). Family-based QTL mapping approaches use existing breeding families and improve both precision and accuracy in QTL detection, compared with the single-family approach (Rosyara, Gonzalez-Hernandez, Glover, Gedye, & Stein, 2009). GWAS, on the other hand, uses correlations between molecular polymorphism and phenotypic variation in assemblages of individuals drawn from natural populations (Yu & Buckler, 2006). In contrast to traditional mapping approaches, GWAS uses linkage disequilibrium between markers and traits of interests and therefore improves mapping resolution by capturing many generations of historical recombination. Moreover, GWAS removes limitations frequently encountered in breeding of species with delayed maturity and long generation times and is therefore a valuable approach to mapping in perennial species of plants. In recent years, genome-wide association approaches have been shown to be an especially powerful tool for identifying the genetic contributors to complex traits.

In plant genomics, numerous other tools have been developed to map quantitative traits. For example, nested association mapping (NAM) utilizes a combination of linkage and association analyses of a large multifamily recombinant mapping population that takes advantage of both historical and recent recombination events to improve resolution and statistical power (Yu, Holland, McMullen, & Buckler, 2008). This strategy has been deployed in maize, where efforts over the last decade have led to the development of robust NAM approaches (Buckler et al., 2009; Tian et al., 2011). Another area of development has been to use multiple generations of intercrossing between several individuals, which can both increase the number of recombination events and broaden the base of genetic diversity for mapping traits (Liu, Kowalski, Lan, Feldmann, & Paterson, 1996). These have led to the development of multiple advanced generation intercrossed lines in *A. thaliana* (Kover et al., 2009) and wheat (Huang, George, et al., 2012; Huang, Kurata, et al., 2012). These and other approaches (such as bulk segregant analysis mapping/extreme QTL mapping) (Ehrenreich et al., 2009) promise to increase the ease and resolution of QTL mapping.

9. ADVANCES THROUGH WHOLE GENOME RESEQUENCING

The previous sections have described areas of research that were already well developed from both theoretical and empirical perspectives prior to the emergence of population genomics as a field. However, the ability to analyze entire populations of genomes not only has facilitated refined analysis of the evolutionary dynamics at work in plant genome evolution but also has ushered in an era of discovery in areas of investigation that were virtually impossible prior to the new era of whole genome sequencing (Table 12.1). For example, population genomic analyses facilitated the resolution of a long-standing riddle in the study of sex determination in date palm (*Phoenix dactylifera*). Al-Dous et al. (2011) used whole genome sequencing of male and female date palm cultivars to identify a small region of the genome (~600 kb) with SNP genotypes consistent with males being the heterogametic sex. Additional research has found this to be a nonrecombining XY-like region (Cherif et al., 2013). Now, breeders who formerly had to wait up to 6 years to determine the sex of seedlings now have the ability to do so with a simple PCR assay.

In other examples, whole genome resequencing efforts have illuminated important details about the domestication process through comparative analysis of domesticated and progenitor populations. A striking example is the study of Guo et al. (2013), which resequenced domesticated watermelons and their wild progenitor and discovered that disease resistance genes had been lost during domestication. They reported that members of the NBS-LRR class of resistance genes showing presence/absence variation among genomes in the progenitor population were completely absent in domesticated cultivars. This could indicate that resistance alleles have a cost in domesticated species, potentially through their effects on fruit quality or yield.

Finally, whole genome sequencing of entire populations of genomes has been extended beyond conventional analysis of DNA sequence variation. For example, Schmitz et al. (2013) resequenced the epigenomes (i.e. genome-wide characterization of methylated cytosines) of 152 *A. thaliana*

Table 12.1 Whole genome sequencing of populations of plant genomes (excluding studies using genotyping arrays)

Species	Number of genomes	Interesting finding
Arabidopsis thaliana (thale cress)	80	Accumulation of deleterious mutations at higher rates in marginal populations (Cao et al., 2011)
Citrullus lanatus (watermelon)	20	Loss of disease resistance genes during domestication (Guo et al., 2013)
Phoenix dactylifera (date palm)	9	Determination of males as the heterogametic sex (Al-Dous et al., 2011)
Oryza sativa	1083	Single origin of domestication in southern China (Huang, Kurata, et al., 2012)
Oryza rufipogon	446	Limited number of individuals used during domestication (loss of genetic diversity) and mapping of two traits leaf sheath colour and tiller angle (Huang, Kurata, et al., 2012)
Oryza sativa	40	Identification of candidate regions selected during domestication (Xu et al., 2011)
Oryza rufipogon	5	A complex evolutionary history between *O. indica* and *O. nivara* (Xu et al., 2011)
Oryza nivara	5	Independent domestication of *O. japonica* from the Chinese *O. rufipogon* population (Xu et al., 2011)
Zea mays (domesticated maize)	58	Recovery of diversity after domestication and important number of genes under strong selection for important phenotypes (Hufford et al., 2012)
Zea parviglumis (teosinte)	14	Introgression into the domesticated maize (Hufford et al., 2012)
Prunus persica (peach)	11	Evidence for selection for different maturation times among varieties (International Peach Genome Initiative, 2013)
Glycine max (soybean)	31	Mapping soybean agronomic traits as well as cloning devastating resistance gene in Asian soybean rust (Schmutz et al., 2010)
Cicer arietinum (chickpea)	90	Candidate genes for disease resistance genes and agronomic traits highlighted (Varshney et al., 2013)
Capsella rubella	2	Genomewide relaxation of purifying selection in *C. rubella* and a shift in expression of flowering genes in the transition to selfing (Slotte et al., 2013)

accessions and generated a genomic map of SMPs (single methylation polymorphisms). From this map of SMPs, the authors characterized the extent of natural epigenomic variation in a population, quantified the rate of decay of linkage disequilibrium among SMPs, and mapped methylation quantitative loci (i.e. genes controlling variation in differentially methylated regions). These and many other examples of whole genome resequencing (Table 12.1) illustrate the diversity of applications and questions to which population genomic approaches are now being applied.

10. CONCLUSION

The ability to explore the diversity of genomes in species and their evolutionary relatives has led to a greater understanding of the evolutionary forces that drive genome variation and genome structure. Population genomics provides a synthesis of genomics, molecular evolution, population genetics, and even molecular biology and ecology. As genomics technology continues to improve, we can look forward to providing in-depth analysis in unprecedented detail.

As the field of population genomics advances, there continues to be a need to explore several areas. First, we need to apply population genomic analyses to a wider array of taxonomically diverse species. Currently, genome-wide information is available primarily for members of the Brassicaceae and cereal crop species. By increasing the diversity of taxa, comparative analysis that consider phylogenetic context will become tractable that will greatly improve our understanding of the mechanisms of evolution in plant species. Second, integrating functional (e.g. transcriptomic, proteomic, and epigenomic) and structural (i.e. sequence) information in population genomic studies will provide opportunities to study functional polymorphism at a genome-wide scale and facilitate characterization of the mechanisms of evolution in entire networks of genes. These are exciting possibilities that we expect will attract much attention in the coming years.

REFERENCES

Ai, B., Wang, Z. S., & Ge, S. (2012). Genome size is not correlated with effective population size in the Oryza species. *Evolution*, *66*(10), 3302–3310.

Al-Dous, E. K., George, B., Al-Mahmoud, M. E., Al-Jaber, M. Y., Wang, H., Salameh, Y. M., et al. (2011). De novo genome sequencing and comparative genomics of date palm (Phoenix dactylifera). *Nature Biotechnology*, *29*(6), 521–527.

Alonso-Blanco, C., Aarts, M. G., Bentsink, L., Keurentjes, J. J., Reymond, M., Vreugdenhil, D., et al. (2009). What has natural variation taught us about plant development, physiology, and adaptation? *Plant Cell*, *21*(7), 1877–1896.

Anderson, J. T., Willis, J. H., & Mitchell-Olds, T. (2011). Evolutionary genetics of plant adaptation. *Trends in Genetics*, *27*(7), 258–266.

Baudry, E., Kerdelhué, C., Innan, H., & Stephan, W. (2001). Species and recombination effects on DNA variability in the tomato genus. *Genetics*, *158*(4), 1725–1735.

Beavis, W. D. (1998). QTL analyses: Power, precision, and accuracy. In A. H. Paterson (Ed.), *Molecular analysis of complex traits* (pp. 145–161). Boca Raton: CRC Press.

Begun, D. J., & Aquadro, C. F. (1992). Levels of naturally occurring DNA polymorphism correlate with recombination rates in D. melanogaster. *Nature*, *356*(6369), 519–520.

Bergelson, J., Kreitman, M., Stahl, E. A., & Tian, D. (2001). Evolutionary dynamics of plant R-genes. *Science*, *292*(5525), 2281–2285.

Bergelson, J., & Roux, F. (2010). Towards identifying genes underlying ecologically relevant traits in Arabidopsis thaliana. *Nature Reviews Genetics*, *11*(12), 867–879.

Buckler, E. S., Holland, J. B., Bradbury, P. J., Acharya, C. B., Brown, P. J., Browne, C., et al. (2009). The genetic architecture of maize flowering time. *Science*, *325*(5941), 714–718.

Burke, J. M., Tang, S., Knapp, S. J., & Rieseberg, L. H. (2002). Genetic analysis of sunflower domestication. *Genetics*, *161*(3), 1257–1267.

Bustamante, C. D., Nielsen, R., Sawyer, S. A., Olsen, K. M., Purugganan, M. D., & Hartl, D. L. (2002). The cost of inbreeding in Arabidopsis. *Nature*, *416*(6880), 531–534.

Cao, J., Schneeberger, K., Ossowski, S., Günther, T., Bender, S., Fitz, J., et al. (2011). Whole-genome sequencing of multiple Arabidopsis thaliana populations. *Nature Genetics*, *43*(10), 956–963.

Castric, V., Bechsgaard, J., Schierup, M. H., & Vekemans, X. (2008). Repeated adaptive introgression at a gene under multiallelic balancing selection. *PLoS Genetics*, *4*(8), e1000168.

Castric, V., & Vekemans, X. (2007). Evolution under strong balancing selection: How many codons determine specificity at the female self-incompatibility gene SRK in Brassicaceae? *BMC Evolutionary Biology*, 7, 132.

Charlesworth, D. (2006). Balancing selection and its effects on sequences in nearby genome regions. *PLoS Genetics*, *2*(4), e64.

Charlesworth, B. (2009). Fundamental concepts in genetics: Effective population size and patterns of molecular evolution and variation. *Nature Reviews Genetics*, *10*(3), 195–205.

Charlesworth, B., & Barton, N. (2004). Genome size: Does bigger mean worse? *Current Biology*, *14*(6), R233–R235.

Charlesworth, D., Lyons, E. E., & Litchfield, L. B. (1993). Inbreeding depression in two highly inbreeding populations of Leavenworthia. *Proceedings of the Royal Society of London B: Biological Sciences*, *258*, 209–214.

Cherif, E., Zehdi, S., Castillo, K., Chabrillange, N., Abdoulkader, S., Pintaud, J. C., et al. (2013). Male-specific DNA markers provide genetic evidence of an XY chromosome system, a recombination arrest and allow the tracing of paternal lineages in date palm. *New Phytologist*, *197*(2), 409–415.

Chia, J. M., Song, C., Bradbury, P. J., Costich, D., de Leon, N., Doebley, J., et al. (2012). Maize HapMap2 identifies extant variation from a genome in flux. *Nature Genetics*, *44*(7), 803–807.

Clark, R. M., Linton, E., Messing, J., & Doebley, J. F. (2004). Pattern of diversity in the genomic region near the maize domestication gene tb1. *Proceedings of the National Academy of Sciences of the United States of America*, *101*(3), 700–707.

Clark, R. M., Schweikert, G., Toomajian, C., Ossowski, S., Zeller, G., Shinn, P., et al. (2007). Common sequence polymorphisms shaping genetic diversity in Arabidopsis thaliana. *Science*, *317*(5836), 338–342.

Cutter, A. D., & Payseur, B. A. (2013). Genomic signatures of selection at linked sites: Unifying the disparity among species. *Nature Reviews Genetics*, *14*(4), 262–274.

Doebley, J., Stec, A., Wendel, J., & Edwards, M. (1990). Genetic and morphological analysis of a maize-teosinte F2 population: Implications for the origin of maize. *Proceedings of the National Academy of Sciences of the United States of America, 87*(24), 9888–9892.

Ehrenreich, I. M., Hanzawa, Y., Chou, L., Roe, J. L., Kover, P. X., & Purugganan, M. D. (2009). Candidate gene association mapping of Arabidopsis flowering time. *Genetics, 183*(1), 325–335.

Eyre-Walker, A. (2006). The genomic rate of adaptive evolution. *Trends in Ecology & Evolution, 21*(10), 569–575.

Fay, J. C., Wyckoff, G. J., & Wu, C. I. (2002). Testing the neutral theory of molecular evolution with genomic data from Drosophila. *Nature, 415*(6875), 1024–1026.

Fishman, L., Kelly, A. J., & Willis, J. H. (2002). Minor quantitative trait loci underlie floral traits associated with mating system divergence in Mimulus. *Evolution, 56*(11), 2138–2155.

Flowers, J. M., Molina, J., Rubinstein, S., Huang, P., Schaal, B. A., & Purugganan, M. D. (2012). Natural selection in gene-dense regions shapes the genomic pattern of polymorphism in wild and domesticated rice. *Molecular Biology and Evolution, 29*(2), 675–687.

Flowers, J. M., & Purugganan, M. D. (2008). The evolution of plant genomes: Scaling up from a population perspective. *Current Opinion in Genetics and Development, 18*(6), 565–570.

Fournier-Level, A., Korte, A., Cooper, M. D., Nordborg, M., Schmitt, J., & Wilczek, A. M. (2011). A map of local adaptation in Arabidopsis thaliana. *Science, 334*(6052), 86–89.

Foxe, J. P., Dar, V. U., Zheng, H., Nordborg, M., Gaut, B. S., & Wright, S. I. (2008). Selection on amino acid substitutions in Arabidopsis. *Molecular Biology and Evolution, 25*(7), 1375–1383.

Foxe, J. P., Slotte, T., Stahl, E. A., Neuffer, B., Hurka, H., & Wright, S. I. (2009). Recent speciation associated with the evolution of selfing in Capsella. *Proceedings of the National Academy of Sciences of the United States of America, 106*(13), 5241–5245.

Gaut, B. S., Morton, B. R., McCaig, B. C., & Clegg, M. T. (1996). Substitution rate comparisons between grasses and palms: Synonymous rate differences at the nuclear gene Adh parallel rate differences at the plastid gene rbcL. *Proceedings of the National Academy of Sciences of the United States of America, 93*(19), 10274–10279.

Gaut, B. S., Wright, S. I., Rizzon, C., Dvorak, J., & Anderson, L. K. (2007). Recombination: An underappreciated factor in the evolution of plant genomes. *Nature Reviews Genetics, 8*(1), 77–84.

Glémin, S. (2007). Mating systems and the efficacy of selection at the molecular level. *Genetics, 177*(2), 905–916.

Glémin, S., Bazin, E., & Charlesworth, D. (2006). Impact of mating systems on patterns of sequence polymorphism in flowering plants. *Proceedings of the Biological Sciences, 273*(1604), 3011–3019.

Goodwillie, C., Ritland, C., & Ritland, K. (2006). The genetic basis of floral traits associated with mating system evolution in Leptosiphon (Polemoniaceae): An analysis of quantitative trait loci. *Evolution, 60*(3), 491–504.

Gos, G., Slotte, T., & Wright, S. I. (2012). Signatures of balancing selection are maintained at disease resistance loci following mating system evolution and a population bottleneck in the genus Capsella. *BMC Evolutionary Biology, 12*, 152.

Gossmann, T. I., Keightley, P. D., & Eyre-Walker, A. (2012). The effect of variation in the effective population size on the rate of adaptive molecular evolution in eukaryotes. *Genome Biology and Evolution, 4*(5), 658–667.

Gossmann, T. I., Song, B. H., Windsor, A. J., Mitchell-Olds, T., Dixon, C. J., Kapralov, M. V., et al. (2010). Genome wide analyses reveal little evidence for adaptive evolution in many plant species. *Molecular Biology and Evolution, 27*(8), 1822–1832.

Grillo, M. A., Li, C., Fowlkes, A. M., Briggeman, T. M., Zhou, A., Schemske, D. W., et al. (2009). Genetic architecture for the adaptive origin of annual wild rice, Oryza nivara. *Evolution*, *63*(4), 870–883.

Grotkopp, E., Rejmánek, M., Sanderson, M. J., & Rost, T. L. (2004). Evolution of genome size in pines (Pinus) and its life-history correlates: Supertree analyses. *Evolution*, *58*(8), 1705–1729.

Guo, Y. L., Bechsgaard, J. S., Slotte, T., Neuffer, B., Lascoux, M., Weigel, D., et al. (2009). Recent speciation of Capsella rubella from Capsella grandiflora, associated with loss of self-incompatibility and an extreme bottleneck. *Proceedings of the National Academy of Sciences of the United States of America*, *106*(13), 5246–5251.

Guo, S., Zhang, J., Sun, H., Salse, J., Lucas, W. J., Zhang, H., et al. (2013). The draft genome of watermelon (Citrullus lanatus) and resequencing of 20 diverse accessions. *Nature Genetics*, *45*(1), 51–58.

Hamrick, J. L., & Godt, M. J. W. (1996). Effects of life history traits on genetic diversity in plant species. *Philosophical Transactions: Biological Sciences*, *351*, 1291–1298.

Hancock, A. M., Brachi, B., Faure, N., Horton, M. W., Jarymowycz, L. B., Sperone, F. G., et al. (2011). Adaptation to climate across the Arabidopsis thaliana genome. *Science*, *334*(6052), 83–86.

Hazzouri, K. M., Escobar, J. S., Ness, R. W., Killian Newman, L., Randle, A. M., Kalisz, S., et al. (2013). Comparative population genomics in Collinsia sister species reveals evidence for reduced effective population size, relaxed selection, and evolution of biased gene conversion with an ongoing mating system shift. *Evolution*, *67*(5), 1263–1278.

Huang, B. E., George, A. W., Forrest, K. L., Kilian, A., Hayden, M. J., Morell, M. K., et al. (2012). A multiparent advanced generation inter-cross population for genetic analysis in wheat. *Plant Biotechnology Journal*, *10*(7), 826–839.

Huang, X., Kurata, N., Wei, X., Wang, Z. X., Wang, A., Zhao, Q., et al. (2012). A map of rice genome variation reveals the origin of cultivated rice. *Nature*, *490*(7421), 497–501.

Hufford, M. B., Xu, X., van Heerwaarden, J., Pyhäjärvi, T., Chia, J. M., Cartwright, R. A., et al. (2012). Comparative population genomics of maize domestication and improvement. *Nature Genetics*, *44*(7), 808–811.

Jiao, Y., Zhao, H., Ren, L., Song, W., Zeng, B., Guo, J., et al. (2012). Genome-wide genetic changes during modern breeding of maize. *Nature Genetics*, *44*(7), 812–815.

Kondrashov, A. S. (1984). Deleterious mutations as an evolutionary factor. 1. The advantage of recombination. *Genetics Research*, *44*(2), 199–217.

Kover, P. X., Valdar, W., Trakalo, J., Scarcelli, N., Ehrenreich, I. M., Purugganan, M. D., et al. (2009). A multiparent advanced generation inter-cross to fine-map quantitative traits in Arabidopsis thaliana. *PLoS Genetics*, *5*(7), e1000551.

Lee, C. R., & Mitchell-Olds, T. (2012). Environmental adaptation contributes to gene polymorphism across the Arabidopsis thaliana genome. *Molecular Biology and Evolution*, *29*(12), 3721–3728.

Leffler, E. M., Bullaughey, K., Matute, D. R., Meyer, W. K., Ségurel, L., Venkat, A., et al. (2012). Revisiting an old riddle: What determines genetic diversity levels within species? *PLoS Biology*, *10*(9), e1001388.

Lin, J. Z., & Ritland, K. (1997). Quantitative trait loci differentiating the outbreeding Mimulus guttatus from the inbreeding M. platycalyx. *Genetics*, *146*(3), 1115–1121.

Liu, F., Charlesworth, D., & Kreitman, M. (1999). The effect of mating system differences on nucleotide diversity at the phosphoglucose isomerase locus in the plant genus Leavenworthia. *Genetics*, *151*(1), 343–357.

Liu, S. C., Kowalski, S. P., Lan, T. H., Feldmann, K. A., & Paterson, A. H. (1996). Genome-wide high-resolution mapping by recurrent intermating using Arabidopsis thaliana as a model. *Genetics*, *142*(1), 247–258.

Liu, X. Q., Wang, L., Liu, X. D., Liu, X. Q., Wang, D. B., Wang, C. T., et al. (2010). The molecular evolution of the rice blast resistance gene Pi36. *International Journal of Plant Sciences, 171*(3), 235–243.

Liu, F., Zhang, L., & Charlesworth, D. (1998). Genetic diversity in Leavenworthia populations with different inbreeding levels. *Proceedings of the Biological Sciences, 265*(1393), 293–301.

Lockton, S., Ross-Ibarra, J., & Gaut, B. S. (2008). Demography and weak selection drive patterns of transposable element diversity in natural populations of Arabidopsis lyrata. *Proceedings of the National Academy of Sciences of the United States of America, 105*(37), 13965–13970.

Lu, J., Tang, T., Tang, H., Huang, J., Shi, S., & Wu, C. I. (2006). The accumulation of deleterious mutations in rice genomes: A hypothesis on the cost of domestication. *Trends in Genetics, 22*(3), 126–131.

Lynch, M., & Conery, J. S. (2003). The origins of genome complexity. *Science, 302*(5649), 1401–1404.

Mandel, J. R., Nambeesan, S., Bowers, J. E., Marek, L. F., Ebert, D., Rieseberg, L. H., et al. (2013). Association mapping and the genomic consequences of selection in sunflower. *PLoS Genetics, 9*(3), e1003378.

Maynard Smith, J., & Haigh, J. (1974). The hitch-hiking effect of a favourable gene. *Genetical Research, 23*(1), 23–35.

Meyer, S. E., Nelson, D. L., Clement, S., & Ramakrishnan, A. (2010). Ecological genetics of the Bromus tectorum (Poaceae)-Ustilago bullata (Ustilaginaceae) pathosystem: A role for frequency-dependent selection? *American Journal of Botany, 97*(8), 1304–1312.

Mitchell-Olds, T., & Schmitt, J. (2006). Genetic mechanisms and evolutionary significance of natural variation in Arabidopsis. *Nature, 441*(7096), 947–952.

Morris, G. P., Ramu, P., Deshpande, S. P., Hash, C. T., Shah, T., Upadhyaya, H. D., et al. (2013). Population genomic and genome-wide association studies of agroclimatic traits in sorghum. *Proceedings of the National Academy of Sciences of the United States of America, 110*(2), 453–458.

Muse, S. V. (2000). Examining rates and patterns of nucleotide substitution in plants. *Plant Molecular Biology, 42*(1), 25–43.

Muse, S. V., & Gaut, B. S. (1997). Comparing patterns of nucleotide substitution rates among chloroplast loci using the relative ratio test. *Genetics, 146*(1), 393–399.

Ness, R. W., Morgan, A. D., Colegrave, N., & Keightley, P. D. (2012). Estimate of the spontaneous mutation rate in Chlamydomonas reinhardtii. *Genetics, 192*(4), 1447–1454.

Ness, R. W., Wright, S. I., & Barrett, S. C. (2010). Mating-system variation, demographic history and patterns of nucleotide diversity in the Tristylous plant Eichhornia paniculata. *Genetics, 184*(2), 381–392.

Nordborg, M. (2000). Linkage disequilibrium, gene trees and selfing: An ancestral recombination graph with partial self-fertilization. *Genetics, 154*(2), 923–929.

Nordborg, M., Hu, T. T., Ishino, Y., Jhaveri, J., Toomajian, C., Zheng, H., et al. (2005). The pattern of polymorphism in Arabidopsis thaliana. *PLoS Biology, 3*(7), e196.

Olsen, K. M., Caicedo, A. L., Polato, N., McClung, A., McCouch, S., & Purugganan, M. D. (2006). Selection under domestication: Evidence for a sweep in the rice waxy genomic region. *Genetics, 173*(2), 975–983.

Olsen, K. M., & Wendel, J. F. (2013). A bountiful harvest: Genomic insights into crop domestication phenotypes. *Annual Review of Plant Biology, 64*, 47–70.

Olson-Manning, C. F., Wagner, M. R., & Mitchell-Olds, T. (2012). Adaptive evolution: Evaluating empirical support for theoretical predictions. *Nature Reviews Genetics, 13*(12), 867–887.

Ossowski, S., Schneeberger, K., Lucas-Lledó, J. I., Warthmann, N., Clark, R. M., Shaw, R. G., et al. (2010). The rate and molecular spectrum of spontaneous mutations in Arabidopsis thaliana. *Science*, *327*(5961), 92–94.

Pérusse, J. R., & Schoen, D. J. (2004). Molecular evolution of the GapC gene family in Amsinckia spectabilis populations that differ in outcrossing rate. *Journal of Molecular Evolution*, *59*(4), 427–436.

Phillips, P. C., Otto, S. P., & Whitlock, M. C. (2000). Beyond the average; the evolutionary importance of gene interactions and variability of epistatic effects. In J. B. Wolf, E. D. Brodie, & M. J. Wade (Eds.), *Epistasis and the evolutionary process*. New York, NY: Oxford University Press.

Pollak, E. (1987). On the theory of partially inbreeding finite populations. I. Partial selfing. *Genetics*, *117*(2), 353–360.

Pritchard, J. K., Pickrell, J. K., & Coop, G. (2010). The genetics of human adaptation: Hard sweeps, soft sweeps, and polygenic adaptation. *Current Biology*, *20*(4), R208–R215.

Qiu, S., Zeng, K., Slotte, T., Wright, S., & Charlesworth, D. (2011). Reduced efficacy of natural selection on codon usage bias in selfing Arabidopsis and Capsella species. *Genome Biology and Evolution*, *3*, 868–880.

Rose, L. E., Michelmore, R. W., & Langley, C. H. (2007). Natural variation in the Pto disease resistance gene within species of wild tomato (Lycopersicon). II. Population genetics of Pto. *Genetics*, *175*(3), 1307–1319.

Rosyara, U. R., Gonzalez-Hernandez, J. L., Glover, K. D., Gedye, K. R., & Stein, J. M. (2009). Family-based mapping of quantitative trait loci in plant breeding populations with resistance to Fusarium head blight in wheat as an illustration. *Theoretical and Applied Genetics*, *118*(8), 1617–1631.

Roux, C., Pauwels, M., Ruggiero, M. V., Charlesworth, D., Castric, V., & Vekemans, X. (2013). Recent and ancient signature of balancing selection around the S-locus in Arabidopsis halleri and A. lyrata. *Molecular Biology and Evolution*, *30*(2), 435–447.

Savolainen, O., Langley, C. H., Lazzaro, B. P., & Fr, H. (2000). Contrasting patterns of nucleotide polymorphism at the alcohol dehydrogenase locus in the outcrossing Arabidopsis lyrata and the selfing Arabidopsis thaliana. *Molecular Biology and Evolution*, *17*(4), 645–655.

Schierup, M. H., Mable, B. K., Awadalla, P., & Charlesworth, D. (2001). Identification and characterization of a polymorphic receptor kinase gene linked to the self-incompatibility locus of Arabidopsis lyrata. *Genetics*, *158*(1), 387–399.

Schmitz, R. J., Schultz, M. D., Urich, M. A., Nery, J. R., Pelizzola, M., Libiger, O., et al. (2013). Patterns of population epigenomic diversity. *Nature*, *495*(7440), 193–198.

Schmutz, J., Cannon, S. B., Schlueter, J., Ma, J., Mitros, T., Nelson, W., et al. (2010). Genome sequence of the palaeopolyploid soybean. *Nature*, *463*(7278), 178–183.

Schultz, S. T., & Lynch, M. (1997). Mutation and extinction: The role of variable mutational effects, synergistic epistasis, beneficial mutations, and degree of outcrossing. *Evolution*, *51*, 1363–1371.

Slotte, T., Bataillon, T., Hansen, T. T., St Onge, K., Wright, S. I., & Schierup, M. H. (2011). Genomic determinants of protein evolution and polymorphism in Arabidopsis. *Genome Biology and Evolution*, *3*, 1210–1219.

Slotte, T., Foxe, J. P., Hazzouri, K. M., & Wright, S. I. (2010). Genome-wide evidence for efficient positive and purifying selection in Capsella grandiflora, a plant species with a large effective population size. *Molecular Biology and Evolution*, *27*(8), 1813–1821.

Slotte, T., Hazzouri, K. M., Agren, J. A., Koenig, D., Maumus, F., Guo, Y. L., et al. (2013). The Capsella rubella genome and the genomic consequences of rapid mating system evolution. *Nature Genetics*, *45*(7), 831–835.

Slotte, T., Hazzouri, K. M., Stern, D., Andolfatto, P., & Wright, S. I. (2012). Genetic architecture and adaptive significance of the selfing syndrome in Capsella. *Evolution, 66*(5), 1360–1374.
Smith, N. G., & Eyre-Walker, A. (2002). Adaptive protein evolution in Drosophila. *Nature, 415*(6875), 1022–1024.
Stahl, E. A., Dwyer, G., Mauricio, R., Kreitman, M., & Bergelson, J. (1999). Dynamics of disease resistance polymorphism at the Rpm1 locus of Arabidopsis. *Nature, 400*(6745), 667–671.
Strasburg, J. L., Kane, N. C., Raduski, A. R., Bonin, A., Michelmore, R., & Rieseberg, L. H. (2011). Effective population size is positively correlated with levels of adaptive divergence among annual sunflowers. *Molecular Biology and Evolution, 28*(5), 1569–1580.
Strasburg, J. L., Scotti-Saintagne, C., Scotti, I., Lai, Z., & Rieseberg, L. H. (2009). Genomic patterns of adaptive divergence between chromosomally differentiated sunflower species. *Molecular Biology and Evolution, 26*(6), 1341–1355.
Sung, W., Ackerman, M. S., Miller, S. F., Doak, T. G., & Lynch, M. (2012). Drift-barrier hypothesis and mutation-rate evolution. *Proceedings of the National Academy of Sciences of the United States of America, 109*(45), 18488–18492.
Sweigart, A. L., & Willis, J. H. (2003). Patterns of nucleotide diversity in two species of Mimulus are affected by mating system and asymmetric introgression. *Evolution, 57*(11), 2490–2506.
The International Peach Genome Initiative (2013). The high-quality draft genome of peach (*Prunus persica*) identifies unique patterns of genetic diversity, domestication and genome evolution. *Nature Genetics, 45*, 487–494.
Tian, F., Bradbury, P. J., Brown, P. J., Hung, H., Sun, Q., Flint-Garcia, S., et al. (2011). Genome-wide association study of leaf architecture in the maize nested association mapping population. *Nature Genetics, 43*(2), 159–162.
Tian, D., Traw, M. B., Chen, J. Q., Kreitman, M., & Bergelson, J. (2003). Fitness costs of R-gene-mediated resistance in Arabidopsis thaliana. *Nature, 423*(6935), 74–77.
Varshney, R. K., Song, C., Saxena, R. K., Azam, S., Yu, S., Sharpe, A. G., et al. (2013). Draft genome sequence of chickpea (Cicer arietinum) provides a resource for trait improvement. *Nature Biotechnology, 31*(3), 240–246.
Vinogradov, A. E. (2003). Selfish DNA is maladaptive: Evidence from the plant Red List. *Trends in Genetics, 19*(11), 609–614.
Wang, E., Wang, J., Zhu, X., Hao, W., Wang, L., Li, Q., et al. (2008). Control of rice grain-filling and yield by a gene with a potential signature of domestication. *Nature Genetics, 40*(11), 1370–1374.
Whitney, K. D., Baack, E. J., Hamrick, J. L., Godt, M. J., Barringer, B. C., Bennett, M. D., et al. (2010). A role for non adaptive processes in plant genome size evolution? *Evolution, 64*(7), 2097–2109.
Wright, S. (1939). The distribution of self-sterility alleles in populations. *Genetics, 24*(4), 538–552.
Wright, S. I., & Andolfatto, P. (2008). The impact of natural selection on the genome: Emerging patterns in Drosophila and Arabidopsis. *Annual Review of Ecology, Evolution, and Systematics, 39*, 193–213.
Wright, S. I., Lauga, B., & Charlesworth, D. (2002). Rates and patterns of molecular evolution in inbred and outbred Arabidopsis. *Molecular Biology and Evolution, 19*(9), 1407–1420.
Wright, S. I., Ness, R. W., Foxe, J. P., & Barrett, S. C. H. (2008). Genomic consequences of outcrossing and selfing in plants. *International Journal of Plant Sciences, 169*, 105–118.

broad sense, i.e. including researchers seeking entry into this new area) develop their own conceptual frameworks so they can recognize and keep up with advances in this important area. The work and thought of Lewis, Ohno, Goldschmidt and more modern thinkers/researchers on duplication and selection, polyploidy, subfunctionalization, the gene balance hypothesis, balanced gene drive and genome dominance are covered. Most importantly, this lesson plan—with its 34 references—is based on the assumption that research in any area is a network of thought, not a timeline of facts. Since almost everything is unknown, what exactly do we students need to know urgently to approach intellectual satisfaction? This course should help students stay up-to-date with the bits of data that—presented with a "good idea"—should naturally fall into place within their own prepared minds.

1. FIRST DAY OF CLASS

This short course is designed for upper division or graduate students of evolution who have had at least one rigorous course in molecular genetics or molecular biology with a genetic emphasis. It is similarly applicable to researchers from other areas of science who see genome-based studies as an important new dimension of making research progress towards answers to long-standing questions in their area(s) and who have some background in molecular genetics and/or molecular biology. A total of thirteen 1.5 h sessions, in addition to this opening day, should be adequate, with up to 26 students, each responsible for presenting two or more of the key papers assigned with each lesson. Unfortunately, assigned papers—focal papers—are often published in elite journals that are not open access.

Here is a glossary of essential terms specific to any discussion of any sort of DNA duplication: http://genomevolution.org/wiki/index.php/Glossary.

This seminar course has no textbook (although one review is emphasized), Google is often recommended and students will frequently be directed to CoGepedia (http://genomevolution.org/wiki/index.php/Main_Page), the wiki attached to the comparative genomics toolbox I will use in this course. This toolbox is CoGe (http://genomevolution.org/CoGe/index.pl), a public resource running on servers at the US National Science Foundation-funded plant cyberinfrastructure project called iPlant.

There have been many reviews on the importance of DNA, gene, chromosomal segment and whole-genome duplication to some aspect of evolving organisms or populations. Here are four of my favourites for reference

(Comai, 2005; Freeling, 2009[1]; Semon & Wolfe, 2007; Van de Peer, Maere, & Meyer, 2009).

For further thought (today's assignment). In addition to testing yourself with the glossary in CoGepedia, you should become familiar with the several ways that DNA, genes (both tandem and dispersed), chromosomal segments and whole genomes can duplicate. For DNA, research the phrase "simple sequence" and understand conceptually the sort of sequences that are hardmasked (sequence turned into "n") by the application WindowMasker. How does WindowMasker work? One useful mechanism that is likely to produce simple sequence repeats (duplications and iterations) is slipped-strand mispairing during replication (Levinson & Gutman, 1987).

It is well known that some transposons pick up pieces of genes and move them around the genome. This too is a form of duplication and—like slipped-strand mispairing—forms part of the mutagenic background of all genomes (Bennetzen, 2005).

Focal paper. Bennetzen, J. L. (2005). Transposable elements, gene creation and genome rearrangement in flowering plants. *Current Opinion in Genetics & Development, 15*(6), 621–627. What do transposons have to do with duplication of genetic material? Clue: "PackMules".

2. LESSON 1: DUPLICATION OF GENES AND THE RELAXATION OF SELECTION

One sort of duplication, tandem or local gene duplication, is a regular occurrence in all eukaryotic genomes. It is easy to "see" tandem duplications if genomes and genomic regions are properly displayed. Even at the level of comparing one entire genome to another in a dot plot, the shape of the orthologous line of synteny might indicate a patch of local repeats; why is this? For example, here is a gene–gene syntenic dot plot of the genomes of two species in the same genus: *Arabidopsis thaliana* and *Arabidopsis lyrata* rendered in CoGe's SynMap application: http://genomevolution.org/r/99yj. This analysis was simplified because the comparison algorithm Quota-Align was set to 1:1, so only the single best syntenic line will be shown, the line of orthology. By clicking on any one "dot" or orthologous

[1] This essay/review is the fundamental single paper supporting this course. It is worth the price if you have to pay for it. Use this chapter now to clearly define the different sorts of DNA, gene and genome duplications that exist in genomes. Read and be ready to discuss these duplicate gene categories during this class session. Each sort of duplication changes the balance among genes in the genome in different ways. How so?

A. thaliana–A lyrata pair, the visualization reloads just that syntenic line(s) that represents one chromosome. Clicking on any dot once again brings up a sequence comparison in a new application called GEvo. Here is such an example: http://genomevolution.org/r/99yz. These two stretches of orthologous chromosome are obviously syntenous (click on "hits" to create line of sequence similarity), but there are several *Cytochrome P450* genes on both *A. thaliana* and *A. lyrata*. The patch of tandems is syntenic, but the details of gene number are divergent in the two genomes. Click on gene models to identify genes; Shift–Command click on any HSP to draw all possible lines of sequence similarity at the same time. Play around with SynMap and GEvo; there are tutorials in CoGepedia. SynMap will be used in subsequent lessons.

Studies of mutant gene phenotypes specified by knockouts of genes that were, in fact, tandem repeats are the origin of the single most fundamental idea of gene duplication: if a gene is duplicated, then at least one of the genes can be mutationally altered and survive purifying selection (because the mutant gene is covered). Perhaps the most famous example of tandemly repeated genes is the ANT-C and BX-C clusters of diverged homeobox genes in *Drosophila* and, in one form or another, every animal including humans. Lewis showed that the transcription factors encoded by each of these individual repeats specified the exact identity of a parasegment in the fruit fly. So, one such gene encodes the anterior half of wing identity, for example, research ANT-C and BX-C.

Although the idea that the relaxation of purifying selection on any one copy of a local duplicate fuels/accelerates the evolution of novel gene function may well predate the work of Lewis, it was Lewis who eventually proved that "pseudoalleles" were actually duplicate genes that had diverged to similar, but different, novel functions (Lewis, 1951).

Focal paper #1. Lewis, E. B. (1951). Pseudoallelism and gene evolution. *Cold Spring Harbor Symposia on Quantitative Biology, 16*, 159–174. Read through this quickly. This classic paper is difficult to understand due to vocabulary.

Focal web source #1. See "*HOX* gene" in Wikipedia. If you want more detail on the evolution of this tandem gene array from its inferred "ProtoHox" origin to other end points in the animal phylogenetic tree, there is a clear but necessarily complicated review in which facts are clearly differentiated from inferences (Garcia-Fernandez, 2005).

It is tempting to think that two tandem genes that are very similar in sequence are somehow "younger" than another tandem pair where the two genes are more diverged. In fact, since genes that can pair can "convert" (by genetic recombination) their sequence one to another, this method of ageing is invalid. Tandem repeats are frequent in all eukaryote genomes, including the yeasts. Read the second focal paper to better understand how to measure the age of tandem repeat events and how to avoid overestimating tandem duplication rates in events per generation (Gao & Innan, 2004).

Focal paper#2. Gao, L. Z., & Innan, H. (2004). Very low gene duplication rate in the yeast genome. *Science, 306*(5700), 1367–1370.

Sometimes, the area of conversion of one chromosomal region to another homologous region is very long (Wang, Tang, Bowers, Feltus, & Paterson, 2007). How long is the region in the case of rice chromosomes 11 and 12?

3. LESSON 2: MEASURING GENETIC DIVERGENCE AND ITS MEANING

Hurst (2002) had prepared a very short and useful essay on how to use a pair of homologous open reading frames to estimate both divergence and the overall intensity of selection on the function of the encoded protein. The metrics here are the neutral base substitution rate (Ks) and the base substitution rate of changes that could be under some level of selection (Ka) and how Ks is one metric to quantify divergence and Ka/Ks is one way to estimate whether or not a sequence is under selection. The second focal paper (Liman & Innan, 2003) uses Ka/Ks to provide evidence that there is a vestigial set of genes in Old World monkeys, including ourselves, that once functioned to transduce pheromones, those chemicals that cause stereotypical sexual behaviour.

Focal paper #1. Hurst, L. D. (2002). The Ka/Ks ratio: Diagnosing the form of sequence evolution. *Trends in Genetics, 18*(9), 486.

Focal paper #2. Liman, E. R. & Innan, H. (2003). Relaxed selective pressure on an essential component of pheromone transduction in primate evolution. *Proceedings of the National Academy of Sciences of the United States of America, 100*(6), 3328–3332.

For further thought. How can Ka, Ks and ratios help see whether or not a gene functions and provide an indication of how this function has been selected for or against?

4. LESSON 3: VISUALIZING DNA SEQUENCE DIVERGENCE USING GEvo IN CoGe RUNNING ON iPLANT SERVERS (USE FIREFOX)

There is no focal paper for this lesson; read or watch CoGepedia tutorials in CoGe if you need help learning how to use GEvo.

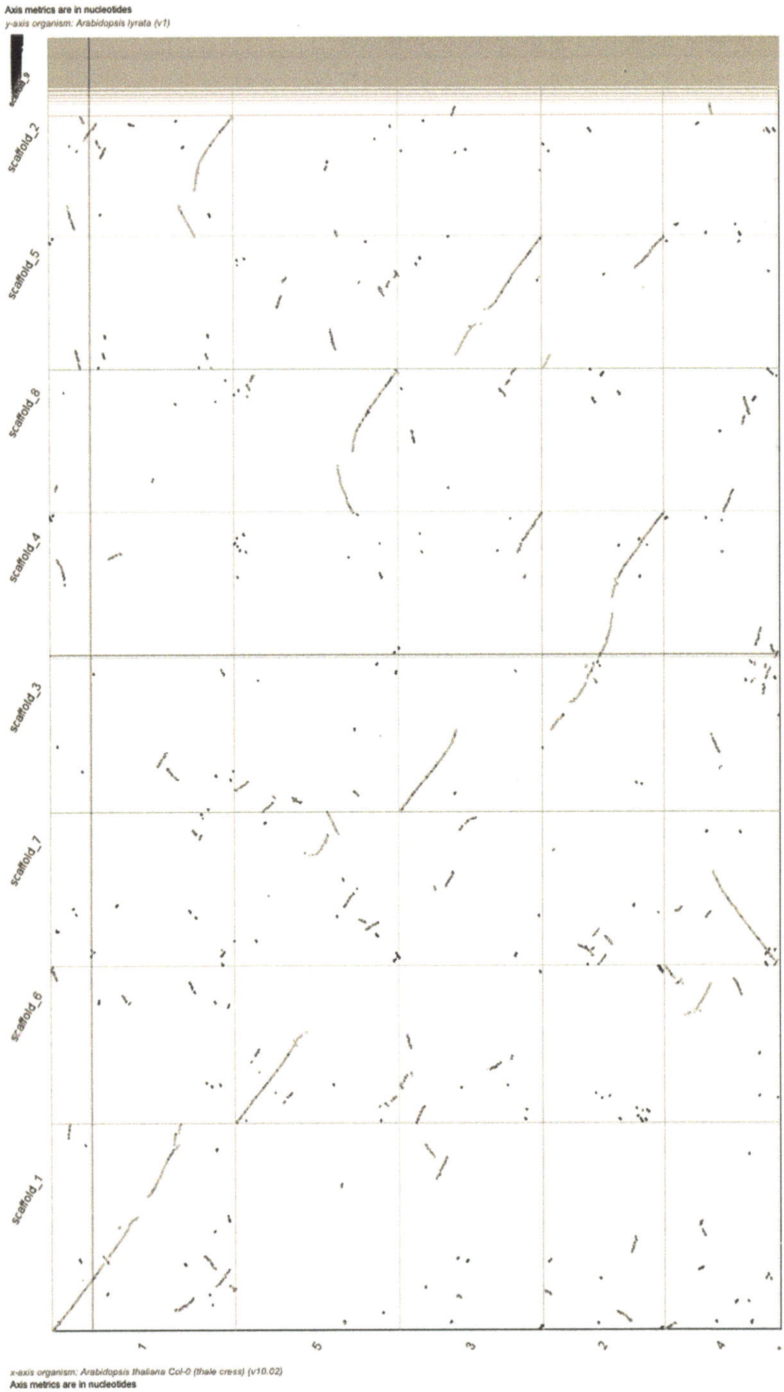

Figure 13.2 (A) This SynMap panel is almost but not quite identical to what you should see when you visit the iPlant servers via the link http://genomevolution.org/r/9a6t and click "OK". The lines comprised of syntenic gene pairs are "naturally" formed because the order of homologous genes in *Arabidopsis thaliana* (*X*-axis) is often similar to the order of homologous genes in *Arabidopsis lyrata*. Each dot is a pair of genes, and the Ks of that gene pair has been calculated automatically and rendered as a colour per (B).

(*Continued*)

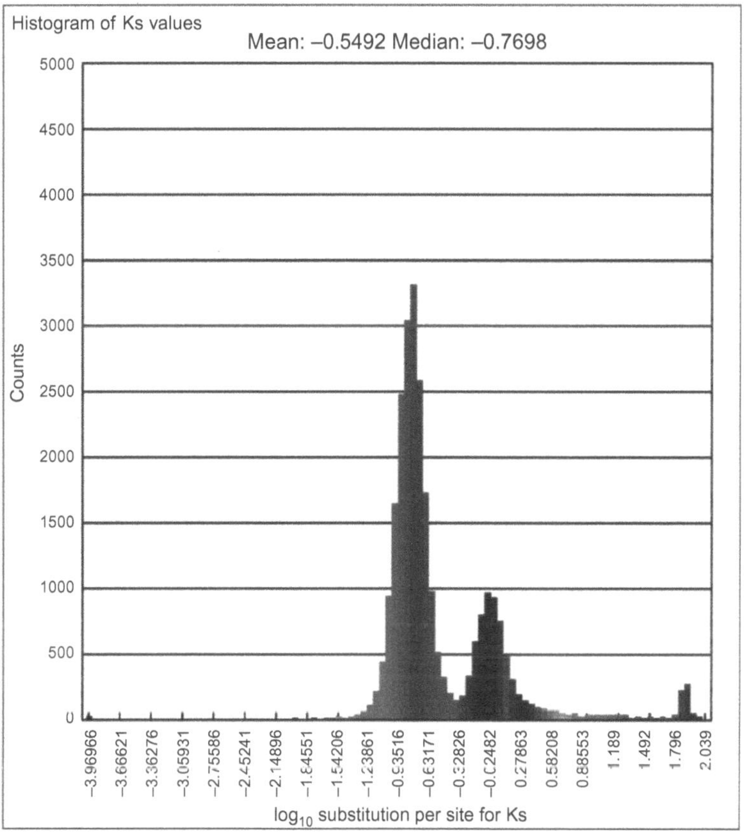

Figure 13.2—Cont'd The orthologous and most recently diverged genes are in light blue. The red syntenic lines are remnants from the most recent tetraploidy shared by all species closely related to *Arabidopsis*. The reason there are no additional syntenic lines or stray dots reflecting more ancient duplication events and noise is because SynMap was told, via the program "Quota-Align", to analyse these two species to a "syntenic depth" of 2:2 and no deeper. If you use the link earlier (not (A)), Quota-Align has been turned "off" and *all* homologous hits are reflected in the dot plots. See (B). Exercise: is it true that adding up the lines of the same colour (the same level of divergence or age) approximately reconstructs the entire genome? (B) This is the log distribution of Ks values (*Y*-axis) for all "dots" or pairs of homologous genes (*X*-axis) between *Arabidopsis thaliana* and *Arabidopsis lyrata*, as derived from the SynMap experiment (http://genomevolution.org/r/9a6t). The modes of Ks defined here are the *A. thaliana*–*A. lyrata* orthology (the divergence time of these two species, this being bluish green), and the red mode reflects the most recent ancient tetraploid shared by these two species. There exist many small blue—the higher Ks shoulder of the "red" mode—lines in the output of the SynMap experiment because Quota-Align was turned "off". These blue lines are thought to reflect more ancient whole-genome duplication events. *Reprinted from Freeling (2009), with permission from Annual Reviews.* (See colour plate.)

6. LESSON 5: POLYPLOIDS

Because there are new species of eukaryotes being sequenced *ab initio* every week or so, the only way to keep up-to-date is to frequently add annotations manually to the phylogenetic trees that interest you or find a reliable website that does this for you. Since our focus is in higher plants, the appropriate section of CoGepedia is informative: as of 29 August 2013, there were 24 ancient polyploidy events affixed to the vascular plant phylogenetic tree, with citations and information. One simple way for two tetraploidies in a row to lead to a hexaploidy (a genomic tripling) is also drawn out. So, counting hexaploidies as two tetraploidies, 10 whole-genome duplications (tetraploidies) separate maize from *Arabidopsis*. Discuss what you have learned.

For further thought. Find lineages in vascular plants that represent the maximum and minimum of tetraploidies. How might knowing these lineages prove to be useful?

In general, a tetraploid is either an autotetraploid or an allotetraploid. This distinction is featured in your focal paper, as are some ideas on why tetraploidies are so common in higher plants (Comai, 2005).

Focal paper #1. Comai, L. (2005). The advantages and disadvantages of being polyploid. *Nature Reviews. Genetics, 6*, 836–846.

Because we researchers are human beings, there is a natural focus on our own lineage, eukaryotic animals. At the base of the vertebrate lineage, after the branch to amphioxus but before the branch to sharks, there is evidence for two consecutive ancient tetraploidies. Between that paired event, called "2R" (approximately 500 MYA) and humans living today, there have been no further whole-genome duplications (although the bony fish lineage has had some). Ohno is generally credited with amassing evidence in support of the importance of whole-genome duplication to understanding the evolution of animals (Ohno, 1970). The essence of Ohno's work is captured in reviews (Meyer & Schartl, 1999; Van de Peer et al., 2009), and Ohno did have one accessible journal article summarizing his data and logic before publishing his book in 1970 (Ohno, Wolf, & Atkin, 1968).

Focal paper #2. Van de Peer, Y., Maere, S., & Meyer, A. (2009). The evolutionary significance of ancient genome duplications. *Nature Reviews. Genetics, 10*, 725–732.

For yet further thought. Post hoc ergo propter hoc. This fallacy of rhetoric means because an action precedes an event in time does not mean that the action caused the event. However, the action could have done so. It is important, when practising critical thinking, to not discard everything illogical, because doing so might well discard the truth. The truth exists apart from all argument. Discuss.

Find some references as to trends in evolution or trends in the "maximums" of ability or complexity, for any lineage, but include both vascular plants and vertebrates. This is a heavy topic, so students with a tolerance or respect for vague, nuanced and sometimes ingenuously bogus arguments should relish this assignment. Please note that most animal comparative developmental biologists agree that the *morphological* complexity—as evidenced in what is often called "the body plan"—of a fish and a man is about the same, and it is not crystal clear that amphioxus, without 2R, has a more or much more primitive body plan than the vertebrate one. The prevertebrate amphioxus certainly has a notochord (positioned as the vertebral axis), mouth, gills, segments and anus (but not jaws or paired fins/limbs). The genome release paper for amphioxus (Putnam et al., 2008), while not a focal paper for this lesson, clarified the placement of 2R in the chordate phylogenetic tree and informs all questions of how gene duplication might "influence" or "fuel" or "drive" increases in the complexity of body plans or, for that matter, any other measurable characteristic of life.

Lesson 11 will introduce a focal paper that attempts to define morphological complexity and goes on to postulate that this sort of complexity has increased its maximums from the first vascular plant in the mid-Silurian (approximately 425 MYA; Google Cooksonia) to this day. However, additional fossils from this same period suggest that early plant morphological complexity may have been underestimated (Kotyk, Basinger, Gensel, & de Freitas, 2002). Finding out whether or not the duplication of genetic material actually "caused" increases in morphological complexity is dependent on this complexity increase being true or at least probable. In fact, there is debate whenever the subject of increasing complexity of most anything is mentioned in an evolutionary context. Achieving intellectually satisfying understanding when it comes to evolutionary trends is challenging. When researching the general area of increasing complexity during evolution, expect a lot of clutter— even metaphysical clutter.

7. LESSON 6: FRACTIONATION

As you should have learned from the previous lesson, saying that a lot of important evolution followed polyploidy does not mean that polyploidy somehow caused anything new and profound or drives any trends.

Posttetraploid fractionation has been put forth as a population genetic-type mechanism that might ensure genetic isolation between descending lineages. This work originated from data following the tetraploidy in the fungal lineage of bread yeast (Scannell, Byrne, Gordon, Wong, & Wolfe, 2006). Describe this scheme. How would one test whether or not this clever scheme accounts for lineage-specific differences in rates of successful speciation? What are some other similar mechanisms that could help explain changes in species richness?

Focal paper #2. Scannell, D. R., Byrne, K. P., Gordon, J. L., Wong, S., & Wolfe, K. H. (2006). Multiple rounds of speciation associated with reciprocal gene loss in polyploid yeasts. *Nature, 440*, 341–345.

For further thought. Given that we are dealing with a plant chromosome, if there is a *cis*-acting unit of smaller features on the chromosome, and one short deletion removes the function of one of these features, what do you expect to happen to the rest of the parts of the *cis*-acting unit? Let us say that this gene is a singlet; that is, there is no possibility for functional redundancy. Discuss.

For additional further thought. Go back to CoGepedia and the phylogenetic tree of plants decorated with known polyploidy events and calculate the expected gene content of a few plants living today if fractionation had not happened.

And more thought. In humans, pheromone receptor network genes are not functional, but they have not been removed as DNA; they are still around the genome as pseudogenes (Liman & Innan, 2003). These pheromone network genes still function in New World monkeys, but do not function in Old World monkeys such as humans. In plants, nonfunctional genes are deleted entirely (unless they are in "heterochromatin" or other recombination-resistant region). So, plants or plant chromosomes must be different from those in mammals, right? Discuss this. What could account for the way plants delete away functionless genes leaving no trace while mammals leave gene corpses around to decompose by point mutations happening at random? The rate of synonymous base substitution (Ks; Lesson 2) is not that different per million years for mammals and plants. So, what could be different? Discuss.

8. LESSON 7: SUBFUNCTIONALIZATION, NONFUNCTIONALIZATION (AND NEOFUNCTIONALIZATION)

In 1999, Force et al. (1999), then a graduate student in the Postlethwait Lab at University of Oregon, published a model that explained how a pair of duplicate genes simply must behave, and in doing so, he and co-workers outlined the subfunctionalization hypothesis to explain why there are "too many" duplicate genes preserved after an ancient tetraploidy. The following must-read paper (Lynch & Force, 2000) follows up on this idea:

Focal paper #1. Lynch, M., & Force, A. (2000). The probability of duplicate gene preservation by subfunctionalization. *Genetics, 154*, 459–473.

Let us say a gene has two *cis*-acting regulatory sites (+ and +) that both must be operative to encode a function in both root and shoot, where one site specifies root and the other shoot. A wild-type gene is a + + gene. After tetraploidy, there are now two of these genes, and thereafter, regulatory double-site losses can happen without losing function in either root or shoot.

+	+	gene	−	+	gene
−	−	gene	+	−	gene

The first is called "nonfunctionalization" and the second "subfunctionalization". These outcomes of double mutation were seen as neutral, since only redundant functions were removed. However, in subsequent generations, both copies of a subfunctionalized gene were necessary to have in the same cells, so this duplicate pair was permanentized while the mutant gene of the nonfunctionalized pair could be lost without selective consequence. Since subfunctionalization is neutral and should occur, these authors postulated that subfunctionalization was a cause of overretention of pairs following ancient tetraploidies.

For further thought. If a gene is particularly subfunctionalizable as a post-tetraploid pair, do you think it would also be subfunctionalizable if the redundancy had been caused by a tandem duplication?

For yet further thought. Consider this diagram of a doublet of genes retained in *Brassica rapa* from a hexaploidy that happened about 13 MYA:

5′ −++++−++−++++−+++−−++−+++ gene −+ 3′

5′ ++++++++++−−+++++++++++++ gene ++ 3′

A "-" in the previously mentioned gene doublet marks the position of a deleted *cis*-acting regulator. In fact, the "regulator" is monitored as a CNS like those studied in Lesson 3. Although this pair of genes is almost nonfunctionalized, it is actually subfunctionalized, and—according to the theory—potentially permanentized. Given this example, do you think that certain sort of genes might be more subfunctionalizable than others? If so, then these should be overretained posttetraploidy. Should a tandem duplicate be just as subfunctionalizable as a duplicate derived from a tetraploidy? Discuss.

Following the tradition of using fancy words with many syllables, two terms, in addition to subfunctionalization and nonfunctionalization, are in common use when cataloguing the possible outcomes of being a duplicate gene: *neofunctionalization* and *afunctionalization*. The former is used when one or both of the duplicates somehow acquire new function, presumably via the addition of new genetic material or by the creation of a new sequence by ligating the ends following a deletion or rearrangement. The second term is when no mutations happened to the duplicate at all.

For yet further thought. Research on how transposons inserted near/in genes have caused new function. Google "transposon insertion position effects" as a start. Present these cases for neofunctionalization. Maybe subfunctionalized or nonfunctionalized genes are more susceptible to becoming neofunctionalized later. There are data and hypotheses that support these possibilities. Clue: here is one of my favourite papers: Lynch, Leclerc, May, and Wagner (2011).

9. LESSON 8: THE GENE BALANCE HYPOTHESIS

Duplication of genetic material has the capacity to change the balance of gene dosage and, in so doing, change the balance of gene products inside the cell. Further, different sorts of duplication change balance in different ways; you should understand these different ways by now. The gene balance hypothesis derives from several lines of inquiry. Although reading reviews is rarely a good way to learn a subject, there are always exceptions. Birchler & Veitia (2007)—two of the three pillars of this theory (the third is Hurst; Papp, Pal, & Hurst, 2003)—wrote a lucid review:

Focal paper #1. Birchler, J. A., & Veitia, R. A. (2007). The gene balance hypothesis: From classical genetics to modern genomics. *The Plant Cell, 19*, 395–402.

Test your understanding of the gene balance hypothesis. "Because some gene products interact with set stoichiometries, there is a tendency for purifying selection to not recognize whole-genome duplications because all genes in the genome change dosage contemporaneously, so balances are unchanged. On the other hand, duplications of individual genes whose products polymerize into multisubunit machines (like ribosomal protein genes) are expected to cause imbalance and be purified from the gene pool. The overall result is that purifying selection reinforces the inheritance of the *status quo* of balance from one generation to the next". If this statement makes no sense, reread Focal paper #1 for understanding. Not every concept is easy to understand.

Yet further thought. Exactly how does this hypothesis predict different fates for different sorts of duplication: tandem, whole-genome, chromosomal segment or dispersed. Again, if an answer to this question seems out of reach, reread Focal paper #1 for understanding.

10. LESSON 9: DIFFERENT SORTS OF GENE DUPLICATION CHANGE THE GENE CONTENT OF THE GENOME IN DIFFERENT WAYS, AND HOW THE SUBFUNCTIONALIZATION IDEA BECAME DOWNGRADED

Your general reviews cite the various laboratories that have documented which sorts of genes tend to be found in tandem arrays, which in retained gene pairs following ancient tetraploidies and which move around during evolution. Tandems and posttetraploidy retention duplicate different sorts of genes than does tetraploidy, as reviewed (Freeling, 2009). Those genes that move around and, for that reason, are easy to duplicate by segregation are a special subset of gene categories as well (Freeling et al., 2008). While my own 2009 review (Freeling, 2009) is not the only or even best review on how genomes have changed their gene contents, I have it as a focal reading because the figures illustrating the message of this lesson are explained there in detail.

Focal reading. Freeling, M. (2009). Bias in plant gene content following different sorts of duplication: Tandem, whole-genome, segmental, or by transposition. *Annual Review of Plant Biology, 60*, 433–453. Figure 13.4 and Tables 13.1 and 13.2 are explained in detail in this chapter. Discuss the meaning of these data.

If you do not know what a gene's "GO term" means, then Google that. If you are concerned about, since a gene can have many GO terms, how did these tables come up with just one per gene, then read your focal paper in detail. The best way to get your questions answered is to answer them yourself.

For further thought. Although subfunctionalization does not explain the gene content changes that occurred following the most recent tetraploidy in the *Arabidopsis* lineage, that does not mean that subfunctionalization is not important. Subfunctionalization is neutral and must happen. That is why the subfunctionalization idea is beautiful. Subfunctionalization is not disproved; it is simply downgraded from being an explanation for the over-retention of genes posttetraploidy, at least for the first 100 million years or so. (How about for the longer term or for special sorts of genes? Is there room for subfunctionalization is such cases?)

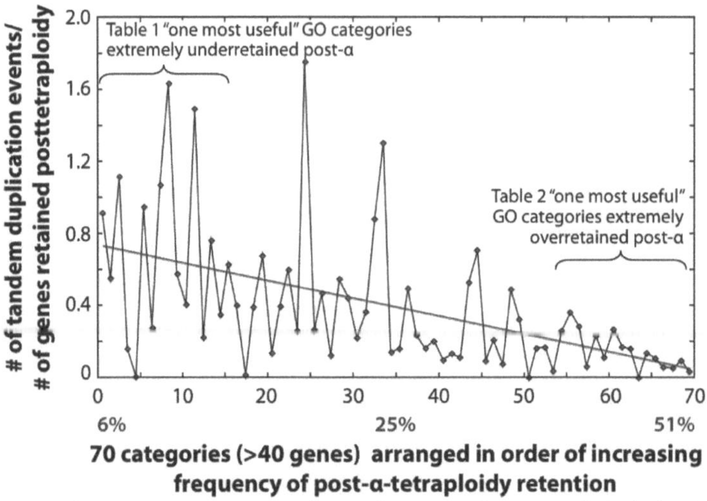

Figure 13.4 There is a negative correlation between the sorts of *Arabidopsis* genes retained following tandem duplication events and those retained following whole-genome duplication. On the *X*-axis are the GO categories of genes retained as pairs increasing from a low of 6% (see Table 13.1 for a list) on the left to 51% on the right (see Table 13.2 for a list). On the *Y* is a metric for the number of tandem duplication events for these same GO categories. *Reprinted from Freeling (2009), with permission from Annual Reviews.*

For yet more thought. Consider the statement: the GO categories in Tables 13.1 and 13.2 are really different. Table 13.2 genes encode products that are "connected" to other gene products, while the proteins encoded by the Table 13.1 genes often do not participate in multisubunit complexes. Does this make sense?

For even more thought. What is to stop repeated tetraploidies in a lineage from accumulating "connected genes" to inelegant levels? Images of "Rube Goldberg machines" are available on the Internet; these define "inelegant".

11. LESSON 10: DRIVE

Using the word "drive" in evolutionary discussions is dangerous, and that is because it has been associated with some ideas that are now thought to have been oversold, especially the "molecular drive" idea of Dover (Google this). "Drive" is a powerful and pretentious word that should be used only when you want "drive" to mean essentially what it means in the phrase "meiotic drive". The chromosomal mechanisms and power sources of

Table 13.1 The 15 least α-retained GO terms from the left side of Fig. 13.3, local (tandem) duplication preferred and listed from least to most extreme

"One most useful" gene ontology (GO) ID #, description	Number of genes	Frequency of α-retention[a]	Frequency of tandem events	Tandem[b]/α
GO:0046872, metal ion binding	79	0.20	0.07	0.35
GO:0016491, oxidoreductase activity	290	0.20	0.15	0.76
GO:0004871, signal transducer activity	80	0.19	0.04	0.22
GO:0004289, subtilase activity	52	0.17	0.26	1.49
GO:0008270, zinc ion binding	174	0.17	0.07	0.41
GO:0004194, pepsin A activity	71	0.17	0.10	0.57
GO:0019825, oxygen binding	208	0.17	0.28	1.63
GO:0004888, transmembrane receptor activity (vague, picked-over)	111	0.16	0.17	1.07
GO:0008168, methyltransferase activity	50	0.16	0.04	0.27
GO:0016740, transferase activity	65	0.15	0.15	0.95
GO:0003755, peptidyl-prolyl *cis–trans* isomerase activity	53	0.15	0.00	0.00
GO:0003899, DNA-directed RNA polymerase activity	47	0.15	0.02	0.16
GO:0005524, ATP binding	171	0.13	0.15	1.12
GO:0008026, ATP-dependent helicase activity	65	0.12	0.07	0.55
GO:0030533, triplet codon-amino acid adaptor activity	631	0.06	0.05	0.91

[a]Data from Figure 13.4 sorted in descending order.
[b]From the y axis of Figure 13.4: third column value over second column value, which equals the # tandem events/# retained genes after α-tetraploidy.

Table 13.2 The 15 most α-retained GO terms from the right side of Fig. 13.3, α-retentions preferred and listed from most to least extreme

"One most useful" gene ontology (GO) ID #, description	Number of genes	Frequency of α-retention[a]	Frequency of tandem events	Tandem[b]/α
GO:0008233, peptidase activity (includes proteasome core genes)	55	0.51	0.02	0.04
GO:0004712, protein serine/threonine/tyrosine kinase activity	47	0.49	0.04	0.09
GO:0003735, structural constituent of ribosome	352	0.42	0.02	0.05
GO:0004722, protein serine/threonine phosphatase activity	141	0.41	0.02	0.05
GO:0003700, transcription factor activity (vague on purpose)	1675	0.41	0.04	0.10
GO:0005509, calcium ion binding	139	0.41	0.05	0.13
GO:0003743, translation initiation factor activity	61	0.41	0.00	0.00
GO:0030599, pectinesterase activity	70	0.40	0.06	0.16
GO:0003779, actin binding	53	0.40	0.07	0.17
GO:0046910, pectinesterase inhibitor activity	64	0.38	0.10	0.26
GO:0005525, GTP binding	134	0.37	0.04	0.11
GO:0016887, ATPase activity (vague, picked-over for specificity)	87	0.37	0.08	0.23
GO:0005516, calmodulin binding	98	0.37	0.02	0.06
GO:0008415, acyltransferase activity	71	0.37	0.10	0.28
GO:0005199, structural constituent of cell wall	41	0.37	0.13	0.36

[a]Data from Figure 13.4 sorted in descending order.
[b]From the y axis of Figure 13.4: third column value over second column value, which equals the # tandem events/# retained genes after α-tetraploidy.

meiotic drive are known, and it is proved that these mechanisms "drive" particular genes, those linked to a particular site on a chromosome, into the population (Zimmering, Sandler, & Nicoletti, 1970). For example, if a region of chromosome carries a site that extends microtubules such that it "leads the way" to the poles during meiotic divisions, and if a megagamete develops only from the haploid cell on the outside of a linear tetrad, then this special region of chromosome, dragging its genes with it, will be preferentially inherited via the female and therefore driven into the population. This is an example of the way meiotic drive should be used in evolutionary biology today, and this is how I will use it.

Thus, the word "drive" connotes a *known* mechanism and a predictable one-way movement of genetic information nonrandomly into the population. So, the drive machine, just by existing, causes particular genes to pass preferentially to the next generation.

Focal assignment. Find any example of meiotic drive and explain; make sure the mechanism and its source of power (something using ATP) are made clear.

For further thought. Find at least one example of the use of the word "drive" in the general area of evolution, genetics or population genetics. Comment.

12. LESSON 11: BALANCED GENE DRIVE

Assuming that the maxima of morphological complexity of plants have increased since plants came out of the water in the mid-Silurian, some mechanism must explain this. There is one paper (Freeling & Thomas, 2006) that not only attempts to support the contention that morphological complexity has risen in plants but also hypothesizes that repeated tetraploidies constitute *drive* to increase morphological complexity. This paper couples two propositions, neither of which is proved to be correct. Although the generally considered "typical" land plant of 430 MYA was the extinct *Cooksonia* sp., and this plant was essentially a root stalk with vasculature going up to a round sporangia (Google this), paleontologists have found additional complexity when more complete fossils are examined (Kotyk et al., 2002).

For further thought. Find examples of mid-Silurian fossil plants and think about modern vascular plants. How could you argue that plants have/have not gained in maxima of morphological complexity from the first land plant until today. Please do this before you read the focal paper.

Focal paper. Freeling, M., & Thomas, B. C. (2006). Gene-balanced duplications, like tetraploidy, provide predictable drive to increase morphological complexity. *Genome Research, 16*, 805–814. It is best to have a testable idea, but not so testable that the idea should be tested before publication. What about this idea?

The evidence that the tetraploidies that happened at the base of the vertebrates, and at two other instances in vertebrate evolution, somehow caused any particular trends in animal phenotype is not strong, mainly because there is no clear trend of rising morphological complexity from fish to man (Donoghue & Purnell, 2005). While the case for rising morphological complexity is better in plants (Focal paper), there is also a question that the first vascular plants might have been more morphologically complex than thought originally (Kotyk et al., 2002) due to incomplete fossil sampling. If you enjoy evolutionary arguments based on inadequate data and know-it-all rhetoric, enter the field of evolution and especially the subfield of biological complexity.

One sort of complexity has certainly risen as a result of repeated tetraploidies: that is the complexity of genetic information.

For yet further thought. During the 1940s and 1950s, the evolutionary genetic community was working out what is now called "the modern synthesis", the merger of Mendelian genetics with Darwinian theory. This is the textbook "Theory of Evolution". The modern synthesis engendered the new field of population genetics. Central to this synthesis was an emphasis on selection and rejection of any sort of mutation—or coordinated barrage of mutations—that, once present, played outsized roles in the direction and speed of evolution. In the arcane terms of evolutionary biology, *selectionist* theory carried the day and *mutationist* theory was rejected. Goldschmidt, with his "hopeful monsters", was one of the most famous proponents of mutationist theory. A particularly lucid paper by Goldschmidt will be a focal paper for the last lesson of this course, so you might want to read it early (Goldschmidt, 1953). Goldschmidt was especially troubled by how little we knew of "the intimate architecture of the chromosome" and was not ready to turn the directionalities of evolution over to random point mutations and small population sizes. Since tetraploidy is a "macromutation" and it may be seen to drive evolution in particular directions, balanced gene drive is one of the best examples of a mutationist mechanism. In Lesson 12, there will be a particular experiment designed to differentiate between selectionist and mutationist explanations for a posttetraploidy problem called "genome dominance". As will be seen, the selectionist alternative was supported by the data. Find a few examples, preferably real, of phenomena that could be explained by either selectionist or mutationist thought and how they turned out. Discuss.

13. LESSON 12: GENOME DOMINANCE

Recall from Lesson 6 that fractionation of the two subgenomes that come together originally to form a wide hybrid and then a tetraploid is rarely even: one homeologous stretch of chromosome has significantly more genes than the other. That is, one subgenome in an ancient tetraploid is overfractionated, while the other is underfractionated (more intact). This uneven aspect of fractionation occurs regularly in all sorts of eukaryotes. The explanation for this typical result of ancient whole-genome duplication was found through studies in maize (Schnable, Springer, & Freeling, 2011).

If you are troubled by the fancy word "homeologous", know that the original word has one more syllable, "homoeologous". They are synonyms, so look either of them up if you need to. Do not bother your instructor.

Focal paper. Schnable, J. C., Springer, N. M., & Freeling, M. (2011). Differentiation of the maize subgenomes by genome dominance and both ancient and ongoing gene loss. *Proceedings of the National Academy of Sciences of the United States of America, 108*, 4069–4074.

The primary conclusion of this chapter is that the underfractionated genome carries genes that are relatively expressed to message levels significantly higher than levels expressed by the homeologous genes. So, the reason that one subgenome loses genes more than the other is that deleting (fractionating) this gene preferentially did not matter too much to the total message level because this gene expressed relatively less.

Test your understanding. There is a control experiment that is the "crown jewel" of this paper. (Often, it is the control of an experiment on which the most profound conclusions rest.) This control was designed to differentiate between selectionist and mutationist explanations for genome dominance. A mutationist explanation would be something like the following: the two posttetraploid genomes are structurally dissimilar in a way that was probably inherited from the parents of the tetraploid. If this were true, then one chromosomal might be less "deletable" than the other. What control tested for this mechanist alternative? Of course, the result of that control was that the selectionist explanation prevailed: deletions actually must occur on both homeologous chromosomes at an equal frequency, but deleting genes from the underexpressed subgenome just did not matter as much, and these sometimes escaped purifying selection. This was not the control experiment result that I had expected. Understand this argument.

For further thought. So, how could a whole genome get tagged to express less—on average—than another whole genome? The answer to this question is ready to be known at the time of this course outline. Such a whole-genome "mutational event" would probably fit Goldschmidt's definition of "systemic mutation". Reading new papers using the query "genome dominance" should quickly discover what really happened after 2014. It is likely that small RNAs and some sorts of epigenetic tags are involved.

For even further thought. Can you think of a way to use Ka and Ks (Lesson 3) to test the hypothesis that the recessive subgenome is under more relaxed selection? Additionally, can you guess whether or not those classical genes of maize, those that have been knocked out to get the most studied phenotypes, tend to reside on the recessive or dominant subgenome and why? The answer to this question is known.

The phenomenon of genome dominance can be discussed within the gene balance hypothesis. Please follow this train of thought.

A romantic hypothesis. It would be romantic if the underexpressed, most "beat up", subgenome were preferentially the origin of novelty. Is this testable? Discuss.

14. LESSON 13: CHIMPS, MAN AND A GOODBYE PRESENT

In case you are thinking that repeated tetraploidies might provide an explanation of most everything newly evolved, think again. There is no doubt that a man is more *capable* than a chimp, a bear or a fish. There can be no doubt that, within vertebrates, there has been an increase in the capability to anticipate and avoid threat, defend oneself in such ways as to prevent recurrence of the threat and—finally—defend oneself with clever combinations of subterfuge, misinformation and surprise attack that only humans could accomplish. Let us call this capability "cleverness". And yet, we must be careful; the average chimp is a far more able gymnast than the most accomplished human, so our definition of "capability" must be precise. Whatever it is that explains the trend of increasing maximums of "cleverness" in vertebrates, it cannot be repeated tetraploidies! There is some evidence that tandem duplications are involved (Geschwind & Konopka,

2012). There will surely be much more data pertaining to this topic in the years to come.

Focal paper #1. Geschwind, D. H., & Konopka, G. (2012). Neuroscience: Genes and human brain evolution. *Nature, 486*(7404), 481–482.

For further thought. According to the gene balance hypothesis, tandem repeats and dispersed (out-of-synteny) repeats should preferentially increase the content of genes that "do not play well with others" because that is thought to be why they do not have haplo- or triploinsufficient, negative phenotypes. This must be a clue towards understanding the evolution of "cleverness". Discuss.

As a goodbye present, here is a short paper by Goldschmidt (1953). This is the paper I suggested previously. I first read this essay over 40 years ago and have read it at least once a year ever since. I still find it fresh and provocative. (I think Goldschmidt fully anticipated the macromutagenic power of tetraploidy and would have loved balanced gene drive.) Perhaps what I find compelling about this essay is not only that I feel "the truth" nearby. I admire how Goldschmidt would not bow to convention until particular sorts of data were explained by convention. Goldschmidt wanted to know "the intimate nature of the chromosome" and epigenetics was not even a word yet. Do you agree that Goldschmidt's questions have not yet been explained? So far, what do you think of Goldschmidt as a thinker? Does his "imperious" personality detract from his ideas? Why do you feel this way? His bullheadedness?

Focal paper #2. Goldschmidt, R. B. (1953). Evolution as viewed by one geneticist. *American Scientist, 40*, 84–98.

A final thought. You may find it positive, negative or neither. Goldschmidt ponders why the origin of all the big phylogenetic clades, the phyla, happened such a long time ago? In more recent times, all we (seem to) find are tiny fine-tunings (microevolution) around themes seemingly set in stone, with what Goldschmidt saw as waning potential to ever get beyond being a species or genus, but never a tribe, family, order, etc. It is as if the potential of life on Earth has diminished over time. Are the rules governing evolution as fueled or driven by DNA, gene and/or genome duplication the same today as they were for early eukaryotes, for the creatures of the Cambrian and similar explosions hundreds of millions years ago? How about for those creatures evolving well into our future? Tomorrow? The next epoch? 3.5 billion years from today, the Earth will be so hot that no life could survive and we are halfway there.

Until then, good luck today and tomorrow.

ACKNOWLEDGEMENT

Thanks to Andrew Paterson for his input on the typescript.

REFERENCES

Bennetzen, J. L. (2005). Transposable elements, gene creation and genome rearrangement in flowering plants. *Current Opinion in Genetics and Development, 15*(6), 621–627. http://dx.doi.org/10.1016/j.gde.2005.09.010.

Birchler, J. A., & Veitia, R. A. (2007). The gene balance hypothesis: From classical genetics to modern genomics. *Plant Cell, 19*(2), 395–402. http://dx.doi.org/10.1105/tpc.106.049338, tpc.106.049338 [pii].

Comai, L. (2005). The advantages and disadvantages of being polyploid. *Nature Reviews Genetics, 6*(11), 836–846. http://dx.doi.org/10.1038/nrg1711, nrg1711 [pii].

Donoghue, P. C., & Purnell, M. A. (2005). Genome duplication, extinction and vertebrate evolution. *Trends in Ecology & Evolution, 20*(6), 312–319. http://dx.doi.org/10.1016/j.tree.2005.04.008.

Force, A., Lynch, M., Pickett, F. B., Amores, A., Yan, Y. L., & Postlethwait, J. (1999). Preservation of duplicate genes by complementary, degenerative mutations. *Genetics, 151*(4), 1531–1545.

Freeling, M. (2009). Bias in plant gene content following different sorts of duplication: Tandem, whole-genome, segmental, or by transposition. *Annual Review of Plant Biology, 60*, 433–453. http://dx.doi.org/10.1146/annurev.arplant.043008.092122.

Freeling, M., Lyons, E., Pedersen, B., Alam, M., Ming, R., & Lisch, D. (2008). Many or most genes in Arabidopsis transposed after the origin of the order Brassicales. *Genome Research, 18*(12), 1924–1937. http://dx.doi.org/10.1101/gr.081026.108.

Freeling, M., & Thomas, B. C. (2006). Gene-balanced duplications, like tetraploidy, provide predictable drive to increase morphological complexity. *Genome Research, 16*(7), 805–814. http://dx.doi.org/10.1101/gr.3681406.

Gao, L. Z., & Innan, H. (2004). Very low gene duplication rate in the yeast genome. *Science, 306*(5700), 1367–1370.

Garcia-Fernandez, J. (2005). Hox, ParaHox, ProtoHox: Facts and guesses. *Heredity, 94*(2), 145–152.

Geschwind, D. H., & Konopka, G. (2012). Neuroscience: Genes and human brain evolution. *Nature, 486*(7404), 481–482. http://dx.doi.org/10.1038/nature11380 (News).

Goldschmidt, R. B. (1953). Evolution as viewed by one geneticist. *American Scientist, 40*, 84–98.

Hurst, L. D. (2002). The Ka/Ks ratio: Diagnosing the form of sequence evolution. *Trends in Genetics, 18*(9), 486 (Review).

Kotyk, M. E., Basinger, J. F., Gensel, P. G., & de Freitas, T. A. (2002). Morphologically complex plant macrofossils from the Late Silurian of Arctic Canada. *American Journal of Botany, 89*(6), 1004–1013. http://dx.doi.org/10.3732/ajb.89.6.1004.

Levinson, G., & Gutman, G. A. (1987). Slipped-strand mispairing: A major mechanism for DNA sequence evolution. *Molecular Biology and Evolution, 4*(3), 203–221.

Lewis, E. B. (1951). Pseudoallelism and gene evolution. In *Cold Spring Harbor Symposia on Quantitative Biology*: Vol. 16. (pp. 159–174).

Liman, E. R., & Innan, H. (2003). Relaxed selective pressure on an essential component of pheromone transduction in primate evolution. *Proceedings of the National Academy of Sciences of the United States of America, 100*(6), 3328–3332. http://dx.doi.org/10.1073/pnas.0636123100, 0636123100 [pii].

Lynch, M., & Force, A. (2000). The probability of duplicate gene preservation by subfunctionalization. *Genetics, 154*(1), 459–473.

Lynch, V. J., Leclerc, R. D., May, G., & Wagner, G. P. (2011). Transposon-mediated rewiring of gene regulatory networks contributed to the evolution of pregnancy in mammals. *Nature Genetics*, *43*(11), 1154–1159. http://dx.doi.org/10.1038/ng.917.

Meyer, A., & Schartl, M. (1999). Gene and genome duplications in vertebrates: The one-to-four (-to-eight in fish) rule and the evolution of novel gene functions. *Current Opinion in Cell Biology*, *11*(6), 699–704.

Ohno, S. (1970). *Evolution by gene duplication*. New York: Springer-Verlag.

Ohno, S., Wolf, U., & Atkin, N. B. (1968). Evolution from fish to mammals by gene duplication. *Hereditas*, *59*(1), 169–187.

Papp, B., Pal, C., & Hurst, L. D. (2003). Dosage sensitivity and the evolution of gene families in yeast. *Nature*, *424*(6945), 194–197.

Putnam, N. H., Butts, T., Ferrier, D. E., Furlong, R. F., Hellsten, U., Kawashima, T., et al. (2008). The amphioxus genome and the evolution of the chordate karyotype. *Nature*, *453*(7198), 1064–1071. http://dx.doi.org/10.1038/nature06967.

Scannell, D. R., Byrne, K. P., Gordon, J. L., Wong, S., & Wolfe, K. H. (2006). Multiple rounds of speciation associated with reciprocal gene loss in polyploid yeasts. *Nature*, *440*(7082), 341–345.

Schnable, J. C., Springer, N. M., & Freeling, M. (2011). Differentiation of the maize subgenomes by genome dominance and both ancient and ongoing gene loss. *Proceedings of the National Academy of Sciences of the United States of America*, *108*(10), 4069–4074. http://dx.doi.org/10.1073/pnas.1101368108.

Semon, M., & Wolfe, K. H. (2007). Consequences of genome duplication. *Current Opinion in Genetics & Development*, *17*(6), 505–512. http://dx.doi.org/10.1016/j.gde.2007.09.007, S0959-437X(07)00172-4 [pii].

Thomas, B. C., Pedersen, B., & Freeling, M. (2006). Following tetraploidy in an Arabidopsis ancestor, genes were removed preferentially from one homeolog leaving clusters enriched in dose-sensitive genes. *Genome Research*, *16*(7), 934–946. http://dx.doi.org/10.1101/gr.4708406.

Van de Peer, Y., Maere, S., & Meyer, A. (2009). The evolutionary significance of ancient genome duplications. *Nature Reviews Genetics*, *10*, 725–732. http://dx.doi.org/10.1038/nrg2600, nrg2600 [pii].

Wang, X., Tang, H., Bowers, J. E., Feltus, F. A., & Paterson, A. H. (2007). Extensive concerted evolution of rice paralogs and the road to regaining independence. *Genetics*, *177*(3), 1753–1763. http://dx.doi.org/10.1534/genetics.107.073197, 177/3/1753 [pii].

Woodhouse, M. R., Schnable, J. C., Pedersen, B. S., Lyons, E., Lisch, D., Subramaniam, S., et al. (2010). Following tetraploidy in maize, a short deletion mechanism removed genes preferentially from one of the two homologs. *PLoS Biology*, *8*(6), e1000409. http://dx.doi.org/10.1371/journal.pbio.1000409.

Zimmering, S., Sandler, L., & Nicoletti, B. (1970). Mechanisms of meiotic drive. *Annual Review of Genetics*, *4*, 409–436. http://dx.doi.org/10.1146/annurev.ge.04.120170.002205 (Review).

CHAPTER FOURTEEN

Synthetic Haplotypes, Species, Karyotypes, and Protoorganisms?

Andrew H. Paterson[1]
Plant Genome Mapping Laboratory, Department of Crop and Soil Sciences, Department of Plant Biology, and Department of Genetics, University of Georgia, USA
[1]Corresponding author: e-mail address: paterson@plantbio.uga.edu

Contents

Abstract

How might the abundant new knowledge from genome sequences be applied? A wide range of fundamental scientific discoveries have been articulated in the forgoing chapters. Most of the herbaceous plant genomes that have been sequenced are those of major crops, and this knowledge naturally opens doors to a range of opportunities to improve the productivity and sustainability of agriculture, as well as the quality and supply of its products. Some applications of genomic information to crop improvement are straightforward, building naturally on the long and productive history of the science of plant breeding. Others, when coupled with independent technological innovations, may offer qualitatively different approaches to address long-standing needs and/or challenges. In this closing chapter, a few examples of such needs and/or challenges are considered, together with both 'natural extensions' and 'new directions' that might be motivated by findings to date from the era of plant genome sequencing.

1. SYNTHETIC HAPLOTYPES

An aspect of crop improvement that has always been intriguing to the author is the use of natural processes to create haplotypes that do not occur in

Advances in Botanical Research, Volume 69
ISSN 0065-2296
http://dx.doi.org/10.1016/B978-0-12-417163-3.00014-7

nature, for example, by making crosses between divergent species and transferring particular desirable traits. Made by a process commonly called 'introgression', such 'synthetic haplotypes' might be thought of as a 'baby step' toward the synthetic genomes that have recently been produced by engineering novel microbes (Gibson et al., 2008). In plants, with basic gene sets at least 40 × greater than those of microbes (see Chapters 8 and 9), fully engineered genomes seem improbable any time soon. However, there may be tremendous opportunities to achieve both new research discoveries and practical benefits from more focused development and manipulation of synthetic haplotypes, guided by the knowledge of genomes and genes that we are quickly gaining as illustrated in the forgoing chapters.

It has long been widely recognized that wild relatives of crops often harbour alleles that are absent from the crop gene pool but that can provide intrinsic genetic solutions to challenges affecting crops, such as disease resistance. Extensive collections including many thousands of accessions of wild, exotic, and obsolete cultivated germplasm have been made for many crops, thanks to investments by governments and nongovernment organizations such as the Consultative Group on International Agricultural Research and its network of agricultural research centres. The Svalbard Global Seed Vault now provides a 'safety net' against accidental loss of more than 400,000 samples from approximately one-third of the world's most important food crop varieties (as of this writing). Genotypes comprising early samples of the diversity within several crop gene pools are already being sequenced, and more are likely to be soon.

However, the utilization of this germplasm for practical improvements has been uneven across crops. For example, in the 20 years prior to 2007, it is estimated that 55 traits were introgressed into tomato from wild relatives; however, many other major crops only received 1–3 new traits (Hajjar & Hodgkin, 2007) and some received none.

There are good reasons why addressing agricultural challenges by introgression of genes from exotic germplasm is often a 'choice of last resort' among crop breeders. First and foremost, introgression usually occurs by reciprocal chromosome segment exchange between the wild relative and the crop. The exchanged chromosome segment typically spans millions of nucleotides of DNA and includes hundreds, perhaps even thousands, of genes. Frequently, only *one* of the transferred genes is sufficient to confer the desirable trait. However, many of the others may reduce the commercial value of the resulting genotype by 'linkage drag', conferring traits that are adaptive in the wild but not in agriculture such as slow maturity, poor quality, low yield, indeterminacy, or other traits that have been largely eliminated from the elite crop gene pool through domestication and improvement.

What can genomics offer? It has been appreciated for decades, well prior to genomics, that genetic markers would be useful for minimizing linkage drag by selecting introgressed chromosomal segments of minimal size. In a 'natural extension' of crop improvement approaches, a virtually unlimited supply of DNA markers together with efficient techniques for screening large plant populations for rare recombinants now permits industrial-level searches for preferred genotypes. Indeed, one widely practised application of DNA markers in crop improvement is to quickly eliminate chromosome segments from a donor genotype, whether it is a wild relative used as a source of disease resistance (or some other trait absent from the elite gene pool) or a transformation-efficient genotype used to introduce novel genes into elite germplasm by backcrossing.

The ability to introduce exogenous genes into plants by recombinant DNA-based 'transformation' mitigated the challenge of separating desirable alleles from undesirable ones (and dramatically broadened the range of accessible genes) but imposed some different constraints. First, to efficiently make a transgenic plant that has a new trait, one must know the exact gene that confers the trait—information that is often not known and can be costly and difficult to obtain. Further, producing and screening transgenic plants is costly, and their commercialization involves regulatory hurdles. Finally, transgenics often involve complex intellectual property issues. *In sum, the cost of bringing a transgenic plant to commercialization was estimated to average $136 million in the 2008–2012 timeframe* (McDougall, 2011).

Might naturally occurring processes that comparative genomics studies have shown to be active in plant genomes be emulated to offer a 'new direction' in crop improvement, rapidly extracting desirable alleles from an undesirable genotype with the precision of transformation? It has long been known that the same reciprocal exchanges of DNA between homologous chromosomes (crossing over) that create the mosaic genotypes of central importance to conventional crop improvement are often accompanied by short tracts of unidirectional, local DNA exchanges known as 'gene conversion' (Holliday, 1964). Gene conversions are much shorter than reciprocal recombinations, often confined to single genes or even single exons (Chen, Cooper, Chuzhanova, Ferec, & Patrinos, 2007). Thus, in principle, gene conversion offers a mechanism of gene transfer *with the precision of transformation but by a natural noninvasive means*. DNA recombination is also implicated in the repair of highly deleterious DNA double-strand breaks (DSBs) that are associated with 'genome stress' following events such as hybridization and polyploidy (Fontdevila, 2005; McClintock, 1984). Most models proposed to account for homologous DSB repair (synthesis-dependent strand annealing, classic DSB repair, and break-induced replication, although not single-strand annealing (Helleday, Lo, van Gent, & Engelward, 2007)) also predict the occurrence of tracts of unidirectional gene conversion, even when reciprocal crossing over occurs.

To be of practical value in crop improvement, gene conversion events would need to be sufficiently abundant to identify at manageable cost on a timescale consistent with scientific crop improvement. In the past, gene conversion events were only detectable in cases for which a phenotypic assay was available—however, the advent of massively parallel DNA sequencing methods now makes it feasible to search for them (e.g. Paterson et al., 2012). Building on rich evidence of concerted evolution in tandemly repeated sequences such as ribosomal RNA genes (Eickbush & Eickbush, 2007), and multigene families such as primate olfactory receptor genes (Sharon et al., 1999), we have shown that gene conversion has occurred in the past ~400,000 years between duplicated rice genes that diverged from a common ancestor 70 million years ago (mya) (Wang, Tang, Bowers, & Paterson, 2009). While this is an extreme case, nonrandom similarity between duplicated genes widely distributed across other genomes (Chapman, Bowers, Feltus, & Paterson, 2006; Wang, Tang, Bowers, Feltus, & Paterson, 2007) suggests the phenomenon to be widespread.

A promising discovery derives from the comparison of genomic sequences from several tetraploid (A_tD_t) *Gossypium* (cotton) species and genotypes tracing to a common polyploidization event 1–2 mya, with their putative diploid A and D genome progenitor species (Paterson et al., 2012). Among the 40% of 'At' (A-genome-derived tetraploid alleles) and 'Dt' alleles in an elite tetraploid cotton genotype, 'Acala Maxxa', that differ in sequence from their diploid progenitors, the vast majority of changes are convergent, that is, inferred to be a result of gene conversion rather than random mutation. It remains to be determined whether gene conversion is occurring at a steady-state rate or in bursts, for example, shortly following polyploid formation. A richer understanding of the conditions that favour gene conversion may lead to its practical utilization—for example, the ability to target double-stranded breaks to specific genetic loci (Christian et al., 2010) also may be a means to enrich the frequency of gene conversions at predetermined sites. Likewise, wide hybridization (such as crosses between species, as can easily be practised in breeding many crops) or formation of synthetic polyploids (feasible in a few crops, with varying degrees of difficulty) might stimulate a 'burst' of conversion events (Fontdevila, 2005; McClintock, 1984) that could be screened for a subset of interest, using efficient genomics procedures.

2. NOVEL POLYPLOIDS: SYNTHETIC PLANT SPECIES

Many crop species for which the elite gene pools are genetically depauperate enjoy a host of wild relatives that harbour rich genetic variation that is not accessible by conventional breeding. For example, many of the world's most important food, fibre, and forage crops are polyploids, formed in rare events such as crosses between species with chromosomes that do not normally pair and recombine—followed by formation of unreduced gametes that may restore a stable, doubled, chromosome set. Hybridization of diploid relatives with a polyploid crop, in the infrequent cases that survive, generally results in genetically unstable triploid genotypes that are difficult to maintain sexually (i.e. by seed), although this avenue has been very successful for breeding some vegetatively propagated crops such as Bermuda grass and Miscanthus, and food crops in which seed production is undesirable such as watermelon and banana.

For example, in the cotton genus (Gossypium), much of the extant genetic variation is among eight 'genome types' (A, B, C, D, E, F, G, and K) classified based on the inability of their chromosomes to pair and recombine (Beasley, 1941, 1942). The various species that comprise these genome types are found on six continents, in a wide range of climates, and harbour rich variation in morphology, biochemistry, and development. However, the overwhelming majority of cultivated cotton is tetraploid, resulting from joining of A and D genome progenitors about 1–2 mya (Beasley, 1940; Wendel, 1989). No cotton polyploids involving other genomes are known in nature, and even if they exist, they have not been domesticated; hence, the veritable treasure trove of variation among the remaining six genome types is presently exceedingly difficult to access for crop improvement. Cotton is not unusual in this inability to benefit from much of the variation available in its relatives—many crops share this constraint, to varying degrees. The effects of domestication superimposed on those of polyploid formation have further narrowed the genetic diversity in the gene pools of many major crops. Opportunities to benefit from the 'stacking' of transgenes have been suggested to contribute to focusing of improvement efforts on only a small subset of elite cultivars, further narrowing the genetic diversity that is actually being deployed in the field (Van Esbroeck, Bowman, Calhoun, & May, 1998; Van Esbroeck, Bowman, May, & Calhoun, 1999). In such cases, expanding the germplasm base is an important means of providing ecologically benign low-cost intrinsic genetic solutions to many challenges facing producers, processors, and consumers.

In a 'natural extension' of crop improvement approaches, many scientists have invested decades of effort in laborious and tedious intercrossing methods, coupled with chromosome doubling using colchicine (a secondary metabolite originally extracted from plants of the genus *Colchicum*, which inhibits microtubule polymerization and functions as a 'mitotic poison'), to produce polyploids that are not known in nature. Many of these 'hopeful monsters' are unable to mate with related crops, or perhaps are even sterile, and quickly go extinct. A rare success, triticale (a hybrid between wheat and rye), now contributes about 15 million tons of grain to the annual world food supply (FAO Statistics Division, 2010).

The greatest benefits from the production of synthetic polyploids have not been from the creation of new crops, however, but as sources of genes for introgression into closely related crops. While many cases are in the literature, one example that is particularly well known to the author is the introgression of nematode resistance into cultivated peanut. In this case, a synthetic tetraploid peanut, TxAG6 (Simpson, Nelson, Starr, Woodard, & Smith, 1993), was produced by crossing one species to a second, crossing the F1 hybrid to a third species, and then doubling the chromosomes—thus incorporating elements of three different diploids that would not normally be compatible with cultivated peanut but were made so by joining their genomes in this synthetic polyploid. TxAG6 harbours many attributes that are absent from the elite peanut gene pool—along with many undesirable traits from its wild progenitors—and resistance to the root-knot nematode was chosen as a priority for introgression. Resistance was initially transferred from TxAG6 to a peanut cultivar (Florunner) by conventional backcrossing, using laborious empirical phenotyping methods (counting nematode eggs on the roots of infested plants). As DNA marker analysis became feasible in peanut, it was quickly determined that the introgression event involved a single chromosome segment, also identifying DNA markers diagnostic of the segment (Burow, Simpson,

Paterson, & Starr, 1996), that led to the development of the first DNA marker-based commercial peanut germplasm (Simpson, Starr, Church, Burow, & Paterson, 2003).

Might the nearly complete information about gene and genome composition available from genome sequences shed light on the incompatibilities that hinder crosses between species, opening a 'new direction' in crop improvement by strategies for mitigating these incompatibilities? Having raised this possibility, it is important to first acknowledge that it is an exceedingly great challenge and likely to be a long time coming, if ever. However, research progress in several directions indicates the general feasibility of identifying and mitigating species incompatibilities. It has been suggested that the Bateson–Dobzhansky–Muller model for hybrid incompatibility between species might be mediated by autoimmunity as a pleiotropic effect of evolution of genes that are involved in pathogen response (Bomblies et al., 2007). Such genes are well known, with a small number of highly conserved functional domains, and one could envision various approaches for suppressing their function (albeit with risk of nontarget effects such as disease susceptibility). Likewise, comparative analysis of independent genome duplications in highly divergent taxa has revealed a modest population of genes that are consistently restored to the 'singleton state' (Paterson et al., 2006), in some instances rather quickly (Rong, Feltus, Liu, Lin, & Paterson, 2010), and suggests the hypothesis that these genes may reduce fitness when present in multiple copies. Again, a variety of established strategies might be employed for suppressing individuals or groups of these genes in a newly polyploid nucleus.

Indeed, genomic information might provide some data toward predicting the feasibility of making polyploids from particular species combinations. One could envision that sequencing of a wide range of species pairs for which quantitative information exists (or could be obtained) about the ease (or not) of obtaining a polyploid might permit one to identify some predictors, such as the degree of low-copy, repetitive (particularly centromeric), or organellar DNA divergence among the candidates; degree of gene loss or gene duplication since their divergence; extent of structural rearrangement; or other factors. Likewise, massively parallel profiling (by sequencing) of gene expression in response to pollination (Xu, Wang, Chen, Sun, & Zhang, 2013) with 'compatible' versus incompatible pollen may dramatically expand on knowledge of the factors that act to inhibit fertilization across species or genome types.

3. SYNTHETIC KARYOTYPES: INTRODUCING ARTIFICIAL CHROMOSOMES INTO A GENOME

The packaging of eukaryotic genomes into chromosomes, and a long history of research into chromosome behaviour (see Chapter 2), has long suggested the notion that naturally occurring karyotypes might be modified with the addition of novel chromosomes containing genes or other elements conferring beneficial traits. The identification of centromeric and telomeric sequences in a number of organisms, together with the advent recombinant DNA technology itself (Cohen, Chang, Boyer, & Helling, 1973), laid the groundwork for engineering of 'artificial chromosomes', that is, non-integrating DNA vectors that are transmissible and have the ability to harbour large quantities of DNA (Gaeta, Masonbrink, Krishnaswamy, Zhao, & Birchler, 2012).

Artificial chromosomes were first created in yeast (Murray & Szostak, 1983), with subsequent 'yeast artificial chromosome' (YAC) cloning vectors (Burke, Carle, & Olson, 1987) becoming important to early characterization of the genomes of human and several plants. YACs were ultimately supplanted by the production of artificial chromosomes in *E. coli* (O'Connor, Peifer, & Bender, 1989). Bacterial artificial chromosome vectors, while harbouring less DNA than YACs, were more stable and grew more rapidly (Shizuya et al., 1992) and became the backbones of several very high-quality plant genome sequences, while contributing contiguity to many additional genome sequences via paired-end sequencing of their termini (see Chapters 3 and 4).

Early successes in the development of mammalian artificial chromosomes (Harrington, Van Bokkelen, Mays, Gustashaw, & Willard, 1997) provided conceptual support for the notion that engineering of plant karyotypes might be a viable improvement strategy, a notion that was borne out with the recovery of engineered chromosomes in maize generated by adding plant telomere sequences to transgene cassettes (Yu, Han, Gao, Vega, & Birchler, 2007). Together with various options available for use as selectable markers, the rich set of plant genome sequences aid in the identification of the minimal requirements for creating plant minichromosomes, including functional centromeres and telomeres, and enough endogenous chromatin to contain origins of replication (Gaeta et al., 2012).

Plant minichromosomes offer several advantages over alternative approaches for genetic engineering (Gaeta et al., 2012). Minichromosomes offer a means to introduce multiple genes into a genome, obviating the need for many generations of crossing to 'pyramid' desirable genes and in principle facilitating the engineering of entire biochemical pathways rather than merely specific enzymes (genes). Location on the same minichromosome inherently places genes in close linkage to one another, conferring on the assembled genes transmission between generations as essentially a single genetic locus. The minichromosome approach also avoids 'linkage drag' associated with random integration of transgenes into a genotype that may be transformation-efficient but commercially noncompetitive (see in the preceding text, this chapter). Once a new karyotype is established and stabilized, strategies for targeted modification of plant DNA (Shukla et al., 2009; Townsend et al., 2009) provide a means to further amend a minichromosome, adding new genes or replacing existing ones without disturbing the function or organization of the remainder of the genome.

Significant obstacles remain, however, toward routine practice of the plant minichromosome strategy. In particular, it has been realized for nearly a century that small chromosomes transmit poorly through meiosis (Kato et al., 2005), often failing to pair and recombine with homologous partners (Han, Gao, Yu, & Birchler, 2007; McClintock, 1938). Several potential solutions to this problem have been suggested, including the use of selectable markers (either natural or engineered) or modified minichromosomes (Gaeta et al., 2012).

4. PROTOORGANISMS: MITIGATING THE CONSTRAINTS OF A SPOROPHYTIC STAGE OF DEVELOPMENT

Whether at the level of the haplotype, the karyotype, or the genome, selection acts on variation resulting from segregation and recombination that occur during meiosis with the necessity of an intervening sporophytic stage of development often imposing a lengthy and costly delay (!) in obtaining a desired genotype. Imagine the acceleration of breeding progress that might be realized if, for example, long-lived trees could pass through meiotic cycles in a few weeks or even a few days, rather than decades, and the cost efficiency that might be realized if these cycles could take place *in vitro*, rather than in vast forests.

Recently, it was shown that ordinary mouse somatic cells could be reprogrammed into totipotent cells, then specialized into male and female gamete precursors, and matured into fully competent gametes by culturing in testicular or ovarian cell broths, respectively. Such cultured gametes were then united pairwise, and zygotes grown *in vivo* to generate fully functional and fertile mouse offspring (Hayashi et al., 2012).

Cycling of gametes *in vitro* (COGIV) offers 'new directions' across all plant improvement strategies that depend upon meiosis (Murray, Eckhoff, Wood, & Paterson, accepted). COGIV could enable genetic material to be recombined far more rapidly than by conventional breeding and at a comparatively marginal cost because 'protoorganisms' generated by COGIV need not mature into adults. This system would permit selection among vast numbers of genotypes for rare desirable allele combinations based either on diagnostic DNA markers (established in separate studies of sporophytes) or on targeted searches for rare events such as gene conversions. The temporary bypassing of a sporophytic stage may also reduce barriers to hybridization, mitigating some species incompatibilities and thereby providing access to alleles that are beyond the reach of conventional crosses. Remarkably, COGIV coupled with *in vitro* meat production (van Eelen, van Kooten, & Westerhof, 1999), which has progressed in the laboratory although not yet reaching economic feasibility (Post, 2012), would suggest the possibility of a near-total physical separation of germ lines and commercially important somatic tissue for some organisms!

To date only demonstrated in mice, COGIV may relatively quickly be adapted to work in animals, but it remains a subject for further research whether, or how, plant gametes might be produced from somatic cells *in vitro*. It may not be possible to culture desired genotypes *in vitro*, to demonstrate normal recombination and development, or to avoid excessive somatic mutation during COGIV. Nonetheless, such methodology could shorten the breeding cycle so dramatically as to offer a compelling case for further research investment.

5. SYNTHESIS

While we have learned much about, and from, the genomics of herbaceous plants as highlighted in this volume, the true impacts of this era of science are only just beginning. And none too soon, the sharp twenty-first-century rise in food and energy prices may be a harbinger of the need anticipated by many leading agricultural scientists for a new Green Revolution (Tuberosa, Phillips, & Gale, 2005), of broader scope than its predecessor. Today, we need to sustain 7 billion people of whom an estimated 850 million are already 'food-insecure' while also broadening agriculture to be the foundation of a bio-based economy (producing not only food, feed, and fibre but also much more fuel and chemical feedstocks than in the past)—all with the constraint of using less water than in the past. Indeed, the inability of the prior Green Revolution to adequately address food security in regions dependent upon low-input production has been an important factor in asserting the need for a 'second' Green Revolution (Cantrell & Hettel, 2004; Conway, 1997), particularly aimed at rainfed areas of Asia and Africa (Paterson & Li, 2011).

Cataloguing and characterizing the genomes of herbaceous plants is a starting point—we now have a view 'from 30,000 ft.' of the genetic potential of many materials available to us, which will be enriched with essential details as ongoing improvements in sequencing (Chapter 3) and other genomics technologies provide broader and deeper coverage of angiosperm genomic diversity, among taxa as well as within populations (Chapter 12). Of particular interest for ongoing sequencing are 'orphan crops' (National Research Council, 2006) of local importance but which are scientifically underexplored, other plants with novel adaptations of widespread potential value, and the microbiomes that contribute much to plant (and other) biology (Chapter 11), shedding further light on the 200-million-year accumulation of biodiversity that has permitted angiosperms to become the world's dominant vegetation and to sustain humanity by providing a wide range of ecosystem services.

Translating genomic knowledge into improved human wellbeing via increased agricultural productivity will no doubt involve invigorated partnership both with the accumulated wisdom of conventional agricultural sciences such as plant breeding and with technological innovations such as some described in this chapter. Indeed, in major crops that already receive relatively heavy investment from agribusiness, it may prove challenging to even sustain the current rates of genetic gain, yet an aggregate 60% increase in agricultural productivity will be necessary to adequately feed the projected world population of 9.15 billion in 2050 (Alexandratos & Bruinsma, 2012).

Noting that agriculture is chronically underfunded—with a research and development budget of 0.1% of annual sales, paltry in comparison to the 5–15% that is typical of *other* human–health–maintenance industries such as pharma—and noting that agricultural initiatives can take 15–30 years to yield maximal returns (Alston, Chan-Kang, Marra, Pardey, & Wyatt, 2000), the required investments need to be made soon.

REFERENCES

Alexandratos, N., & Bruinsma, J. (2012). World Agriculture Towards 2030/2050. In Food and Agriculture Organization of the United Nations Agricultural Development Economics Division (Ed.), ESA Working Papers.

Alston, J. M., Chan-Kang, C., Marra, M. C., Pardey, P. G., & Wyatt, T. J. (2000). *A meta-analysis of rates of return to agricultural R&D: Ex pede Herculem?* Washington DC: International Food Policy Research Institute.

Beasley, J. O. (1940). The origin of American tetraploid Gossypium species. *American Naturalist, 74*(752), 285–286.

Beasley, J. O. (1941). Hybridization, cytology, and polyploidy of Gossypium. *Chronica Botanica, 6*, 394–395.

Beasley, J. O. (1942). Meiotic chromosome behavior in species hybrids, haploids, and induced polyploids of *Gossypium*. *Genetics, 27*, 25–54.

Bomblies, K., Lempe, J., Epple, P., Warthmann, N., Lanz, C., Dangl, J. L., et al. (2007). Autoimmune response as a mechanism for a Dobzhansky-Muller-type incompatibility syndrome in plants. *PLOS Biology, 5*(9), 1962–1972.

Burke, D. T., Carle, G., & Olson, M. V. (1987). Cloning of large segments of exogenous DNA into yeast by means of artificial chromosome vectors. *Science, 236*, 806–812.

Burow, M. D., Simpson, C. E., Paterson, A. H., & Starr, J. L. (1996). Identification of peanut (Arachis hypogaea L) RAPD markers diagnostic of root-knot nematode (Meloidogyne arenaria (Neal) Chitwood) resistance. *Molecular Breeding, 2*(4), 369–379.

Cantrell, R. P., & Hettel, G. P. (2004). Rice, biofortification, and enhanced nutrition. In *Paper presented at the World Food Prize Symposium*. Iowa: Des Moines.

Chapman, B. A., Bowers, J. E., Feltus, F. A., & Paterson, A. H. (2006). Buffering of crucial functions by paleologous duplicated genes may contribute cyclicality to angiosperm genome duplication. *Proceedings of the National Academy of Sciences of the United States of America, 103*(8), 2730–2735. http://dx.doi.org/10.1073/pnas.0507782103.

Chen, J. M., Cooper, D. N., Chuzhanova, N., Ferec, C., & Patrinos, G. P. (2007). Gene conversion: Mechanisms, evolution and human disease. *Nature Reviews Genetics, 8*(10), 762–775. http://dx.doi.org/10.1038/nrg2193.

Christian, M., Cermak, T., Doyle, E. L., Schmidt, C., Zhang, F., Hummel, A., et al. (2010). Targeting DNA double-strand breaks with TAL effector nucleases. *Genetics, 186*(2), 757–761.

Cohen, S. N., Chang, A. C. Y., Boyer, H. W., & Helling, R. B. (1973). Construction of biologically functional bacterial plasmids *in vitro*. *Proceedings of the National Academy of Sciences of the United States of America, 70*, 3240–3244.

Conway, G. (1997). *The doubly green revolution: Food for all in the twenty-first century*. Ithaca, NY: Cornell University Press.

Eickbush, T. H., & Eickbush, D. G. (2007). Finely orchestrated movements: Evolution of the ribosomal RNA genes. *Genetics, 175*(2), 477–485. http://dx.doi.org/10.1534/genetics.107.071399.

FAO Statistics Division (2010). Food and agricultural commodities production.

Fontdevila, A. (2005). Hybrid genome evolution by transposition. *Cytogenetic and Genome Research, 110*(1–4), 49–55. http://dx.doi.org/10.1159/000084937.

Gaeta, R. T., Masonbrink, R. E., Krishnaswamy, L., Zhao, C., & Birchler, J. A. (2012). Synthetic chromosome platforms in plants. *Annual Review of Plant Biology, 63*, 307–330.

Gibson, D. G., Benders, G. A., Axelrod, K. C., Zaveri, J., Algire, M. A., Moodie, M., et al. (2008). One-step assembly in yeast of 25 overlapping DNA fragments to form a complete synthetic Mycoplasma genitalium genome. *Proceedings of the National Academy of Sciences of the United States of America, 105*(51), 20404–20409. http://dx.doi.org/10.1073/pnas.0811011106.

Hajjar, R., & Hodgkin, T. (2007). The use of wild relatives in crop improvement: A survey of developments over the last 20 years. *Euphytica, 156*(1–2), 1–13.

Han, F., Gao, Z., Yu, W., & Birchler, J. A. (2007). Minichromosome analysis of chromosome pairing, disjunction, and sister chromatid cohesion. *Plant Cell, 19*, 3853–3863.

Harrington, J. J., Van Bokkelen, G., Mays, R. W., Gustashaw, K., & Willard, H. F. (1997). Formation of de novo centromeres and construction of first-generation human artificial microchromosomes. *Nature Genetics, 15*, 345–355.

Hayashi, K., Ogushi, S., Kurimoto, K., Shimamoto, S., Ohta, H., & Saitou, M. (2012). Offspring from oocytes derived from in vitro primordial germ cell-like cells in mice. *Science, 338*(6109), 971–975.

Helleday, T., Lo, J., van Gent, D. C., & Engelward, B. P. (2007). DNA double-strand break repair: From mechanistic understanding to cancer treatment. *DNA Repair (Amst), 6*(7), 923–935. http://dx.doi.org/10.1016/j.dnarep.2007.02.006.

Holliday, R. (1964). Mechanism for gene conversion in fungi. *Genetical Research, 5*(2), 282–304.

Kato, A., Zheng, Y. Z., Auger, D. L., Phelps-Durr, T., Bauer, M. J., Lamb, J. C., et al. (2005). Minichromosomes derived from the B chromosome of maize. *Cytogenetic and Genome Research, 109*(1–3), 156–165.

McClintock, B. (1938). The production of homozygous deficient tissues with mutant characteristics by means of the aberrant mitotic behavior of ring-shaped chromosomes. *Proceedings of the National Academy of Sciences of the United States of America, 18*, 677–681.

McClintock, B. (1984). The significance of responses of the genome to challenge. *Science, 226*(4676), 792–801.

McDougall, P. (2011). The cost and time involved in the discovery, development and authorisation of a new plant biotechnology derived trait—A consultancy study for Crop Life International. www.croplife.org/view_document.aspx?docId=3338.

Murray, A. W., & Szostak, J. W. (1983). Construction of artificial chromosomes in yeast. *Nature, 305*, 189–193.

Murray, S. C., Eckhoff, P., Wood, L., & Paterson, A. H. (2013). A proposal to use cycling of gametes in vitro for improvement of agricultural species. *Nature Biotechnology, 31*, 877–880.

National Research Council, (2006). *Lost Crops of Africa. Volume II: Vegetables.* Washington, DC: National Academies Press.

O'Connor, M., Peifer, M., & Bender, W. (1989). Construction of large DNA segments in Escherichia coli. *Science, 244*, 1307–1312.

Paterson, A. H., Chapman, B. A., Kissinger, J. C., Bowers, J. E., Feltus, F. A., & Estill, J. C. (2006). Many gene and domain families have convergent fates following independent whole-genome duplication events in Arabidopsis, Oryza, Saccharomyces and Tetraodon. *Trends in Genetics, 22*(11), 597–602.

Paterson, Andrew H., & Li, Zhi-Kang (2011). Paleo-Green Revolution for rice. *Proceedings of the National Academy of Sciences of the United States of America, 108*(27), 10931–10932. http://dx.doi.org/10.1073/pnas.1107959108.

Paterson, A. H., Wendel, J. F., Gundlach, H., Guo, H., Jenkins, J., Jin, D., et al. (2012). Repeated polyploidization of Gossypium genomes and the evolution of spinnable cotton fibres. *Nature, 492*(7429), 423–427. http://dx.doi.org/10.1038/nature11798.

Post, M. J. (2012). Cultured meat from stem cells: Challenges and prospects. *Meat Science, 92*(3), 297–301. http://dx.doi.org/10.1016/j.meatsci.2012.04.008.

Rong, J. K., Feltus, F. A., Liu, L., Lin, L., & Paterson, A. H. (2010). Gene copy number evolution during tetraploid cotton radiation. *Heredity, 105*(5), 463–472. http://dx.doi.org/10.1038/hdy.2009.192.

D

E

F

G

H

L

M

N

S

T

U

V

W

SUBJECT INDEX

Note: Page numbers followed by "*f*" indicate figures and "*t*" indicate tables.

N

O

P

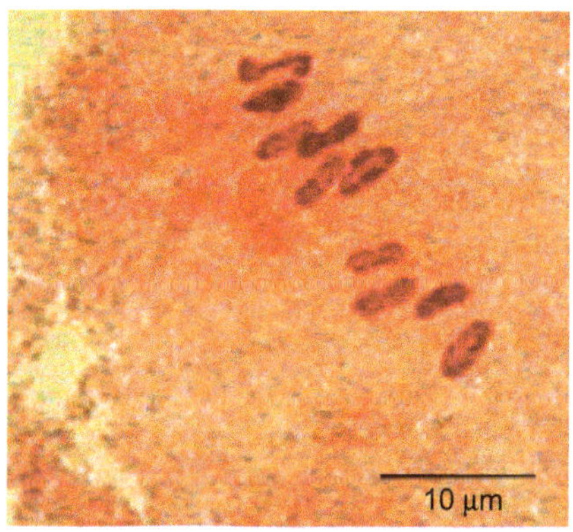

Daniel G. Peterson, Figure 2.2 Light micrograph of early anaphase I chromosomes of *Sorghum bicolor* ($2n = 20$) stained with aceto-orcein. Each chromosome is composed of two homologues attached by chiasmata. Homologues are just beginning to be pulled towards opposite ends of the cell. *Photo by Daniel G. Peterson.*

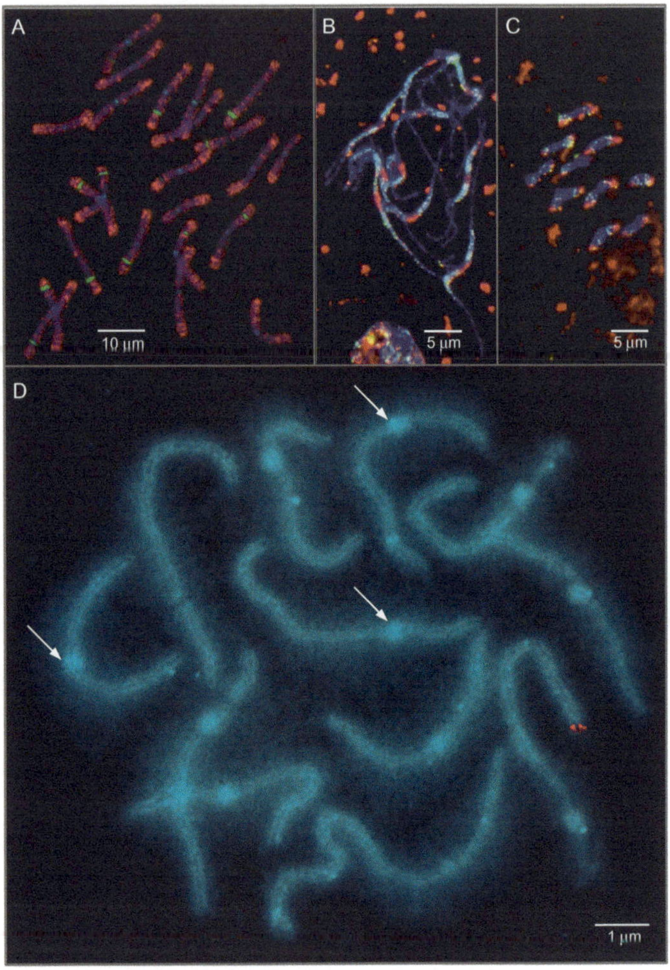

Daniel G. Peterson, Figure 2.5 Fluorescence *in situ* hybridization (FISH) to plant chromosomes. All images are previously unpublished. (A) FISH to mitotic metaphase chromosomes of *Pinus taeda* (loblolly pine; $2n=24$). The chromosomes have been hybridized with a 45S rDNA probe (green/yellow) and a dispersed repeat sequence (red). Image courtesy of Nurul Islam-Faridi. (B) FISH to pachytene chromosomes and (C) anaphase I chromosomes of *Sorghum bicolor* (sorghum; $2n=20$). The pink regions indicate hybridization of a centromeric probe, while the green/yellow signals show hybridization of a pericentromeric repeat sequence to the chromosomes. Note the considerable difference in length between the pachytene and anaphase I chromosomes. (Images by Daniel G. Peterson.) (D) FISH localization of a single-copy sequence (pink signal) on tomato pachytene chromosome *2*. Note that a kinetochore is visible on each chromosome (e.g. arrows) and that chromatin is visible as a halo around each SC. FISH has greatly facilitated assembly of the tomato genome (Stack et al., 2009). *(Image courtesy of Stephen M. Stack.)*

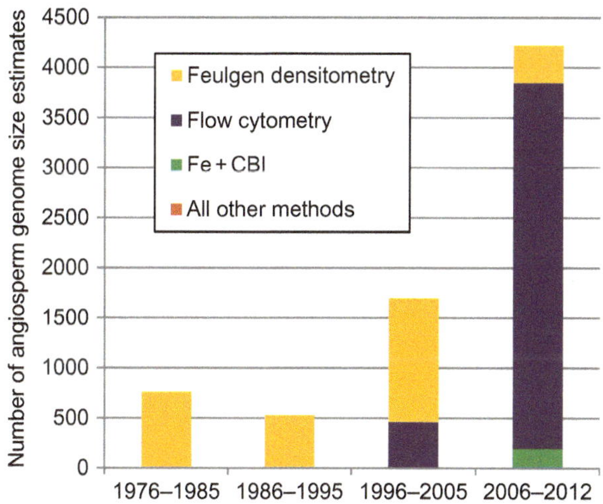

Daniel G. Peterson, Figure 2.7 Methods used in genome size determination. While Feulgen densitometry was the predominant means of determining genome size for 30 years, flow cytometry has emerged as the most common means of estimating genome sizes in plants. All data are from Plant DNA C-values Database (Bennett & Leitch, 2012).

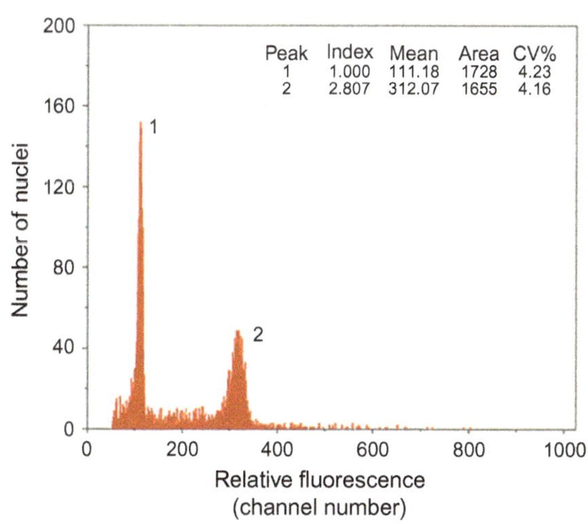

Daniel G. Peterson, Figure 2.8 Histogram showing G_0/G_1 (i.e. 2C) peaks for *Trillium sessile* (peak 1) and *Paris japonica* (peak 2). *P. japonica* has the largest known genome of any plant species (1C = 148,852 Mb). *Image from Pellicer et al. (2010). Used with permission of John Wiley & Sons and the authors.*

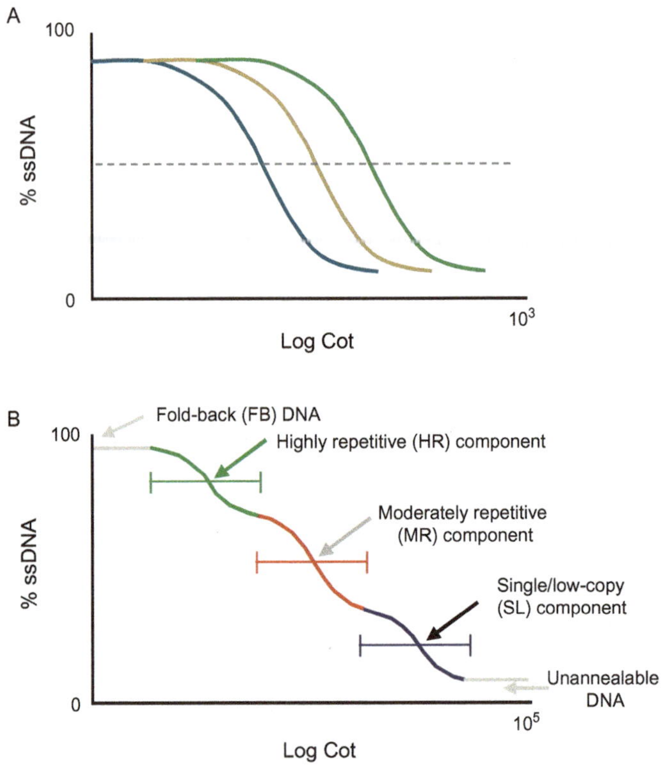

Daniel G. Peterson, Figure 2.9 Cot curves. (A) Cot curves of three bacterial species. The dotted line intersects each curve at its Cot½ value. Each curve is S-shaped and shows 80% reassociation over a 'two Cot decade region' (i.e. if Cot½=*y*, then 80% of DNA reassociates between 0.1*y* and 10*y*). For species with little repetitive DNA (such as those represented by the curves), genome size is directly correlated with Cot½, that is, the species with the right-most curve has the largest genome, while the species with the left-most curve has the smallest genome size. (B) Cot curve of a complex eukaryotic genome. Mathematical analysis of the curve reveals that the genome is composed of three major components. Each component curve has a sigmoid shape and shows 80% reassociation over a two Cot decade region (marked by brackets centred at Cot½ values). The slowest reassociating component represents renaturation of single-/low-copy sequences, while the other components represent reassociation of repetitive DNA sequences. See Peterson, Schulze, et al. (2002) and Peterson, Wessler, Paterson (2002) for further explanation.

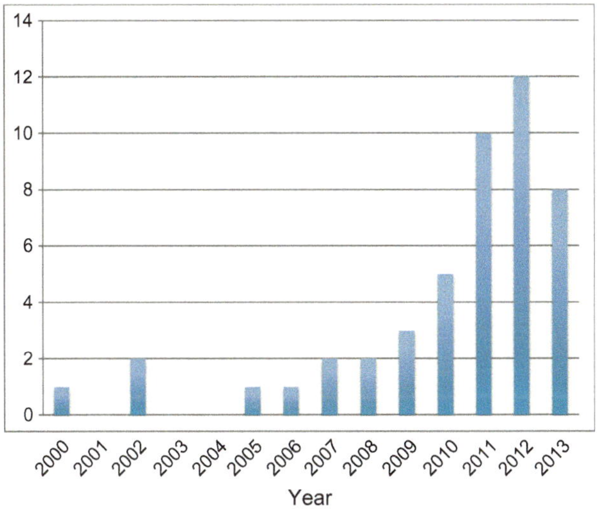

Stephen A. Goff *et al.*, Figure 3.1 A histogram of published plant genomes sequenced over time. Note that there are many other plant genome projects (genomic and transcriptomic) that are available but not published or in progress. See the JGI's 'Phytozome' Website (http://www.phytozome.net/) and the Genomes Online Database (GOLD, http://www.genomesonline.org/cgi-bin/GOLD/index.cgi) for more information on available plant genomes. Genome sequence publications for 2013 are those until April 2013.

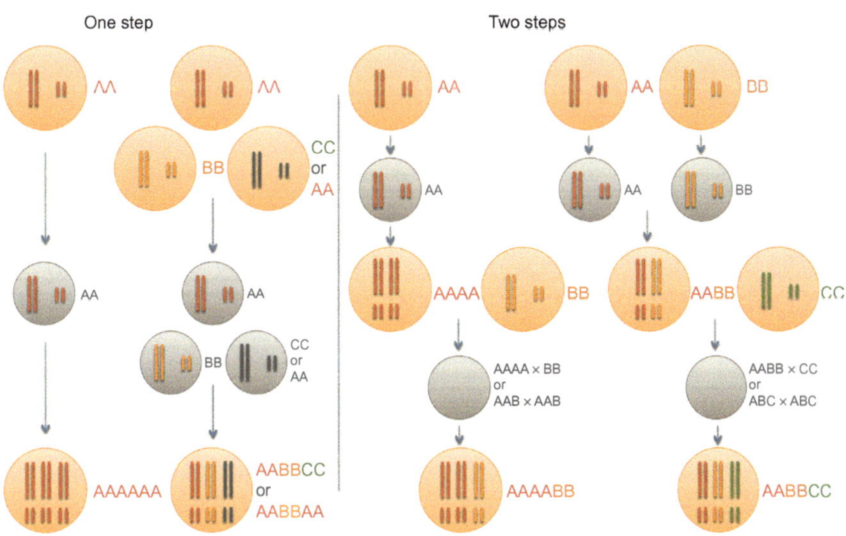

Jingping Li *et al.*, Figure 6.1 Different models to form (paleo)hexaploidy. Panel 1 illustrates one-step autohexaploid formation. Panel 2 illustrates one-step allohexaploid formation. Panel 3 illustrates a two-step autotetraploidy and allotetraploidy hybrid model to form a hexaploidy organism. Panel 4 illustrates a two-step formation of an allohexaploid via two successive allotetraploidizations. The big dark circles represent normal diploid cells (germ-line cells and embryos). The small light circles represent gametes.

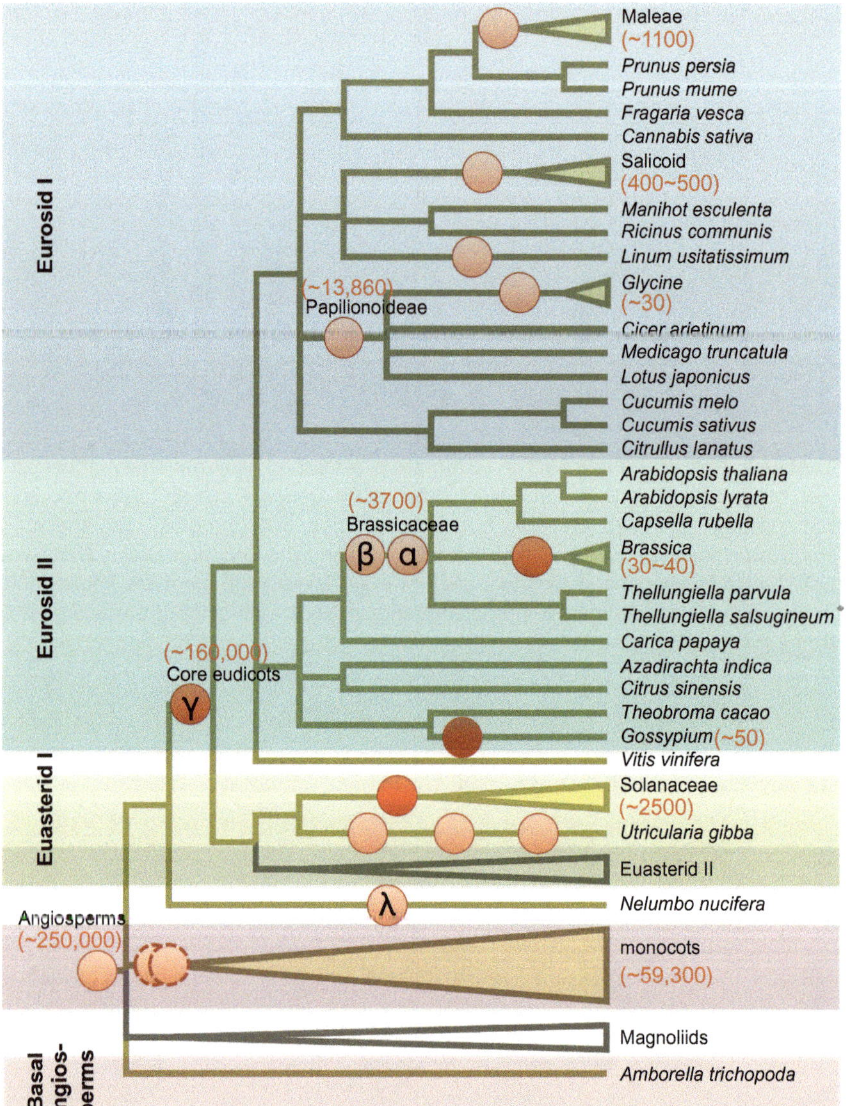

Jingping Li *et al.*, Figure 6.2 Paleopolyploidy events mapped to the phylogeny of major eudicot lineages. Since this chapter focuses on dicots, monocots and other angiosperm clades are abstracted. Inside the parentheses are numbers of known extant species in the respective clades. Light pink circles represent paleotetraploidies. Dark pink circles represent paleohexaploidies. Red circle represents the paleodecaploidy or dodecaploidy in *Gossypium*. The two overlapping light pink circles with dashed outline are an abstract representation of WGD events identified in monocot lineages. Clades that have no genome sequenced yet are omitted or in light grey (euasterid II and magnoliids). The cladogram is based on the NCBI Taxonomy database.

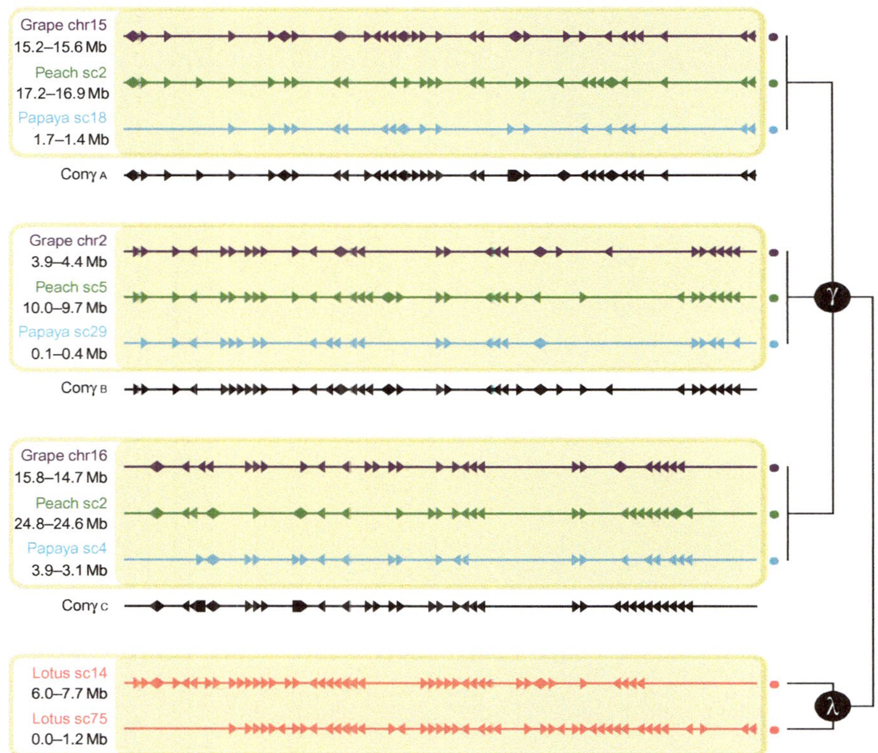

Jingping Li *et al.*, Figure 6.3 Multiple alignments of a set of syntenic regions in papaya, peach, grape and sacred lotus. Triangles represent individual genes and their transcriptional orientations. Genes with no syntenic matches are not plotted. The event γ is the paleohexaploidy that occurred in ancestral eudicots and is shared by the grape, peach, and papaya lineages. The event λ is the paleotetraploidy in the *Nelumbo* (sacred lotus) lineage after it diverged from the rest of eudicot lineages. The two events are also shown in Fig. 6.2. The γ regions are grouped into three γ subgenomes based on parsimony principles. Aligned genes within each γ subgenome are merged into a consensus order (Con γ_A, γ_B, and γ_C, respectively). Ancestral genes with uncertain orientations are represented by squares. The pair of sister λ regions in lotus is displayed at the bottom.

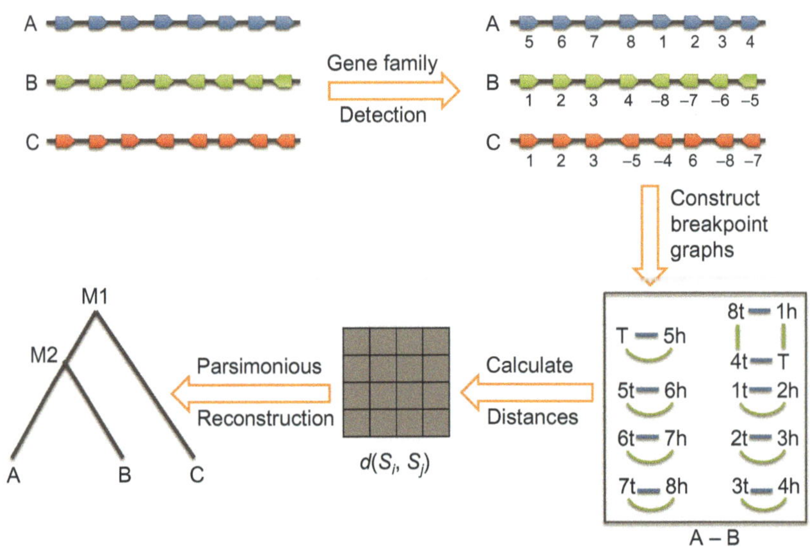

Jingping Li *et al.*, Figure 6.4 Pipeline for computational reconstruction of the ancestral genome. A toy example of three extant genomic regions (A, B, and C) and their ancestral regions (M1 and M2) is used for illustration. Oriented blocks represent genes, which are usually converted to signed integers in computation.

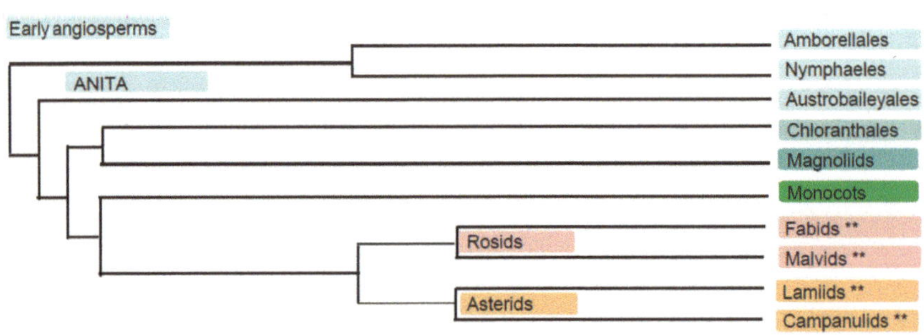

Haibao Tang *et al.*, Figure 8.1 Phylogeny of flowering plants with a focus on early angiosperm evolution. Phylogeny is based on APG III (The Angiosperm Phylogeny Group, 2009). **indicates unranked groups instead of regular taxonomic orders.

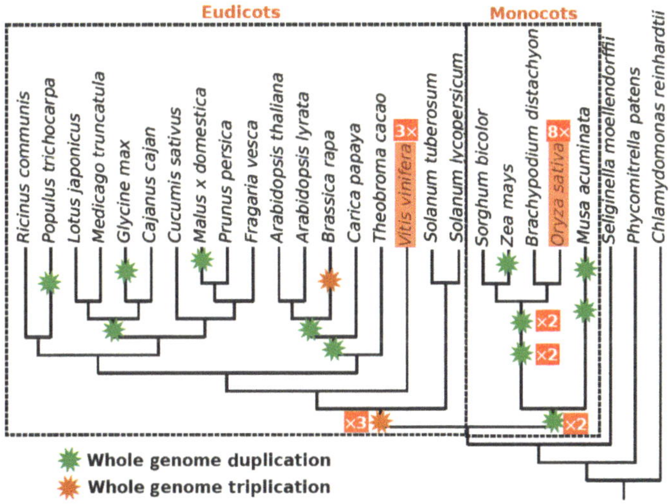

Haibao Tang *et al.*, Figure 8.3 Phylogeny of sequenced plant genomes with known paleopolyploidy events mapped. The stars on this figure represent paleopolyploidy events (Lee et al., 2013). *The polyploidy tree was adapted from 'The duplication history of plants in PGDD' (http://chibba.agtec.uga.edu/duplication/).*

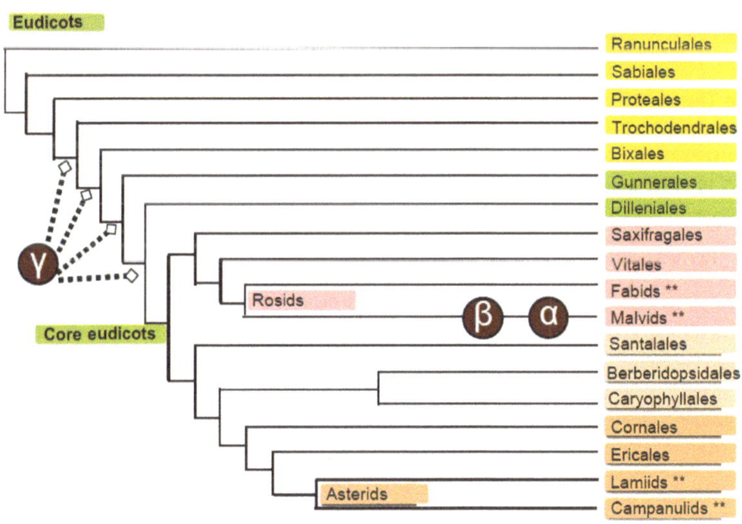

Haibao Tang *et al.*, Figure 8.4 Phylogeny of selected families of eudicots. Circles represent paleopolyploidy events discovered in the Arabidopsis lineage (Bowers et al., 2003; Jaillon et al., 2007; Tang, Wang, et al., 2008). Arrows indicate possible placement for certain events. Phylogeny is based on APG III (The Angiosperm Phylogeny Group, 2009). **indicates unranked groups instead of regular taxonomic orders.

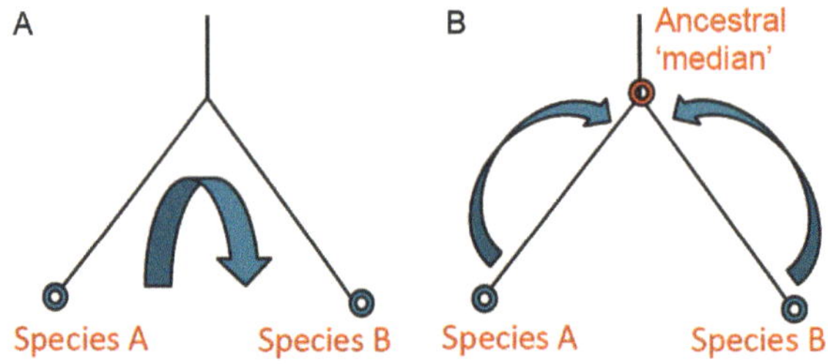

Haibao Tang *et al.*, Figure 8.7 Illustration of how 'ancestral median' genome can cut down the comparison distance between the extant genomes. (A) Direct comparison; (B) Comparison mediated through a reconstructed ancestral genome.

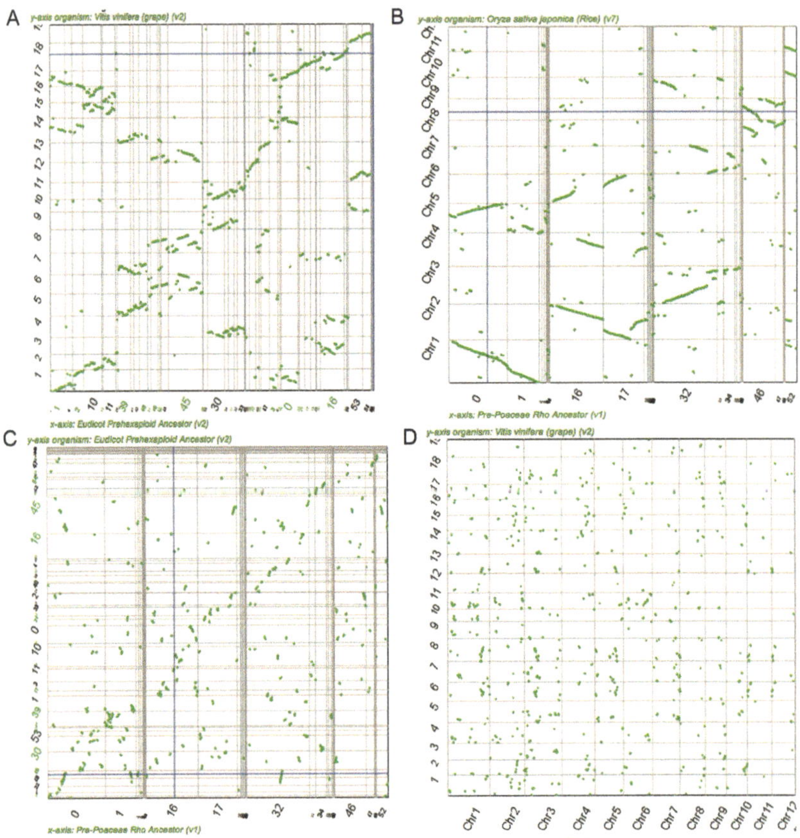

Haibao Tang *et al.*, Figure 8.8 Dot plots between reconstructed genomes and extant genomes using CoGe SynMap (http://genomevolution.org). (A) PreEudicot versus grape; (B) PreGrass versus rice; (C) PreEudicot versus PreGrass; (D) grape versus rice.

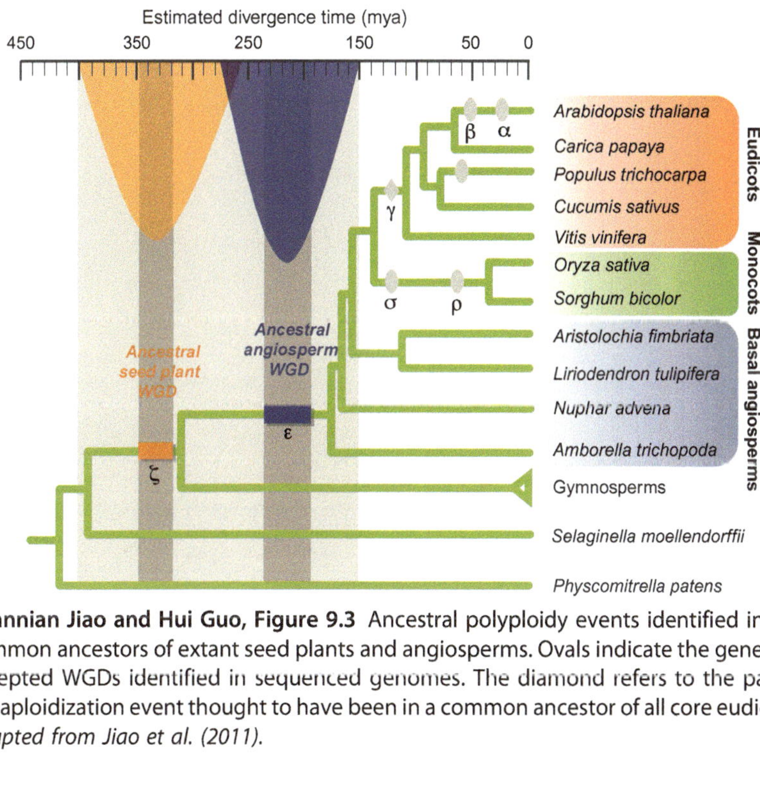

Yuannian Jiao and Hui Guo, Figure 9.3 Ancestral polyploidy events identified in the common ancestors of extant seed plants and angiosperms. Ovals indicate the generally accepted WGDs identified in sequenced genomes. The diamond refers to the paleo-hexaploidization event thought to have been in a common ancestor of all core eudicots. *Adapted from Jiao et al. (2011).*

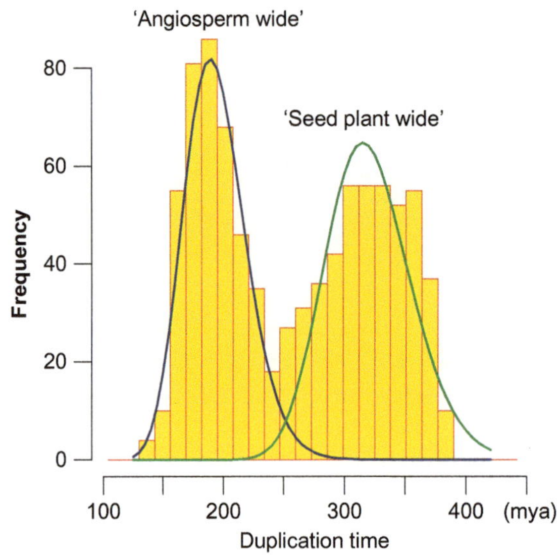

Yuannian Jiao and Hui Guo, Figure 9.4 Age distribution of ancient duplications. Two statistically significant peaks suggest two ancestral polyploidy events shared by seed plants and angiosperms, respectively (Jiao et al., 2011).

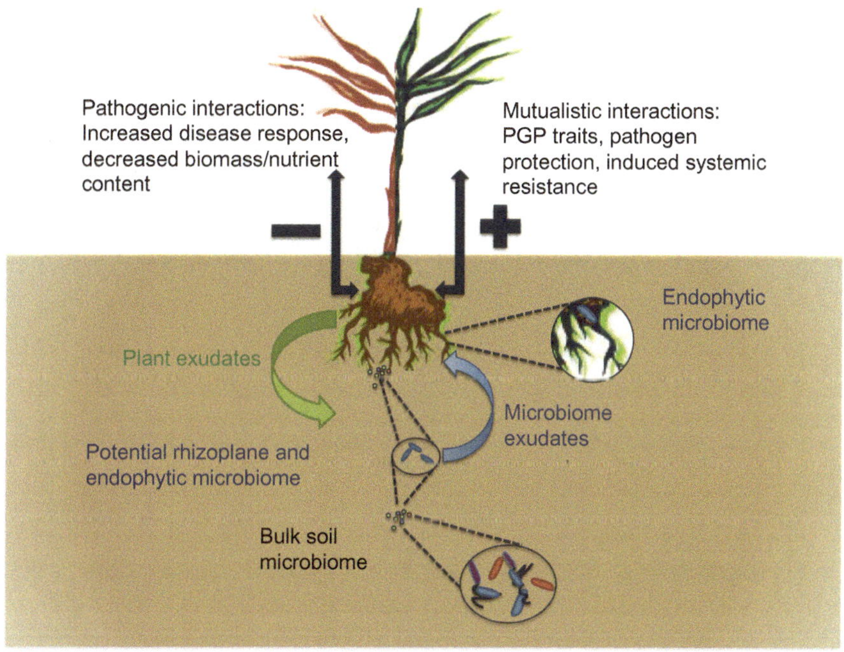

Marnie E. Rout, Figure 11.1 Model of the plant microbiome, with emphasis on the belowground habitats. The belowground regions of the plant microbiome include microorganisms inhabiting areas surrounding the plant roots at the root–soil interface or rhizosphere, those that are adherent to the root surface referred to as the rhizoplane, and those that colonize the internal root tissues that are known as endophytes. The microbial community residing in the bulk soil is primarily under the influence of environmental factors and is not under the direct influence of the plant (roots and exudates).

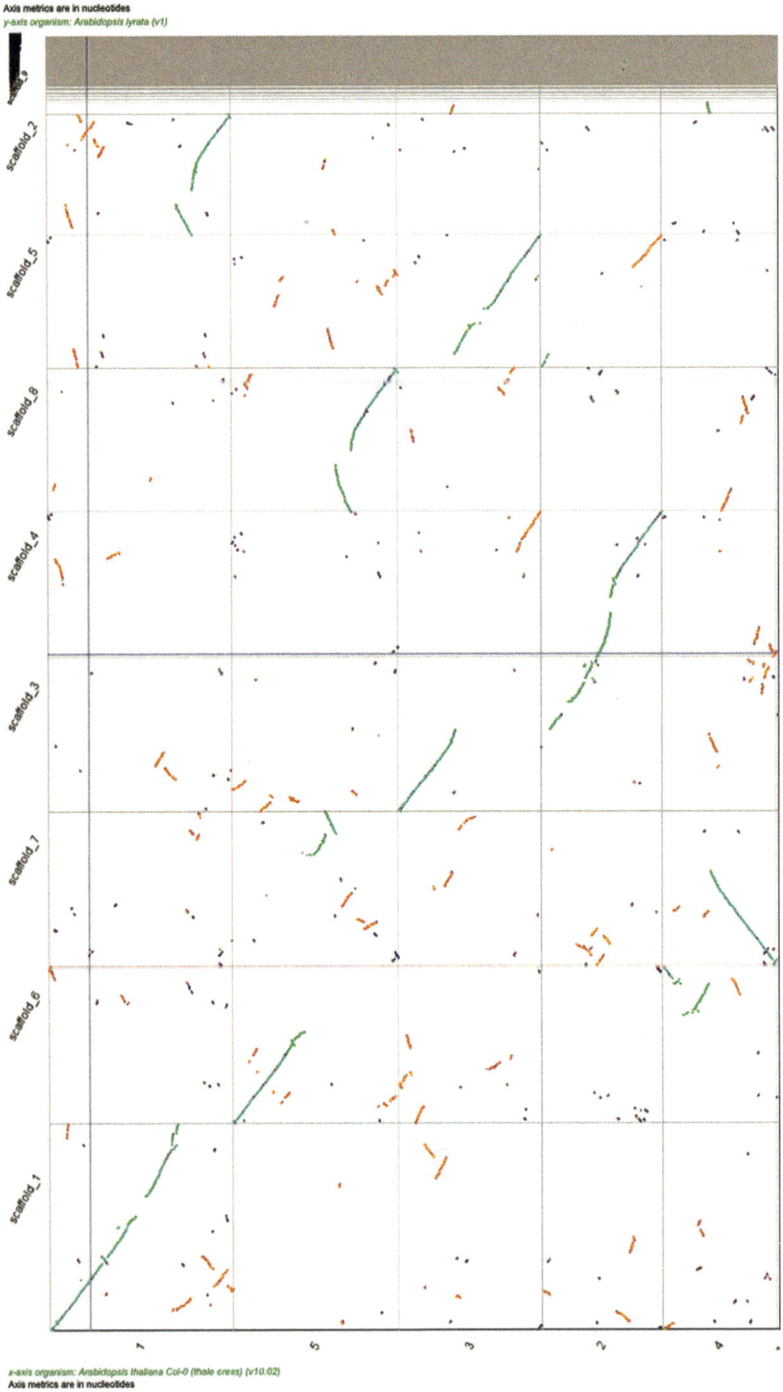

Axis metrics are in nucleotides
y-axis organism: Arabidopsis lyrata (v1)
scaffold_2
scaffold_5
scaffold_8
scaffold_4
scaffold_3
scaffold_7
scaffold_6
scaffold_1
1
5
3
2
4
x-axis organism: Arabidopsis thaliana Col-0 (thale cress) (v10.02)
Axis metrics are in nucleotides

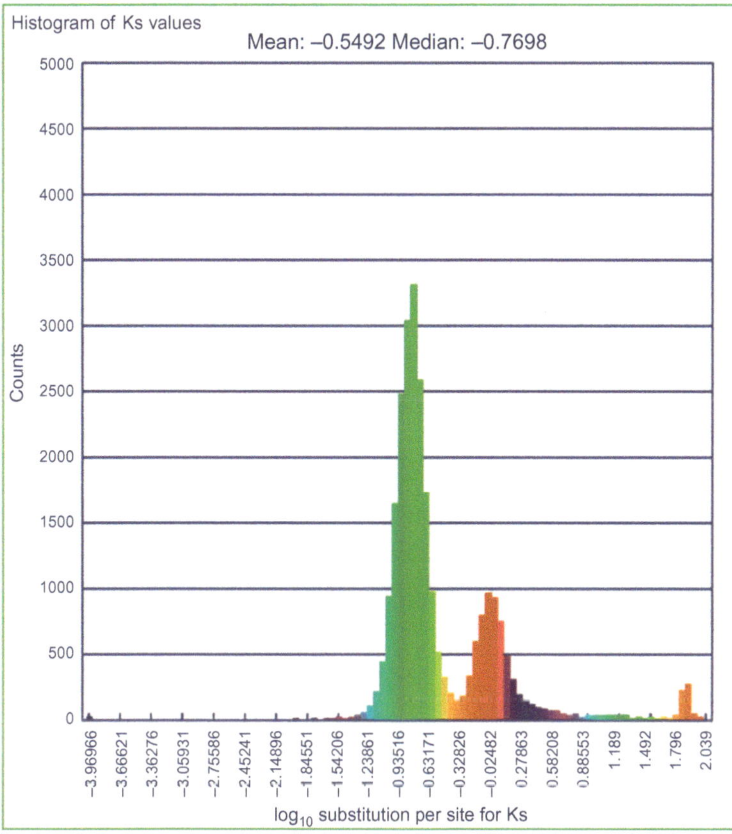

Michael Freeling, Figure 13.2 (A) This SynMap panel is almost but not quite identical to what you should see when you visit the iPlant servers via the link http://genomevolution.org/r/9a6t and click "OK". The lines comprised of syntenic gene pairs are "naturally" formed because the order of homologous genes in *Arabidopsis thaliana* (*X*-axis) is often similar to the order of homologous genes in *Arabidopsis lyrata*. Each dot is a pair of genes, and the Ks of that gene pair has been calculated automatically and rendered as a colour per (B). The orthologous and most recently diverged genes are in light blue. The red syntenic lines are remnants from the most recent tetraploidy shared by all species closely related to *Arabidopsis*. The reason there are no additional syntenic lines or stray dots reflecting more ancient duplication events and noise is because SynMap was told, via the program "Quota-Align", to analyse these two species to a "syntenic depth" of 2:2 and no deeper. If you use the link earlier (not (A)), Quota-Align has been turned "off" and *all* homologous hits are reflected in the dot plots. See (B). Exercise: is it true that adding up the lines of the same colour (the same level of divergence or age) approximately reconstructs the entire genome? (B) This is the log distribution of Ks values (*Y*-axis) for all "dots" or pairs of homologous genes (*X*-axis) between *Arabidopsis thaliana* and *Arabidopsis lyrata*, as derived from the SynMap experiment (http://genomevolution.org/r/9a6t). The modes of Ks defined here are the *A. thaliana–A. lyrata* orthology (the divergence time of these two species, this being bluish green), and the red mode reflects the most recent ancient tetraploid shared by these two species. There exist many small blue—the higher Ks shoulder of the "red" mode—lines in the output of the SynMap experiment because Quota-Align was turned "off". These blue lines are thought to reflect more ancient whole-genome duplication events. *Reprinted from Freeling (2009), with permission from Annual Reviews.*

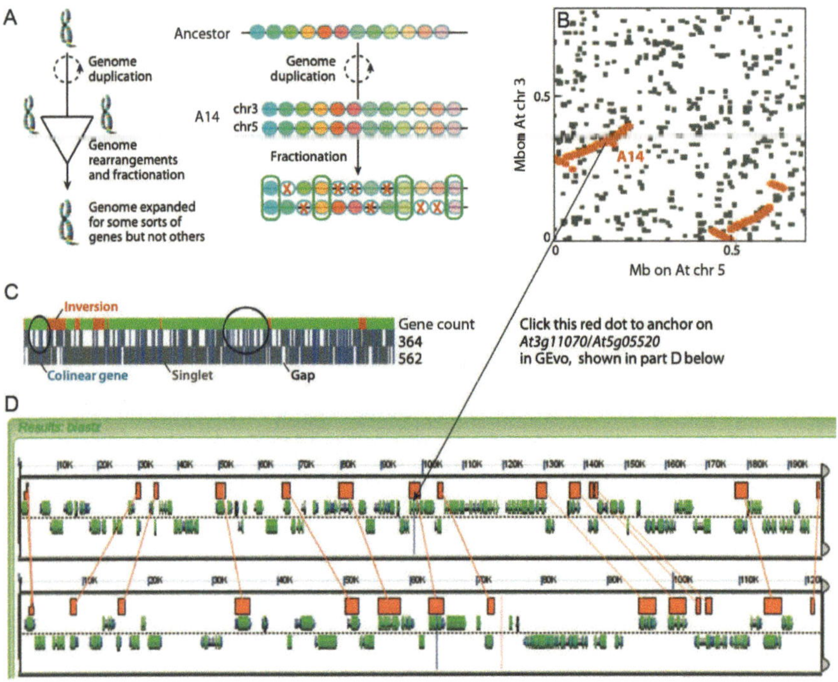

Michael Freeling, Figure 13.3 An example of fractionation of a region of paired chromosome retained from the most recent, whole-genome duplication in the *Arabidopsis* lineage. (A) Fractionation illustrated from chromosome to gene. Each coloured ball is a transcriptional unit or gene. (B) This same region of homeologous chromosome, called "A14", as seen in SynMap (Lesson 4). (C) A diagram of A14 (Thomas, Pedersen, & Freeling, 2006) showing that there are a lot more genes on one homeolog than the other, the consequence of fractionation bias. White spaces in this diagram are gaps that needed to be added to the more fractionated chromosome in order to diagram syntenic alignments. (D) A GEvo graphic (Lesson 3) of A14 showing that this region can be studied right down to nucleotide sequence. *Reprinted with permission from Freeling (2009).*